T0323880

Global One Health and Infectious Diseases

While many terms relate to One Health, the idea remains the same: to think outside a chosen area of specialty and work collaboratively as part of a team to improve health status around the world. This involves the collective effort of physicians, veterinarians, public health practitioners, ecologists, anthropologists, social workers, economists, and many others. Collectively, these are the Global One Health practitioners.

Through the lens of infectious disease, this book brings together the diverse range of topics necessary to be an effective global health practitioner at the intersection of human and animal health, particularly in developing countries. It explores what an aspiring or mid-career practitioner should be aware of when working with infectious diseases, including technical skills, cultural competency, capacity building, big data, and understanding the landscape and history of global health. Each chapter focuses on a specific area of necessary knowledge with background information, case examples, and resources to use moving forward.

An important reference for upper-level undergraduate students, graduate students, and early practitioners in human, animal, and public health, this text highlights the competencies rather than focusing on the problems in Global One Health. It provides a blueprint of areas that the reader should pay attention to, particularly in the realm of infectious diseases.

CRC One Health One Welfare

Learning from Disease in Pets: A 'One Health' Model for Discovery
Edited by Rebecca A. Krimins

Animals, Health and Society: Health Promotion, Harm Reduction and Equity in a One Health World
Edited by Craig Stephen

One Welfare in Practice: The Role of the Veterinarian
Edited by Tanya Stephens

Climate Change and Animal Health
Edited by Craig Stephen, Colleen Duncan

Animal Welfare in a Pandemic: What Does COVID-19 Tell us for the Future?
John T. Hancock, Ros C. Rouse, Tim J. Craig

Global One Health and Infectious Diseases: An Interdisciplinary Practitioner's Guide
Edited by William Sander

For more information about this series, please visit https://www.routledge.com/CRC-One-Health-One-Welfare/book-series/CRCOHOW

Global One Health and Infectious Diseases

An Interdisciplinary Practitioner's Guide

Edited by

William E. Sander

Designed Cover Image: ShutterStock ID: 663481099 (Front Cover), 1626532720 (Back Cover)

First edition published 2025
by CRC Press
2385 NW Executive Center Drive, Suite 320, Boca Raton FL 33431

and by CRC Press
4 Park Square, Milton Park, Abingdon, Oxon, OX14 4RN

CRC Press is an imprint of Taylor & Francis Group, LLC

ISBN: 9781032140742 (hbk)
ISBN: 9781032140674 (pbk)
ISBN: 9781003232223 (ebk)

DOI: 10.1201/9781003232223

Typeset in Times
by codeMantra

This book is dedicated to my supporting family including my superstar wife, Sam, and my two amazing daughters, Delaney and Shae.

Contents

Preface

The aspiring global health practitioner has been appealing for decades and even centuries long before the term "global health" came into being. Missionaries dating back to the 13th and 14th centuries, as part of the Crusades and many other Western explorations, not only brought religion but also made attempts at improving livelihoods, often in caring for those who were sick. This drive has shaped much of how the world is today both in some of the improvements as well as in some of the large disparities. While the idea of international health or global health often originates from those of developed or Western countries bringing expertise and solutions to those of other areas of the world, this idea has shifted more recently. The recognition of imperialism still existing through medicine and health has sparked a change in mindset. However, this shift has been variable across initiatives in the global health field and much of what is called "helicopter medicine" is often practiced.

At the same time, the term "One Health," or the integration of human, animal, and environmental health, has grown with the recognition that complex problems in health require multidisciplinary efforts. Additionally, these problems are often driven by many different factors from across sectors. While many terms relate to One Health (e.g., planetary health, one medicine), the idea to think outside a chosen area of specialty and be able to work collaboratively as part of a team to better solve and improve health status around the world remains the same. This involves the collective effort of physicians, veterinarians, public health practitioners, ecologists, anthropologists, social workers, economists, and many others. Collectively, these are the global one health practitioners.

Within the realm of global one health, infectious diseases are a key component that draws from all of these disciplines and the focus of this book. Often an expert in one area that touches global one health does not consider the other components necessary for successful programs, initiatives, and outcomes in the targeted areas. One can not only focus on technical aspects like surveillance, metrics, investment, and training, but must also consider the upstream and downstream components like ecosystem services, nutrition, climate change, and migration as well as the interactive and relationship components like the role of gender, cultural competency, and consensus building. Without an understanding of these pieces prior to engagement, the chances of success diminish as do the chances of sustainability after a program or intervention concludes or funding ends.

Through the lens of infectious disease, this book brings together the components necessary to be a global one health practitioner. Each chapter focuses on a specific area of knowledge needed with background information, case examples, and resources to use moving forward that not only build your understanding but also can be applied in your health endeavors. Each chapter provides sufficient working knowledge to apply the concept in practice while allowing space for the aspiring practitioner to explore more deeply through additional training and resources. The text is especially applicable for those in graduate and professional programs that touch on the field of global one health, which spans many disciplines, and is still applicable for those who have worked in the field of global health for several years.

About the Editor

William E. Sander is an Assistant Professor for Preventive Medicine and Public Health and Director of the DVM/MPH Joint Degree program at the University of Illinois – College of Veterinary Medicine. His research and teaching focus on the interface across professions of One Health from the local to the global level, leading efforts on sustainability, veterinary and interprofessional education, human-wildlife interactions, emerging infectious diseases, and antimicrobial resistance. Prior to Illinois, he spent 6 years in Washington, D.C. During that time, he spent 2 years at the U.S. EPA Office of Water as an AAAS Science and Technology Fellow and 3 years supporting the Defense Threat Reduction Agency's Cooperative Biological Engagement Program. Additionally, he was a staff veterinarian for 5 years at City Wildlife, Washington, D.C.'s only wildlife rehabilitation center, and practiced for 4 years at VCA small animal practices. He's a diplomate of the American College of Veterinary Preventive Medicine and previously served on their Executive Board. He serves on the board of directors for the Illinois State Veterinary Medical Association, American Association for Food Safety and Public Health, and the Veterinary Sustainability Alliance. Additionally, he is the alternate Delegate for Illinois to the AVMA House of Delegates. He is a 2009 graduate of the University of Wisconsin School of Veterinary Medicine and earned an MPH from Yale University in 2011. His Bachelor of Arts is from Colby College in 2004.

Contributors

A. Alonso Aguirre
Warner College of Natural Resources
Colorado State University
Fort Collins, Colorado

Robyn G. Alders
Development Policy Centre
Australian National University
Fullerton, New South Wales, Australia

Brian F. Allan
University of Illinois
Urbana, Illinois

Sarah E. Baum
School of Public Health
Harvard University
Boston, Massachusetts

Jane Blake
Jacobs
Alexandria, Virginia

D. Katterine Bonilla-Aldana
College of Medicine
Korea University
Seoul, Republic of Korea

John P. Bourgeois
Cummings School of Veterinary Medicine
Tufts University
North Grafton, Massachusetts

Sulagna Chakraborty
University of Illinois
Urbana, Illinois

Helena J. Chapman
Department of Environmental and Occupational Health
Milken Institute School of Public Health
George Washington University
Washington, District of Columbia

Rosa Costa
Kyeema Foundation
Brisbane, Australia

Julia De Bruyn
World Vegetable Center
Strathmore, Victoria, Australia

Evelyne de Leeuw
École de Santé Publique
Université de Montréal
Montréal, Quebec, Canada

Manisha Dhakal
Blue Diamond Society
Kathmandu, Bagamati, Nepal

Katherine A. Feldman
Maryland Department of Health
Ellicott City, Maryland

Leilani V. Francisco
Social and Scientific Systems
DLH Corporation
Silver Spring, Maryland

Leah Goodman
Jacobs
Alexandria, Virginia

Jennie Gordon
School of Medicine and Psychology Rural Clinical School
College of Health and Medicine
Australian National University
Canberra, Australia

Heather Kristen Grieve
Independent Consultant
Armidale, New South Wales, Australia

Kristen T. Honey
U.S. Department of Health and Human Services
Washington, District of Columbia

Susan Horton
University of Waterloo
Scarborough, Ontario, Canada

Laura H. Kahn
One Health Initiative
Bethesda, Maryland

Elsie Kiguli-Malwadde
African Centre for Global Health and Social Transformation
Kampala, Uganda

Jennifer Lane
School of Veterinary Medicine
University of California, Davis
Davis, California

Warren G. Lavey
College of Law
University of Illinois
Urbana, Illinois

J. Austin Lee
Indiana University School of Medicine Department of Emergency Medicine
Bloomington, Indiana

Joann M. Lindenmayer
School of Medicine
Tufts University
Uxbridge, Massachusetts

Catherine Machalaba
EcoHealth Alliance
New York, New York

Shannon Mesenhowski
Bill and Melinda Gates Foundation
Minneapolis, Minnesota

Tonya Nichols
U.S. Environmental Protection Agency
Washington, District of Columbia

Mariam Reda
Personal capacity
Dearborn Heights, Michigan

Alfonso J. Rodriguez-Morales
Grupo de Investigación Biomedicina
Fundación Universitaria Autónoma de las Américas
Pereira, Colombia

and

Universidad Científica del Sur
Lima, Peru

and

Gilbert and Rose-Marie Chagoury School of Medicine
Lebanese American University
Beirut, Lebanon

and

School of Medicine
Universidad Privada Franz Tamayo
Cochabamba, Bolivia

Jonathan Rushton
University of Liverpool
Liverpool, United Kingdom

Frances Siobhán Ryan
The Royal (Dick) School of Veterinary Studies
University of Edinburgh
Edinburgh, United Kingdom

Gareth Richard Salmon
The University of Edinburgh
Edinburgh, United Kingdom

Stephanie J. Salyer
Global Health Center
Division of Global Health Protection
U.S. Centers for Disease Control and Prevention
Atlanta, Georgia

Lindsey McCrickard Shields
PATH
Washington, District of Columbia

Tara E. Stewart Merrill
Cary Institute of Ecosystem Studies
Millbrook, New York

Cheryl Stroud
One Health Commission
Apex, North Carolina

Stewart Sutherland
School of Medicine and Psychology Rural Clinical School
College of Health and Medicine
Australian National University
Canberra, Australia

Deborah Thomson
One Health Lessons
Arlington, Virginia

Neil Vezeau
Personal capacity
Kula, Hawaii

Christopher A. Whittier
Cummings School of Veterinary Medicine
Tufts University
North Grafton, Massachusetts

Johanna Tsing-Min Wong
The University of Edinburgh
Edinburgh, United Kingdom

Jackson Tzu-ie Zee
Vier Pfoten International
Vienna, Austria

1 COVID-19 – A Catalyst for the One Health Movement

D. Katterine Bonilla-Aldana and Alfonso J. Rodriguez-Morales

INTRODUCTION

As not observed for more than a century, the world faced, in December 2019–March 2020, the beginning of an unprecedented pandemic caused by a new species of coronavirus, a Betacoronavirus, the Severe Acute Respiratory Syndrome coronavirus type 2 (SARS-CoV-2), which caused the coronavirus disease 2019 (COVID-19) (Bonilla-Aldana, Cardona-Trujillo, et al., 2020; Rabaan et al., 2020; Rodriguez-Morales, Bonilla-Aldana, et al., 2020; Rodriguez-Morales, Balbin-Ramon, et al., 2020). Although other pandemics have been declared before, e.g., Influenza H1N1 in 2009, COVID-19 has been the most significant in human history. This coronavirus is closely related to the SARS-CoV-1, a Sarbecovirus (subgenus) in the genus Betacoronavirus. From the end of 2019 through the end of 2024, this new coronavirus led to 776.6 million cases and 7.1 million deaths worldwide. Consequently, more than 13.64 billion doses of anti-SARS-CoV-2 vaccines have been administered globally through October 30, 2024. The pandemic has affected more than 250 countries and territories, but the impact has been significantly higher in low- and middle-income countries (Azhar et al., 2014; Dagan et al., 2021; Halfmann et al., 2020; Holshue et al., 2020; Ksiazek et al., 2003; Lipsitch et al., 2020; RECOVERY Collaborative Group, 2021; Rodriguez-Morales, Gallego, et al., 2020; Rodriguez-Morales, MacGregor, et al., 2020; Sah et al., 2021; Schlagenhauf et al., 2021; Vasquez-Chavesta et al., 2020; Wilder-Smith, 2006; Zhao et al., 2020; Zhu et al., 2020).

Like other coronaviruses, especially SARS-CoV-1 and the Middle East Respiratory Syndrome coronavirus (MERS-CoV), SARS-CoV-2 is a zoonotic pathogen (Bonilla-Aldana, Holguin-Rivera, et al., 2020; Bonilla-Aldana, Rodriguez-Morales, et al., 2020; Dhama et al., 2013; Faburay, 2015; Hemida & Ba Abduallah, 2020), detected in a wide range of animals, including domestic cats and dogs (Figure 1.1). In nature, several species of bats are likely involved in enzootic cycles of coronaviruses serving as hosts, where other sylvatic animals may interact and can be infected (Figure 1.1). These interactions can lead to spillover including in humans, with the most significant events including SARS-CoV-1

DOI: 10.1201/9781003232223-1

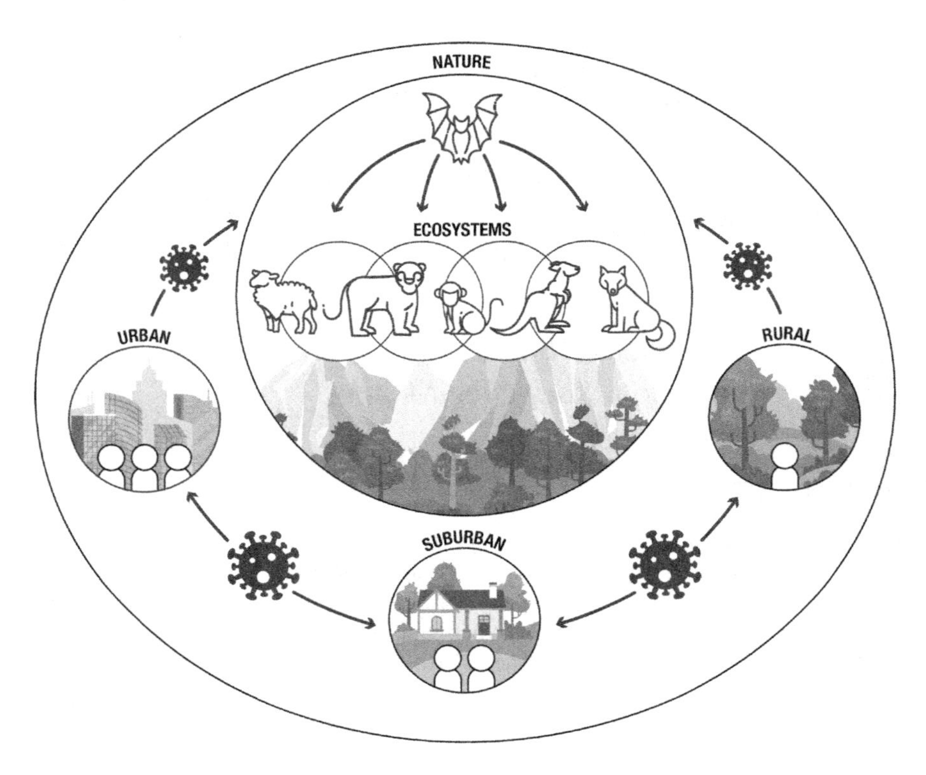

FIGURE 1.1 Interactions and spillover of SARS-CoV-2 and other viral zoonotic pathogens may be related to the importance of One Health.

in China and other countries in 2002, MERS-CoV primarily in the Middle East in 2012, and SARS-CoV-2/COVID-19 leading to a pandemic affecting all continents (Choudhary et al., 2024; Contini et al., 2020).

The spillover and emergence of SARS-CoV-2 was a multifactorial event, where different epidemiological, social, animal, and environmental factors, among others, interplayed in varying roles. SARS-CoV-2/COVID-19 is a disease, a medical condition, and a societal problem where multidisciplinary approaches and strategies were necessary to understand its complexity and prevent and mitigate future similar situations. In that context, considering human, animal, and environmental health, the One Health approach is critical. Climate change, deforestation, landscape use and degradation, illegal wildlife trade, and urbanization of rural areas, among other factors, have been suggested and analyzed in the context of the SARS-CoV-2 emergence (Sparrer McKenzie et al., 2023).

For Global Health, promoting health, keeping the world safe, and serving the vulnerable are vital to understanding the One Health approach and its relevance in this pandemic. COVID-19, even in its first weeks, rapidly became a global problem with a declaration by the World Health Organization as a Public Health Emergency of International Concern (PHEIC) on January 30, 2020.

The COVID-19 pandemic is a prime example of global One Health, both in the efforts to prepare for a pandemic and the failings to do so. Consequently, multiple components must be improved and applied to future health threats in months, years, or even decades. More epidemics are coming, and the recent example of the multi-country outbreak of a zoonotic viral disease caused by a virus discovered in 1958, the monkeypox (Mpox) virus (an Orthopoxvirus of the family Poxviridae), is just one recent example, also declared twice as PHEIC, in 2022 and in 2024 (Antinori et al., 2023; Taha et al., 2023).

The focus on the origin of SARS-CoV-2 from animals, with a subsequent respiratory human-to-human transmission (spillover), is essential. The transmission from humans to animals (i.e., reverse zoonoses), mainly to dogs, cats, lions, tigers, and minks, among others, is essential to understand that multiple components may contribute to the continuation of this pandemic even as it goes to endemicity. Similarly, the recent transmission of Mpox from humans to dogs reminds us of the multiple directions of spillovers (Abdel-Moneim & Abdelwhab, 2020).

In addition, cultural and political understanding of those interactions, determinants, and the relevance of One Health is critical to taking actions that may include better surveillance of humans, animals, and the environment to detect possible biohazards and risks early. More research and funding will be essential in this effort.

CASE EXAMPLE: BATS AS BIOLOGICAL SENTINELS FOR VIRAL BIOTHREATS

Bats are considered hosts of multiple infectious pathogens, including bacteria, parasites, fungi, and numerous viruses. In addition, bats are involved in the persistence of many viral diseases through enzootic cycles of transmission involving other animals, leading to spillover events. Rabies is a perfect example of a One Health disease where this occurs. In addition, multiple zoonotic coronaviruses have been detected in bats, and it has been suggested that they may contribute to the spillover to other animals, including pangolins, camels, other mammals, and human beings.

Detection of zoonotic pathogens in bats through sentinel surveillance is vital for understanding the biothreats and possible spillovers to other animals and humans. In addition, current biotechnological tools, such as molecular biology and particularly metagenomics, may contribute to screening viral biothreats in different regions of the world.

COMPETENCY – APPROACHES AND IMPACTS ON ONE HEALTH

For One Health and Global Health practitioners, understanding the role of environmental health in animal and human health is essential, particularly for the risk of emerging zoonotic pathogens. We need to increase the development of evidence through multiple studies, with multi-, inter-, and transdisciplinary approaches, including medical, animal, environmental, and social sciences, interacting proactively to generate critical information for stakeholders and funders. Dissemination of such data at a global level is urgently needed, as well as proper translation and social appropriation in local instances and authorities, understanding that One Health and Global Health need to be more actively promoted at all levels. Integrating the work on public health with animal health and environmental conservation will lead to applications of One Health in integrative systems composed of multidisciplinary teams.

Setting up One Health approaches may prevent, reduce, and mitigate the impacts of emerging and reemerging diseases. Consequently, improving the health of the environment, animals, and humans with sustainable development and direct relationships with more reachable Sustainable Development Goals (SDGs), as dreamed by the world society, will help reduce the risk or mitigate future zoonotic outbreaks.

APPLICATION OF COMPETENCY AS A GLOBAL HEALTH PRACTITIONER

In education, it is critical to include disciplines and content that bring together and deliver information about One Health in primary, secondary, and post-secondary venues, especially in medicine, veterinary medicine, biology, environmental sciences, and social sciences. Furthermore, community education and the promotion of One Health are needed to stimulate improved multidisciplinary approaches and collaborations. In these actions, Global Health practitioners are essential. Additionally, professionals from the areas mentioned above should be continually reminded about the importance of One Health, including promoting the potential interaction between multiple related disciplines for better approaches and solutions.

Also, for governance and health diplomacy, in the context of zoonoses, it is critical to raise awareness of multisectoral approaches that consider One Health as key in managing and preventing outbreaks of emerging pathogens. At local and global levels, such interactions with health authorities are vital in controlling and providing solutions to emerging infectious diseases.

CONCLUSIONS

Due to the COVID-19 pandemic, multiple actors in different sectors started to understand better the existence and the concept of One Health. For example, the number of articles about One Health in bibliographical databases such as PubMed has significantly grown, especially after 2019. Historically, more than 45% of the articles about One Health have been published between 2020 and 2022, while the first appeared in 1953, with a total of just 2,089.

In conclusion, the One Health approach can no longer be delayed at different levels. One Health is vital for all the potential actors and sectors that need urgent interaction and the development of strategies that consider all the components involved as part of a whole, providing a holistic approach to public health problems that are related to animal and environmental health.

REFERENCES

Abdel-Moneim, A. S., & Abdelwhab, E. M. (2020). Evidence for SARS-CoV-2 infection of animal hosts. *Pathogens*, 9(7). https://doi.org/10.3390/pathogens9070529

Antinori, S., Casalini, G., Giacomelli, A., & Rodriguez-Morales, A. J. (2023). Update on Mpox: A brief narrative review. *Le Infezioni in Medicina*, *31*(3), 269–276.

Azhar, E. I., El-Kafrawy, S. A., Farraj, S. A., Hassan, A. M., Al-Saeed, M. S., Hashem, A. M., & Madani, T. A. (2014). Evidence for camel-to-human transmission of MERS coronavirus. *New England Journal of Medicine*, *370*(26), 2499–2505.

Bonilla-Aldana, D. K., Cardona-Trujillo, M. C., García-Barco, A., Holguin-Rivera, Y., Cortes-Bonilla, I., Bedoya-Arias, H. A., Patiño-Cadavid, L. J., Tamayo-Orozco, J. D., Paniz-Mondolfi, A., Zambrano, L. I., Dhama, K., Sah, R., Rabaan, A. A., Balbin-Ramon, G. J., & Rodriguez-Morales, A. J. (2020). MERS-CoV and SARS-CoV infections in animals: A systematic review and meta-analysis of prevalence studies. *Le Infezioni in Medicina*, *28*(suppl 1), 71–83.

Bonilla-Aldana, D. K., Holguin-Rivera, Y., Perez-Vargas, S., Trejos-Mendoza, A. E., Balbin-Ramon, G. J., Dhama, K., Barato, P., Lujan-Vega, C., & Rodriguez-Morales, A. J. (2020). Importance of the One Health approach to study the SARS-CoV-2 in Latin America. *One Health*, *10*, 100147. https://doi.org/10.1016/j.onehlt.2020.100147

Bonilla-Aldana, D. K., Rodriguez-Morales, A. J., & Dhama, K. (2020). Revisiting the one health approach in the context of COVID-19: A look into the ecology of this emerging disease. *Advances in Animal and Veterinary Sciences*, *8*(3), 234–237. https://doi.org/10.17582/journal.aavs/2020/8.3.234.237

Choudhary, P., Shafaati, M., Abu Salah, M. A. H., Chopra, H., Choudhary, O. P., Silva-Cajaleon, K., Bonilla-Aldana, D. K., & Rodriguez-Morales, A. J. (2024). Zoonotic diseases in a changing climate scenario: Revisiting the interplay between environmental variables and infectious disease dynamics. *Travel Medicine and Infectious Disease*, *58*, 102694. https://doi.org/10.1016/j.tmaid.2024.102694

Contini, C., Di Nuzzo, M., Barp, N., Bonazza, A., De Giorgio, R., Tognon, M., & Rubino, S. (2020). The novel zoonotic COVID-19 pandemic: An expected global health concern. *The Journal of Infection in Developing Countries*, *14*(3), 254–264. https://doi.org/10.3855/jidc.12671

Dagan, N., Barda, N., Kepten, E., Miron, O., Perchik, S., Katz, M. A., Hernán, M. A., Lipsitch, M., Reis, B., & Balicer, R. D. (2021). BNT162b2 mRNA Covid-19 vaccine in a nationwide mass vaccination setting. *New England Journal of Medicine*, *384*(15), 1412–1423.

Dhama, K., Chakraborty, S., Kapoor, S., Tiwari, R., Kumar, A., Deb, R., Rajagunalan, S., Singh, R., Vora, K., & Natesan, S. (2013). One World, One Health—Veterinary perspectives. *Advances in Animal and Veterinary Sciences*, *1*, 5–13.

Faburay, B. (2015). The case for a "one health" approach to combating vector-borne diseases. *Infection Ecology & Epidemiology*, *5*, 28132. https://doi.org/10.3402/iee.v5.28132

Halfmann, P. J., Hatta, M., Chiba, S., Maemura, T., Fan, S., Takeda, M., Kinoshita, N., Hattori, S., Sakai-Tagawa, Y., & Iwatsuki-Horimoto, K. (2020). Transmission of SARS-CoV-2 in domestic cats. *New England Journal of Medicine*, *383*(6), 592–594.

Hemida, M. G., & Ba Abduallah, M. M. (2020). The SARS-CoV-2 outbreak from a one health perspective. *One Health, 10*, 100127. https://doi.org/10.1016/j.onehlt.2020.100127

Holshue, M. L., DeBolt, C., Lindquist, S., Lofy, K. H., Wiesman, J., Bruce, H., Spitters, C., Ericson, K., Wilkerson, S., & Tural, A. (2020). First case of 2019 novel coronavirus in the United States. *New England Journal of Medicine, 382*(10), 929–936.

Ksiazek, T. G., Erdman, D., Goldsmith, C. S., Zaki, S. R., Peret, T., Emery, S., Tong, S., Urbani, C., Comer, J. A., & Lim, W. (2003). A novel coronavirus associated with severe acute respiratory syndrome. *New England Journal of Medicine, 348*(20), 1953–1966.

Lipsitch, M., Swerdlow, D. L., & Finelli, L. (2020). Defining the epidemiology of Covid-19—Studies needed. *New England Journal of Medicine, 382*(13), 1194–1196.

Rabaan, A. A., Al-Ahmed, S. H., Haque, S., Sah, R., Tiwari, R., Malik, Y. S., Dhama, K., Yatoo, M. I., Bonilla-Aldana, D. K., & Rodriguez-Morales, A. J. (2020). SARS-CoV-2, SARS-CoV, and MERS-COV: A comparative overview. *Le infezioni in medicina, 28*(2), 174–184.

RECOVERY Collaborative Group. (2021). Dexamethasone in hospitalized patients with Covid-19. *New England Journal of Medicine, 384*(8), 693–704.

Rodríguez-Morales, A. J., Balbin-Ramon, G. J., Rabaan, A. A., Sah, R., Dhama, K., Paniz-Mondolfi, A., Pagliano, P., & Esposito, S. (2020). Genomic epidemiology and its importance in the study of the COVID-19 pandemic. *Genomics, 1*(3), 139–142.

Rodriguez-Morales, A. J., Bonilla-Aldana, D. K., Balbin-Ramon, G. J., Rabaan, A. A., Sah, R., Paniz-Mondolfi, A., Pagliano, P., & Esposito, S. (2020). History is repeating itself: Probable zoonotic spillover as the cause of the 2019 novel coronavirus epidemic. *Le infezioni in medicina, 28*(1), 3–5.

Rodriguez-Morales, A. J., Gallego, V., Escalera-Antezana, J. P., Méndez, C. A., Zambrano, L. I., Franco-Paredes, C., Suárez, J. A., Rodriguez-Enciso, H. D., Balbin-Ramon, G. J., & Savio-Larriera, E. (2020). COVID-19 in Latin America: The implications of the first confirmed case in Brazil. *Travel Medicine and Infectious Disease, 35*, 101613.

Rodríguez-Morales, A. J., MacGregor, K., Kanagarajah, S., Patel, D., & Schlagenhauf, P. (2020). Going global–Travel and the 2019 novel coronavirus. *Travel Medicine and Infectious Disease, 33*, 101578.

Sah, R., Khatiwada, A. P., Shrestha, S., Bhuvan, K. C., Tiwari, R., Mohapatra, R. K., Dhama, K., & Rodriguez-Morales, A. J. (2021). COVID-19 vaccination campaign in Nepal, emerging UK variant and futuristic vaccination strategies to combat the ongoing pandemic. *Travel Medicine and Infectious Disease, 41*, 102037.

Schlagenhauf, P., Patel, D., Rodriguez-Morales, A. J., Gautret, P., Grobusch, M. P., & Leder, K. (2021). Variants, vaccines and vaccination passports: Challenges and chances for travel medicine in 2021. *Travel Medicine and Infectious Disease, 40*, 101996.

Sparrer McKenzie N., Hodges Natasha F., Sherman Tyler, VandeWoude Susan, Bosco-Lauth Angela M., & Mayo Christie E. (2023). Role of spillover and spillback in SARS-CoV-2 transmission and the importance of One Health in understanding the dynamics of the COVID-19 pandemic. *Journal of Clinical Microbiology, 61*(7), e01610–22. https://doi.org/10.1128/jcm.01610-22

Taha, A. M., Katamesh, B. E., Hassan, A. R., Abdelwahab, O. A., Rustagi, S., Nguyen, D., Silva-Cajaleon, K., Rodriguez-Morales, A. J., Mohanty, A., Bonilla-Aldana, D. K., & Sah, R. (2023). Environmental detection and spreading of mpox in healthcare settings: A narrative review. *Frontiers in Microbiology*, 14. https://www.frontiersin.org/journals/microbiology/articles/10.3389/fmicb.2023.1272498

Vasquez-Chavesta, A. Z., Morán-Mariños, C., Rodrigo-Gallardo, P. K., & Toro-Huamanchumo, C. J. (2020). COVID-19 and dengue: Pushing the Peruvian health care system over the edge. *Travel Medicine and Infectious Disease*, *36*, 101808.

Wilder-Smith, A. (2006). The severe acute respiratory syndrome: Impact on travel and tourism. *Travel Medicine and Infectious Disease*, *4*(2), 53–60.

Zhao, S., Zhuang, Z., Ran, J., Lin, J., Yang, G., Yang, L., & He, D. (2020). The association between domestic train transportation and novel coronavirus (2019-nCoV) outbreak in China from 2019 to 2020: A data-driven correlational report. *Travel Medicine and Infectious Disease*, *33*, 101568. MEDLINE. https://doi.org/10.1016/j.tmaid.2020.101568

Zhu, N., Zhang, D., Wang, W., Li, X., Yang, B., Song, J., Zhao, X., Huang, B., Shi, W., & Lu, R. (2020). A novel coronavirus from patients with pneumonia in China, 2019. *New England Journal of Medicine*, *382*(8), 727–733.

2 Global One Health History

William E. Sander

INTRODUCTION

The history of human and animal health could and does span multiple books already in print (Packard, 2016; Farmer et al., 2013). However, it is important to highlight where these health sectors have come from, where they are now, and how the history of One Health builds from these histories and where this can lead in the future. History is always an important starting point when trying to change what happens in the present. One would be remiss not to consider history when looking to the future. Without a grounding in what has occurred for infectious disease efforts, the battle to stop future infectious diseases in the most effective and strategic ways may be wasted. This chapter will pull together human health history at a global level in the last two centuries followed by animal health and, finally, One Health, with a much shorter timeframe.

GLOBAL HUMAN HEALTH HISTORY

Much of the early global human health history parallels and is inextricably linked to the imperial system of the 16th and 17th centuries (Humphreys, 1999). Wherever colonists went, typically from western Europe, diseases were introduced to native populations. Conversely, these explorers and colonists were exposed to harsh diseases. The result, on both sides, was highly impactful and, for those being colonized, devasting. Many of the early efforts in disease control centered around the idea of protecting colonists from disease and civilizing native populations (Arnold, 2016). This umbrella of healthcare came to be called tropical medicine as it focused on the disease residing most often in tropical and subtropical environments that caused the greatest problems for the colonists.

Healthcare providers in the 17th and 18th centuries were often tied to the military. The military was ultimately responsible for many health units in colonies. This gave rise to heavily used quarantines that were often involuntary (Scultetus et al., 2006). Additionally, colonists often refused to treat or provide support to many that they colonized demonstrating open disregard.

As the understanding of medicine grew in the 19th century, population medicine became more advanced. This included the rise in what would be epidemiologists. The notable John Snow in 1854 utilized these concepts to stop a local cholera outbreak in London. After charting cases and deaths over a period

DOI: 10.1201/9781003232223-2

of time, a spatial association with the water pumps was devised and he famously tore the handle off the suspected contaminated source at the Broad Street pump (Tulchinsky, 2018).

Throughout that century, the term "tropical medicine" took on other terms like military medicine and medical missions. Missions continued well into the 20th century. A strong, altruistic connotation came with missions as they focused more on individualized medicine based out of Western culture (Worboys, 2000). Although medical missions were meant to be selfless and good willed, they were an extension of soft power and Western influence, often displacing traditional healers. These healers, while not trained in Western medicine, did have a wealth of local knowledge that often provided benefits to those they cared for through their treatments, their comforts, and the cultural importance for which they often held.

One of the most famous missionaries of the early 20th century was Dr. Albert Schweitzer who led mission trips from 1913 into the early 1950s to West Africa (Harris, 2016; Polednak, 1989). Trained in theology, he could not get permission from Paris to go on a medical mission trip until he went back to school at the age of 30 to become a physician. Over the course of his four decades in Gabon, he interacted with and saved countless lives but also never tried to integrate into the culture or language of the region. His colonialist tendencies kept his mission steadfast throughout his time in West Africa but also perpetuated what is now considered international or tropical medicine rather than global health. Despite this, he is the only individual physician to be awarded the Nobel Peace Prize.

At the same time as what was considered medicine was shifting in the 19th and 20th centuries, the idea of coming together across countries to tackle leading diseases coalesced. The International Sanitary Conventions was the first notable global or international effort centered around disease, particularly one disease, cholera. Their first meeting happened in 1851 and included the countries of Austria, France, Great Britain, Italy, Portugal, Russia, and Turkey. Cholera drove this shared interest because of the six major cholera pandemics that occurred between 1816 and 1899. Their main focus was on one disease but it was the first time attempting to create an international health policy. The International Sanitary Conventions continued to meet up until World War 1 (Sealey, 2011).

The early 1900s saw international efforts continue to surge. The Office International d'Hygiene Publique was formed to track epidemiologic info across countries of major human diseases. The Rockefeller Foundation International Board of Health was stood up through philanthropic support much the way the Bill and Melinda Gates Foundation is now. This International Board of Health started in 1913 and became the largest funder of global health through 1950 including starting many public health schools around the globe. Following the catastrophic impact of World War 1, the newly formed League of Nations attempted to set up the Health Committee which worked with the Pan-American Sanitary Bureau, Red Cross, and Rockefeller Foundation. The Pan-American Sanitary Bureau, started in 1924, was the forebearer, in many ways, to the Pan-American Health Organization and focused much of its efforts on yellow fever and immigration (McCarthy, 2002). All these efforts were still centered on the traditional global powers and were often very disease specific.

Following World War 2 and the founding of the United Nations (UN), the World Health Organization (WHO) started in 1946 with 61 initial countries signing on, far more than any previous international efforts. Importantly, the definition of health they utilized shifted the focus to the "state of complete physical, mental, and social well-being and not merely the absence of disease or infirmity." The mandate of WHO had several core areas that included health standards, data collection, epidemiologic surveillance, research, training, and emergency relief. Despite the overall shift in mandate in WHO, the infectious disease effort still targeted disease by disease. Some of that was seen by a dramatic increase in the availability of vaccinations and in reducing child mortality (McCarthy, 2002). This culminated with the biggest disease success, the eradication for the first time in human history of a disease, smallpox, in 1977. Smallpox was specifically targeted because of an effective vaccine, high morbidity and mortality, a singular host species in people, and global impact of the disease. While Europe, the United States, Australia, New Zealand, and USSR eliminated the disease by the early 1940s, the rest of the world only did so in the subsequent two decades – a monumental effort given the accessibility, resources, and infrastructure in place for many of these countries (Fenner, 1993).

Despite targeted disease success, the effort did highlight the resources needed just to get at one disease. It illuminated the complexities of debt and economic development in low-resource countries. Rather than being disease-focused, the Declaration of Alma Ata in 1978 focused on health for all by providing accessibility to primary healthcare. Foreign aid to lower resource countries shifted more toward health infrastructure and training along with accessibility. However, with a brand new pandemic in the 1980s of human immunodeficiency virus (HIV)/acquired immunodeficiency syndrome (AIDS), the disease-focused efforts were still very strong in the international community. The United States, always a major financial contributor to international health, launched an ambitious funding scheme in 2002 with the President's Emergency Plan for AIDS Relief or PEPFAR, which infused billions of dollars since into this one disease and parts of healthcare that touch it. Even in 2000, the G8, a group of highly industrialized countries (France, Germany, Italy, the United Kingdom, Japan, the United States, Canada, and Russia), included HIV/AIDS as a national security priority (Merson et al., 2008).

As foreign aid continued to increase in the 2000s to support international human health in lower-resourced nations, the question of which countries receive foreign aid became a more contentious piece as it was not just based on health metrics (if available), but on political and cultural connections. One might assume that the countries with the least gross national income (GNI) may be the best target for healthcare assistance. However, in Nigeria, for example, the GNI in 2016 per capita was 2970USD while many countries are much lower like Cambodia (1020USD) or Ethiopia (550USD), but Nigeria's deaths per 100 births before age 5 in 2014 was almost twice as high as in Ethiopia (112.5 to 61.8, respectively)

(The World Bank, 2023). Industrialized or higher-resourced countries are also not contributing equally. The UN set a goal of 0.7% ODA/GNI back in 1970 that has been re-endorsed by countries within the UN multiple times since. ODA is the official development assistance, so this parameter means that countries part of the Organisation for Economic Co-operation and Development (OECD) contribute 0.7% of their GNI annually to foreign aid. The OECD is made up of 38 countries and their contributions to ODA are reported every year. In 2018, only four countries met that goal (Denmark, Norway, Luxembourg, and Sweden), while many countries did not, including the world's leading economy, the United States, at 0.17% (OECD, 2024). Of U.S. global health funding, almost half goes to HIV/AIDS, 5.3B USD (KFF, 2024).

Additionally, in the 21st century, there has been a greater return of philanthropic agencies like Rockefeller Foundation of the early 20th century, as well as other non-governmental organizations contributing to health aid globally. These include intergovernmental groups, like the UN, World Organization for Animal Health, and the World Bank, non-profit organizations, for-profit businesses, and philanthropic groups. Much more is discussed around these different stakeholders in Chapter 4. The complex global funding and health assistance scene has many benefits to countries and areas that need it but also adds layers of potential duplication, mixed priorities, and potential blind spots in addressing human healthcare.

As this shift from colonial medicine to international health to global health continues, frameworks that all stakeholders can work under have been key. One main framework has been the work of the UN on the Millenium Development Goals (MDGs) in 2000 and the Sustainable Development Goals (SDGs) in 2015 (Figure 2.1) (Nakatani, 2016). Each set of goals was meant to tackle some of the biggest global challenges in a 15-year timeframe and were mutually agreed upon by all countries in the UN. The most recent of these, the SDGs, had 17 goals, all of which touch on human health needs in some form. While many of the MDGs improved greatly over the 15-year period, the SDGs are not on the same trajectory. As of 2023 in the 140 targets within the 17 goals, half are moderately or severely offtrack and over 30% have seen no movement or gone backward from the 2015 baseline (United Nations, Department of Economic and Social Affairs, 2023). As this more global context takes hold and global health 2.0 is re-envisioned (Figure 2.2), there is a careful balance of moral obligation and national autonomy. The flow of foreign aid has to be carefully monitored and managed as it comes from many stakeholders and for many different reasons (Fransen et al., n.d.).

The old idea of international health was staunchly entrenched in colonial medicine with the consequences of introduction of non-native diseases, facilitation of rapid spread of disease, and extraction of wealth. It was a universal practice focusing on the colonizers for stereotypes and models of disease. That approach legitimized and controlled those being colonized and reinforced many other aspects of colonialism. It also centered on the colonizers' national interests and security.

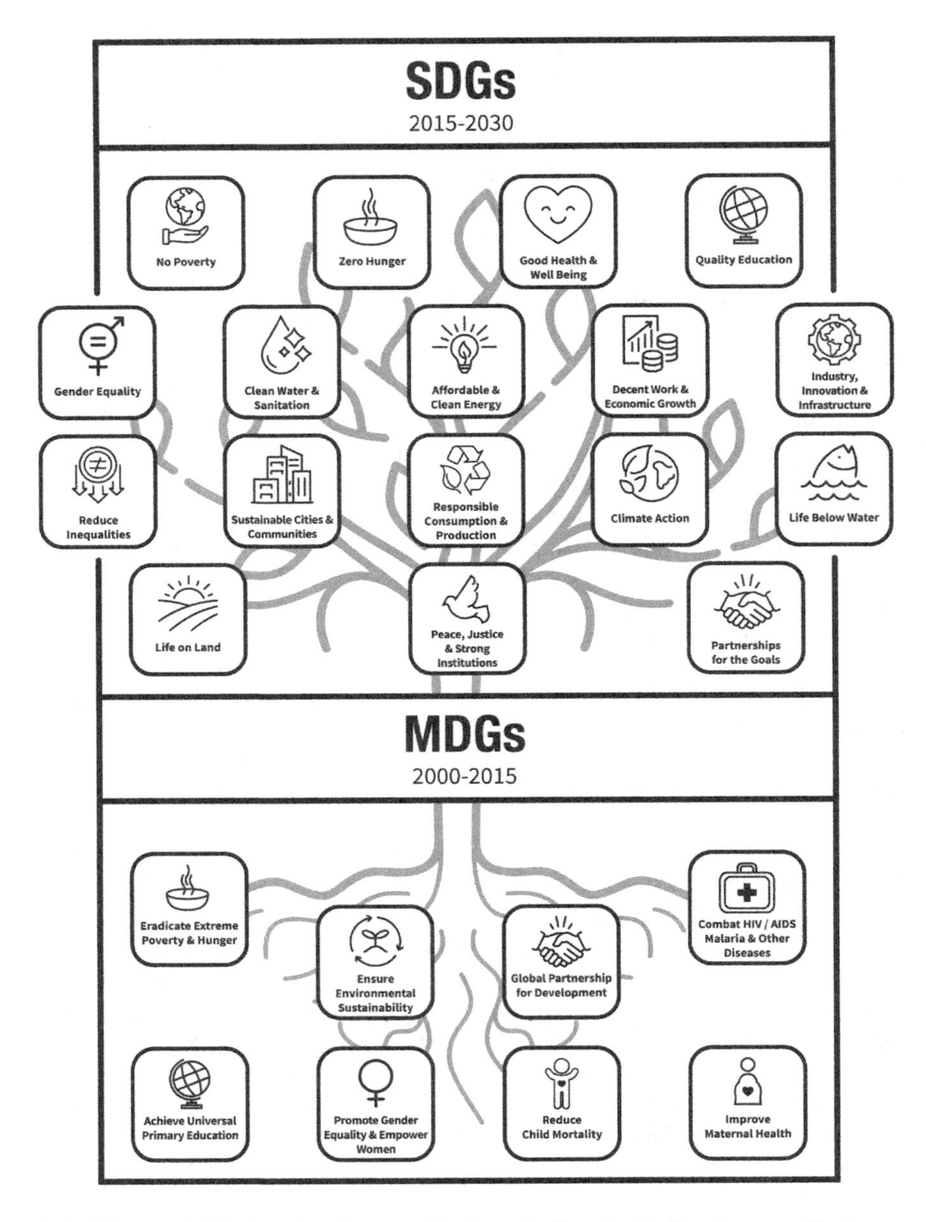

FIGURE 2.1 Millenium Development Goals to the Sustainable Development Goals.

Global health 2.0 seeks to capitalize on the globalization of our societies and focus more on the collective rather than an us-versus-them approach. This framing will be critical as the two biggest drivers of future infectious diseases are urbanization or crowding of people and increased proximity of humans to animals.

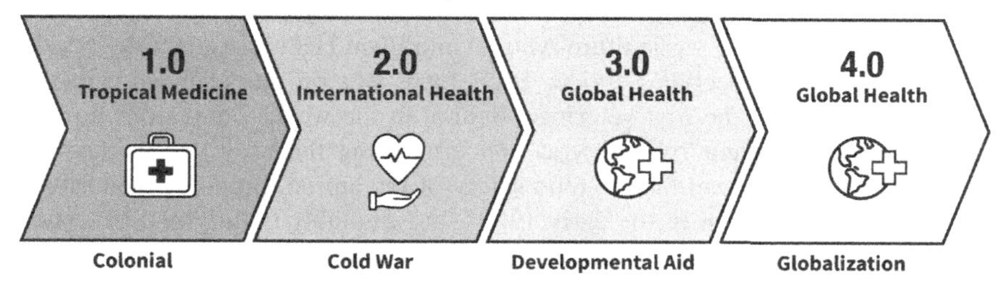

FIGURE 2.2 Evolution of Global Health over time.

This new global health provides many opportunities for improved health and for the global health practitioner but also necessitates that practitioner is aware of the complexities underlying the global context.

GLOBAL ANIMAL HEALTH HISTORY

While formal and informal physicians have existed for hundreds of years, the first formal veterinarians were trained in 1761 at the world's first veterinary school in Lyon, France (Hannaway, 1977). Veterinary medicine was specifically focused on working and production animals and the associated ailments and diseases that inhibited economic benefits from these animals. Until the early 1900s, horses were the primary mode of transportation, if one was not walking, as well as being useful draft animals for work. Livestock served not only for food and fiber but also for labor in plowing fields. The top diseases of concern were rinderpest, anthrax, blackleg, sheep pox, scabies, glanders, contagious bovine pleuropneumonia, and strangles (Woods, 2016). Much like in human medicine, even though the causative agents for these diseases were unknown, the diseases themselves were still a big area of focus in how to contain, reduce, and eliminate them.

Most veterinarians were trained in Europe until the mid-1800s when veterinary schools first opened in the United States (Smith, 2010). From 1866 to 1925, there were 41 veterinary schools founded including institutions like Harvard University, Columbia University, and George Washington University. All of these ceased to exist for a variety of reasons and were primarily created to care for working horses in cities. From World War 2 on, veterinary schools were primarily founded at land-grant institutions as a way to improve agricultural performance of livestock (Smith, 2010). By 1959, 18 of the 33 veterinary schools of today existed followed by another wave of 10 schools in the 1970s and 1980s because of the influence of James Herriot and the increasing importance of companion animals in society in the United States (Fish, 2011).

Although formal veterinary training originated in Europe, the United States has become the world leader in veterinary education in training over the last century, with the American Veterinary Medical Association providing the highest standard for veterinary schools to meet. As such, it is worth looking a little more into the history of the United States animal health sector before looking internationally. The U.S. Bureau of Animal Industry (1884–1942) preceded the U.S. Department of Agriculture Animal and Plant Health Inspection Service and Food Safety Inspection Service. Their focus was on one of those primary diseases included in the first veterinary school in the world, contagious bovine pleuropneumonia. Their role grew as concerns about the health of workers in slaughter facilities as well as the food safety of the animal products themselves was called into question in the early 1900s, most notably highlighted in Upton Sinclair's *The Jungle*. Eggs began to be inspected by veterinarians in the military before expanding to the public (Olmstead & Rhode, 2015). As can be noticed in Figure 2.3, the change in veterinary focus over the last 100 years is highlighted in the last few decades with the emphasis on companion animals, now a part of many U.S. families, and the consolidation of species farmed. Livestock farm sizes have continually increased over the last 30 years with changes in production technologies, increased enterprise specialization, and vertical coordination and integration.

Globally, the World Organisation for Animal Health (WOAH) was founded for much the same reason as the International Sanitary Conventions in trying to combat one disease. Instead of cholera on the human side, rinderpest was the driver for WOAH after rinderpest was introduced into Belgium in 1920. WOAH was founded in 1924 when 28 countries signed an international agreement to combat the disease. Founded in Paris and still where it resides, it was named Office International Epizootic (OIE) but has more recently changed to be entirely in English. With the creation of the UN and its many organizations in 1948, there was initial conflict with the UN Food and Agriculture Organization (FAO) as some overlap existed. The FAO was founded to eliminate hunger and improve nutrition as well as improve standards of living by increasing agricultural productivity. The conflict between WOAH and FAO was eventually resolved through a mutual agreement in 1951 (Otte et al., 2004). WOAH serves as the international animal health guidance organization. Since 1998, the World Trade Organization and the World Bank gave the international agreement powers around animal disease control for animal trade to WOAH (World Trade Organization, 2024).

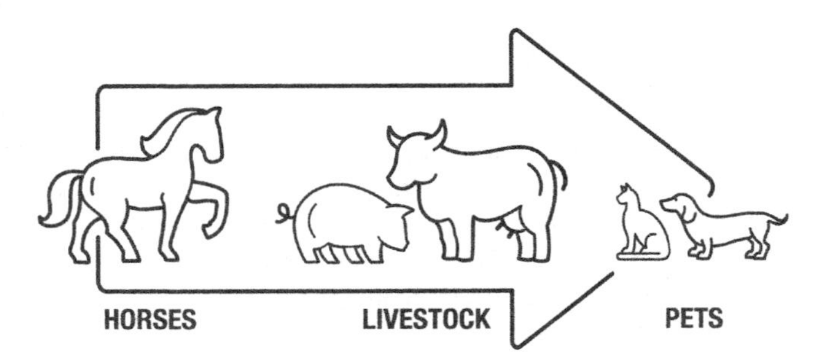

FIGURE 2.3 The change in veterinary focus in the Western world over the last 100 years.

Over the 75 years, the disease list WOAH considers has grown from 9 to 117 spanning aquatic and terrestrial animals of economic importance.

Similar to the concerted efforts of smallpox eradication, the WOAH along with many partners targeted the eradication of rinderpest (Roeder, 2011). As mentioned, rinderpest, a morbillivirus primarily affecting cattle with high morbidity and mortality particularly in naïve populations, was the driving disease for WOAH's creation 100 years ago. The disease has been documented all the way back to the Roman empire; implicated in several wars and natural disasters; directly impacted food, animal labor, and human nutrition; and paved the way for quarantines and movement restrictions in public health. An effective vaccine was developed, but mobilizing resources through a concerted international effort proved to be the most challenging. Finally, with a combined effort of the FAO, WOAH, and the International Atomic Energy Agency in the 1990s, the last recorded incidence was in 2001 with the last vaccine given in 2006. WOAH declared the disease eradicated on May 25, 2011 (Food and Agriculture Organization & World Organisation for Animal Health, 2021).

Infectious disease and the ability to contain and control them has been a major focus in animal health (Hartung, 2013). Over the last 100 years, there have been some dramatic animal disease outbreaks. While rinderpest was introduced into Belgium in 1920, contagious bovine pleuropneumonia was introduced in the Netherlands. The 1950s saw a major outbreak of foot-and-mouth disease in Mexico from Brazil affecting over 1 million cattle, sheep, and goats – large number of cloven-hoofed animals at that time given farming practices. While African swine fever (ASF) is a major recent global concern, a devasting incursion of ASF occurred on the island of Hispaniola in 1978, driving all swine to be destroyed. In 1997, the Netherlands had another major disease outbreak in classical swine fever, leading to the destruction of 11 million pigs. In the last decade, the effects of highly pathogenic avian influenza have been felt particularly in the United States in 2014–2015 and 2022 up to the present, with a total of over 130 domestic poultry (chickens and turkeys) affected (Woods, 2016). ASF has left its mark over the last 15 years from the Caucus region of eastern Europe to southeast Asia and China to the island of Hispaniola again (Rousier & Ntsama, 2019; World Health Organization, 2023).

The underpinning of infectious disease in animals is only a superficial indicator of the larger animal health picture. Over the last 70 years, much work, at the global level, has been tied to developing agricultural programs. This has occurred through partnerships between FAO and the UN Development Programme, investments from the World Bank into small shareholder farming and microfinance schemes, and through a great deal of research in international agricultural research centers. These centers span the range of farmed animals including the International Livestock Research Institute, the International Food Policy Research Institute, the International Institute of Tropical Agriculture, and WorldFish (Voth, 2004). The constraints in animal agriculture fall into two groups: biological and socioeconomic. Biological challenges include, certainly, not only diseases and pests, but also feed supply, livestock management, and genetic potential (National Intelligence

Council, 2012). On the socioeconomic side, the challenges include broad markets, unmanaged human population growth, and government policies.

These constraints that agricultural development groups and research are still trying to address mirror some of the common animal health barriers throughout history. Challenges around transboundary informal movements have always been there but have increased even more with the rapid ability to transit the globe. Vaccination programs have had some success (e.g., rinderpest) but continue to face many problems including funding. New virus lineages continue to emerge coupled with failure of timely outbreak reporting and response. While farmers know a great deal, there is still the challenge of limited farmer knowledge and recognition of certain disease processes and concerns. This is coupled with often low-level technical capacity across countries because of limitations and resources allocated to national programs. Finally, standardizing biosecurity, one of the best preventive measures, continues to be poorly executed globally (Hulme, 2020).

GLOBAL ONE HEALTH HISTORY

As evidenced so far in this chapter, historically, animal and human health have often been in silos. This starts at the structure where most countries have a Ministry of Health and a separate Ministry of Agriculture, or their equivalents. Training of the workforce is separated and has been for the last 100 years, with medical schools and veterinary medical schools in separate curricula. Finally, priority has historically been high for human health and lower for animal health, which reflects on initiatives, resources, and interests. However, when one looks deeper, there have been many overlapping points throughout health history often framed as One Medicine approaches. This overlap exists in the clinical space, the public health space, and in research.

Rudolf Virchow famously coined "between animal and human medicine there are no dividing lines – nor should there be. The object is different but the experience obtained constitutes the basis of all medicine" to highlight the interrelationship of human and veterinary medicine in the 19th century (University of Geneva, 2024). The father of modern medicine, Sir William Osler, who studied with Virchow, held joint appointments in Montreal Veterinary College as well as McGill University in the 1800s, again highlighting the substantial overlap in curricula that used to exist between the two spaces (Bliss, 1999). In the United States, Drs. Theobald Smith and F. L. Kilbourne, a physician and veterinarian, worked together on the challenge of cattle tick fever, *Babesia bigemina*, at the turn of the 20th century (Assadian & Stanek, 2002). More recently, the 1996 Nobel Prize in immunology was given jointly to a physician and veterinarian, Drs. Rolf Zinkernagel and Peter Doherty (Greenspan, 2018).

The WHO included veterinary public health from their founding and developed a joint zoonoses list with FAO that started with 150 diseases in 1967 and has grown to over 200 in 2000 (World Health Organization, 2024). Dr. Calvin Schwabe defined the term "One Medicine" and many state boards of health in the United States were involved in animal disease surveillance up to the mid-20th century (Olmstead & Rhode, 2015; Schwabe, 1984). More recently, the One Health approach has been embraced by FAO, OIE (now WOAH), WHO, national agencies, and professional groups across multiple disciplines (World Health Organization, 2021). The public health overlap for veterinarians is part of their professional obligations around disease surveillance, zoonotic disease reduction, and community health education. The joint FAO/WHO report in 1975 defined veterinary public health as the contributions to the physical, mental, and social well-being of humans through an understanding and application of veterinary science (Joint FAO WHO Expert Committee on Veterinary Public Health & FAO, 1975).

As a term, One Health has become part of the lexicon in the last 20 years starting with the Wildlife Conservation Society publishing the 12 Manhattan Principles in 2004. Internationally, in 2010, the FAO, OIE (now WOAH), and WHO collaborate to sign the Tripartite Concept Note around One Health to address risks at the human-animal-ecosystem interface. The UN with the World Bank that same year recommended adoption of One Health approaches as does the European Union. In 2011, the first International One Health Congress was held in Melbourne, Australia, and has been held every other year since then. The U.S. Centers for Disease Control and Prevention created their One Health office in 2009 as the first U.S. agency or department to do so (Ancheta et al., 2021). The quadripartite was signed in 2021 adding on the UN Environmental Programme. Most recently, in 2021, the One Health High-Level Expert Panel (OHHLEP) was assembled from international experts spanning all disciplines of One Health to develop additional definitions, action plans, and initiatives. The OHHLEP definition of One Health is now accepted to be the standard internationally and the one in which this book is focused around:

> One Health is an integrated, unifying approach that aims to sustainably balance and optimize the health of people, animals and ecosystems.
>
> It recognizes the health of humans, domestic and wild animals, plants, and the wider environment (including ecosystems) are closely linked and inter-dependent.
>
> The approach mobilizes multiple sectors, disciplines and communities at varying levels of society to work together to foster well-being and tackle threats to health and ecosystems, while addressing the collective need for clean water, energy and air, safe and nutritious food, taking action on climate change, and contributing to sustainable development
>
> *One Health High-Level Expert Panel (OHHLEP) et al. (2022)*

All of these developments in the last 20 years have been driven by many emerging infectious disease outbreaks like Severe Acute Respiratory Syndrome coronavirus type 1 (SARS-CoV-1), avian influenza, and Ebola. These outbreaks highlighted the fact that 75% of emerging infectious diseases in humans are zoonotic, five new human diseases appear every year, and 80% of agents with potential bioterrorist use are zoonotic (Ghai & Behravesh, 2024). At the same time, the human population has grown tremendously over the last 100 years. It took all of human history up until 1803 to reach 1 billion people, just over 100 years to reach 2 billion people, and since then, the world has added another billion people every 15–20 years with projections that the human population will reach 11 billion toward the end of this century before slowing in growth (United Nations, Department of Economic and Social Affairs, Population Division, 2015). To meet human demands and shifts in nutrition desires, domestic livestock production continues to increase over the last 50 years, growing from 70 million tons in 1961 to 350 million tons in 2018 with the biggest area of growth in Asia (Ritchie et al., 2023). That growth will only increase over the next 30 years with a projected doubling of demand for poultry meat commodity growth (Alexandratos & Bruinsma, 2012). All of this is on the backdrop of two enormous global issues of antimicrobial resistance (see Chapter 6) and climate change (see Chapter 15).

The challenge throughout health history is how to set up effective collaborations and strategies to tackle continual health concerns. As the One Medicine idea has morphed into One Health and even Planetary Health over the last 30 years, there are three strategies often thought about. One is a level-based collaboration where the strategy is looking at one level at a time such as individual medicine or population health or comparative medicine. A second is solution-based collaboration where the health strategy is geared around one solution (e.g., combined immunization programs). A third is third-party-based collaboration where there is some intermediary between human and animal health systems (Yasobant et al., 2019). These different approaches each have their own merit and lead to health professionals' approaches that build on the successes of health history.

By building on this One Health movement, there are many examples in the last decade illustrating the momentum that will carry One Health into the future. Kenya created their Zoonotic Disease Unit in 2011 (Mbabu et al., 2014), while Rwanda created their One Health Strategic Plan in 2015 (Nyatanyi et al., 2017). The Inter-Sectoral Coordination Committee on Zoonoses was started in Mongolia (Batsukh et al., 2012), while Bangladesh has their own One Health professional association (One Health Bangladesh, 2024). Tanzania, in 2018, drafted their National One Health Strategic Plan and the United States has considered over the last 4 years versions of an Advancing Emergency Preparedness through One Health Act. All that has been illustrated, as previously mentioned, at the international stage with the addition of the environment into the One Health Quadripartite formation in 2021. Many lessons have been learned from the collective health history of the world and all of them help inform where One Health goes into the future recognizing that health across humans, animals, and ecosystems is interconnected.

REFERENCES

Alexandratos, N., & Bruinsma, J. (2012). *World Agriculture towards 2030/2050: The 2012 revision* (Working paper No. 12-03). Food and Agriculture Organization. https://www.fao.org/4/ap106e/ap106e.pdf

Ancheta, J., Fadaak, R., Anholt, R. M., Julien, D., Barkema, H. W., & Leslie, M. (2021). The origins and lineage of One Health, Part II. *The Canadian Veterinary Journal = La Revue Veterinaire Canadienne*, *62*(10), 1131–1133.

Arnold, D. (2016). Introduction: Disease, medicine and empire. In *The rise and fall of modern empires, Volume II* (pp. 343–368). Saul Dubow:

Assadian, O., & Stanek, G. (2002). Theobald Smith—The discoverer of ticks as vectors of disease. *Wiener Klinische Wochenschrift*, *114*(13–14), 479–481.

Batsukh, Z., Tsolmon, B., Otgonbaatar, D., Undraa, B., Dolgorkhand, A., & Ariuntuya, O. (2012). One Health in Mongolia. In J. S. Mackenzie, M. Jeggo, P. Daszak, & J. A. Richt (Eds.), *One Health: The human-animal-environment interfaces in emerging infectious diseases* (Vol. 366, pp. 123–137). Springer Berlin Heidelberg.

Bliss, M. (1999). *William Osler: A life in medicine*. Oxford University Press.

Farmer, P., Kim, JY., Kleinman, A., & Basilico, M. (2013). *Reimaging global health: an introduction* (vol. 26). University of California Press.

Fenner, F. (1993). Smallpox: Emergence, global spread, and eradication. *History and Philosophy of the Life Sciences*, *15*(3), 397–420. JSTOR.

Fish, E. J. (2011). Letters to the editor. *Journal of the American Veterinary Medical Association*, *238*(1), 27–28. https://doi.org/10.2460/javma.238.1.27

Food and Agriculture Organization & World Organisation for Animal Health. (2021). *Joint FAO/OIE committee on global rinderpest eradication: Final report*. FAO and OIE. https://afrohun.org/wp-content/uploads/2021/01/OIE-Joint-FAO-OIE-Committee-on-Global-Rinderpest-Eradication.pdf

Fransen, L., Kuiper, E., & Brady, D. (n.d.). *Global Health 2.0: Paving the way for the future*. European Policy Centre. Retrieved June 15, 2024, from https://epc.eu/en/Publications/Global-Health-20-Paving-the-way-for-the-future~591898

Ghai, R., & Behravesh, C. (2024). Zoonoses—The One Health approach. In J. B. Nemhauser & Centers for Disease Control and Prevention (Eds.), *CDC Yellow Book*. Oxford University Press.

Greenspan, N. S. (2018). Peter Doherty, Nobel Laureate: Questions and reflections concerning MHC restriction and other fruits of a life of biomedical erudition. *Pathogens and Immunity*, *3*(2), 224. https://doi.org/10.20411/pai.v3i2.260

Hannaway, C. C. (1977). Veterinary medicine and rural health care in pre-revolutionary France. *Bulletin of the History of Medicine*, *51*(3), 431–447.

Harris, R. (2016). Schweitzer and Africa. *The Historical Journal*, 59(4), 1107–1132.

Hartung, J. (2013). A short history of livestock production. In A. Aland & T. Banhazi (Eds.), *Livestock housing* (pp. 21–34). Brill | Wageningen Academic. https://doi.org/10.3920/978-90-8686-771-4_01

Hulme, P. E. (2020). One Biosecurity: A unified concept to integrate human, animal, plant, and environmental health. *Emerging Topics in Life Sciences*, *4*(5), 539–549. https://doi.org/10.1042/ETLS20200067

Humphreys, M. (1999). Epidemics and history: Disease, power, and imperialism. *Bulletin of the History of Medicine*, *73*(4), 747–748.

Joint FAO WHO Expert Committee on Veterinary Public Health & FAO (Eds.). (1975). *The veterinary contribution to public health practice: Report of a Joint FAO/WHO Expert Committee on Veterinary Public Health*. World Health Organization.

KFF. (2024, May 24). U.S. Global Health Budget Tracker. *KFF*. https://www.kff.org/interactive/u-s-global-health-budget-tracker/

Mbabu, M., Njeru, I., File, S., Osoro, E., Kiambi, S., Bitek, A., Ithondeka, P., Kairu-Wanyoike, S., Sharif, S., Gogstad, E., Gakuya, F., Sandhaus, K., Munyua, P., Montgomery, J., Breiman, R., Rubin, C., & Njenga, K. (2014). Establishing a One Health office in Kenya. *Pan African Medical Journal, 19*. https://doi.org/10.11604/pamj.2014.19.106.4588

McCarthy, M. (2002). A brief history of the World Health Organization. *The Lancet, 360*(9340), 1111–1112. https://doi.org/10.1016/S0140-6736(02)11244-X

Merson, M. H., O'Malley, J., Serwadda, D., & Apisuk, C. (2008). The history and challenge of HIV prevention. *The Lancet, 372*(9637), 475–488. https://doi.org/10.1016/S0140-6736(08)60884-3

Nakatani, H. (2016). Global strategies for the prevention and control of infectious diseases and non-communicable diseases. *Journal of Epidemiology, 26*(4), 171–178. https://doi.org/10.2188/jea.JE20160010

National Intelligence Council. (2012). *Global food security: Emerging technologies to 2040* (NICR 2012–30). Office of the Director of National Intelligence, United States. https://www.dni.gov/files/documents/nic/NICR%202012-30%20Global%20Food_Security%20Emerging%20Technology%20FINAL.pdf

Nyatanyi, T., Wilkes, M., McDermott, H., Nzietchueng, S., Gafarasi, I., Mudakikwa, A., Kinani, J. F., Rukelibuga, J., Omolo, J., Mupfasoni, D., Kabeja, A., Nyamusore, J., Nziza, J., Hakizimana, J. L., Kamugisha, J., Nkunda, R., Kibuuka, R., Rugigana, E., Farmer, P., … Binagwaho, A. (2017). Implementing One Health as an integrated approach to health in Rwanda. *BMJ Global Health, 2*(1), e000121. https://doi.org/10.1136/bmjgh-2016-000121

OECD. (2024). *Official Development Assistance (ODA)*. https://www.oecd.org/dac/financing-sustainable-development/development-finance-standards/official-development-assistance.htm

Olmstead, A. L., & Rhode, P. W. (2015). *Arresting contagion: Science, policy, and conflicts over animal disease control*. Harvard University Press.

One Health Bangladesh. (2024). *One Health Bangladesh – Multidisciplinary collaborative platform working together to Combat the challenges of emerging infect*. https://onehealthbd.org/

One Health High-Level Expert Panel (OHHLEP), Adisasmito, W. B., Almuhairi, S., Behravesh, C. B., Bilivogui, P., Bukachi, S. A., Casas, N., Cediel Becerra, N., Charron, D. F., Chaudhary, A., Ciacci Zanella, J. R., Cunningham, A. A., Dar, O., Debnath, N., Dungu, B., Farag, E., Gao, G. F., Hayman, D. T. S., Khaitsa, M., … Zhou, L. (2022). One Health: A new definition for a sustainable and healthy future. *PLOS Pathogens*, 18(6), e1010537. https://doi.org/10.1371/journal.ppat.1010537

Otte, M. J., Nugent, R., & McLeod, A. (2004). *Transboundary Animal Diseases: Assessment of socio-economic impacts and institutional responses* (Livestock Policy Discussion Paper No. 9). Food and Agriculture Organization. https://www.fao.org/4/ag273e/ag273e.pdf

Packard, RM. (2016). *A History of Global Health: Interventions into the Lives of Other Peoples*. Johns Hopkins University Press.

Polednak, A. P. (1989). Albert Schweitzer and international health. *Journal of Religion and Health*, 28, 323–329.

Ritchie, H., Rosado, P., & Roser, M. (2023). Meat and dairy production. *Our World in Data*. https://ourworldindata.org/meat-production

Roeder, P. L. (2011). Rinderpest: The end of cattle plague. *Preventive Veterinary Medicine, 102*(2), 98–106. https://doi.org/10.1016/j.prevetmed.2011.04.004

Rousier, A., & Ntsama, F. (2019). Panorama 2019-1: Historical data on animal disease outbreaks: The contribution made by the OIE archives. *Bulletin de l'OIE, 2019*(Panorama-1), 1–3. https://doi.org/10.20506/bull.2019.1.2913

Schwabe, C. W. (1984). *Veterinary medicine and human health* (3rd ed). Williams & Wilkins.

Scultetus, A. H., Villavicencio, L. J., Koustova, E., & Rich, N. M. (2006). To heal and to serve: Military medical education throughout the centuries. *Journal of the American College of Surgeons, 202*(6), 1005–1016.

Sealey, A. (2011). Globalizing the 1926 International Sanitary Convention. *Journal of Global History, 6*(3), 431–455. https://doi.org/10.1017/S1740022811000404

Smith, D. F. (2010). 150th anniversary of veterinary education and the veterinary profession in North America. *Journal of Veterinary Medical Education, 37*(4), 317–327. https://doi.org/10.3138/jvme.37.4.317

The World Bank. (2023, April 7). *World Bank open data*. World Bank Open Data. https://data.worldbank.org/indicator/SH.XPD.CHEX.GD.ZS

Tulchinsky, T. H. (2018). John Snow, cholera, the broad street pump; waterborne diseases then and now. *Case Studies in Public Health, 77*. https://doi.org/10.1016/B978-0-12-804571-8.00017-2

United Nations Department of Economic and Social Affairs. (2023). *The sustainable development goals report 2023: Special edition*. United Nations. https://doi.org/10.18356/9789210024914

United Nations, Department of Economic and Social Affairs, Population Division. (2015). *World population prospects: The 2015 revision, key findings and advance tables* (Working Paper No. ESA/P/WP.241). United Nations. https://population.un.org/wpp/publications/files/key_findings_wpp_2015.pdf

University of Geneva. (2024). *Global health at the human-animal-ecosystem interface*. Coursera. https://www.coursera.org/learn/global-health-human-animal-ecosystem

Voth, D. E. (2004). *An overview of international development perspectives in history: Focus on agricultural and rural development*. https://doi.org/10.22004/AG.ECON.15776

Woods, A. (2016). Patterns of animal disease. In M. Jackson (Ed.), *The Routledge history of disease* (p. 18). Routledge, Taylor & Francis Group.

Worboys, M. (2000). The colonial world as mission and mandate: leprosy and empire, 1900-1940. *Osiris*, 15, 207–218.

World Health Organization. (2021, December 1). *Tripartite and UNEP support OHHLEP's definition of "One Health."* https://www.who.int/news/item/01-12-2021-tripartite-and-unep-support-ohhlep-s-definition-of-one-health

World Health Organization. (2023, July 12). *Ongoing avian influenza outbreaks in animals pose risk to humans*. https://www.fao.org/animal-health/news-events/news/detail/ongoing-avian-influenza-outbreaks-in-animals-pose-risk-to-humans/en

World Health Organization. (2024). *Veterinary public health*. https://www.who.int/teams/control-of-neglected-tropical-diseases/interventions/strategies/veterinary-public-health

World Trade Organization. (2024). *The WTO and WOAH*. https://www.wto.org/english/thewto_e/coher_e/wto_oie_e.htm

Yasobant, S., Bruchhausen, W., Saxena, D., & Falkenberg, T. (2019). One health collaboration for a resilient health system in India: Learnings from global initiatives. *One Health, 8*, 100096. https://doi.org/10.1016/j.onehlt.2019.100096

3 International Health, Commerce, and Collaboration

J. Austin Lee, Rachel Cezar-Martinez, and Evelyne de Leeuw

INTRODUCTION

Humans have been interacting with other animals and their environment for all of human history. Each such interaction allows for the risk of disease transmission. Further, these interactions have often been regulated – from instructions for food storage in the cities of the Levant, sanitation in the Indus Valley, to waste disposal in ancient Rome and China. History shows a deep awareness of the ways that the interface between humans, animals, and the environment can lead to interspecies transmission of disease (de Leeuw & Simos, 2017). And in our increasingly mobile and globalized world, such interactions can subsequently lead to the rapid dissemination of infectious diseases. Current evidence suggests that zoonotic diseases (a cross-species transmission of a disease from an animal to a human), which occur at the interface of these domains (human, animal, environment), account for 60% of all infectious diseases that affect humans, and comprise an estimated 75% of novel and modern infectious disease threats (Bond et al., 2013; CDC, 2024). Even for those infections that do not infect non-human animals (such as smallpox), the environment has played a crucial role in driving human behavior and disease spread.

Over the last several centuries, humans have become increasingly mobile and at risk of disease dissemination, from shipping, railway, and now air travel. Societies began to organize disease control measures starting long before the development of germ theory, and there are numerous examples of practices such as quarantine being utilized as recognized methods for controlling leprosy, bubonic plague, and other communicable infections. Through to the present, disease control begins with recognition and detection. Monitoring for illness has taken many forms: from keeping ships docked outside of a port for a waiting period to ensure no sailors carried yellow fever or cholera, to travelers' recent requirements to have a negative pre-departure and after-arrival tests for coronavirus. As human societies have evolved in their organization and socio-political structures, laws and frameworks have been developed to help detect and mitigate disease spread.

 DOI: 10.1201/9781003232223-3

Regulations and institutional frameworks continue to evolve and have become more sophisticated. There is stronger recognition of "glocal" health, that is, the interaction between global change and contexts, and local causes and responses (Kickbusch & de Leeuw, 1999). Human and animal disease control starts at the local level in our homes, streets, and farms, and at the same time requires planetary vigilance and coordination.

HISTORY OF COLLABORATION

The practice of quarantine, with its root meaning coming from "forty days," was a practice of isolation that initially required ships to remain docked outside of a port. This shipping-based practice began in the northern Mediterranean and its institution came about during outbreaks of bubonic plague, as a means of protecting uninfected populations. One can imagine how such restrictions may help with disease control but can be quite onerous. Geopolitics and tit-for-tat quarantines escalated over many years and expanded the scope to include other diseases such as yellow fever and cholera.

Coordinated modern disease control efforts are often recognized as starting in 1851, with the first of a series of over a dozen conferences over the next century, called the International Sanitary Conventions (Lee et al., 2022). These meetings were initially held among European nations, later expanded to include the United States and other countries, to address the effects of cholera outbreaks and resultant burdensome quarantines and border control measures. In the 1870s, a permanent international health agency was first proposed, which eventually led to the founding in Paris of the Office International d'Hygiène Publique (OIHP) in 1907 (Lloyd, 1930). The OIHP was organized around controlling bubonic plague and cholera, with a focus on regulating disparate quarantine laws. A few years earlier, in 1902, the Pan-American Sanitary Bureau was founded in Washington, DC to coordinate health for North, Central, and South America. The importance of such an agency was stressed with the opening of the Panama Canal. These two organizations, the Pan-American Sanitary Bureau and the OIHP, began working together to share knowledge, experience, and information regarding disease control measures.

After World War 1, the League of Nations established the Health Organization of the League of Nations in 1920, which became another multilateral organization concerned with international health (Sealey, 2011). After World War 2, the launch of the United Nations (UN) system led to the formation of the World Health Organization (WHO) and the merger of the OIHP, the Pan-American Sanitary Bureau (reorganized as the regional office of WHO for the Americas, the Pan American Health Organization), and the Health Organization of the League of Nations.

Further, civil aviation has played a crucial and growing role in global commerce over the past several decades. In 2006, the Collaborative Arrangement for the Prevention and Management of Public Health Events in Civil Aviation (CAPSCA) was founded as a multi-organizational collective of nearly two dozen aviation-related organizations that coordinate planning and response to issues in public health and aviation (Collaborative Arrangement for the Prevention and Management of Public Health Events in *Civil Aviation—CAPSCA*, 2024). CAPSCA is managed by the UN's International Civil Aviation Organization (ICAO) and the WHO.

The World Organisation for Animal Health (WOAH), formerly the Office International des Epizooties (OIE), is a non-UN, Paris-based organization that was founded in 1924. WOAH is frequently still referred to by its historical acronym OIE, and was organized in part as a response to the spread of viral rinderpest. The organization is focused on epizootic disease control, epidemics of disease in non-human animals. The main objectives of WOAH are: transparency, scientific information, international solidarity, sanitary safety, promotion of veterinary services, and food safety and animal welfare.

The World Trade Organization (WTO) is another non-UN organization that plays a central role in international commerce and is focused on promoting free trade through agreements. The WTO has its roots in the post-World War 2 multilateral General Agreement on Tariffs and Trade, that was agreed to in 1947. WTO's work on open trade is focused both on physical goods and services, as well as on protections for intellectual property.

Beyond these key organizations, numerous other entities play substantial roles in supporting disease control efforts. These include other UN agencies (the Food and Agriculture Organization, ICAO, the World Bank and International Monetary Fund, the World Tourism Organization, and the UN International Children's Emergency Fund or UNICEF), non-UN multilateral organizations (Gavi – the Vaccine Alliance, and the Global Fund to Fight AIDS, Tuberculosis and Malaria), and non-governmental organizations around the world.

Modern systems for disease surveillance and outbreak detection have proliferated in the internet age. The Program for Monitoring Emerging Diseases (ProMED) is available online (promedmail.org) through the International Society for Infectious Diseases and uses news reports, public health data, and other available information to collate, map, and organize reports of disease outbreaks among humans, other animals, and plants (*Home—ProMED*, 2024). Similarly, the World Animal Health Information System (WAHIS) is available online (wahis.woah.org) from the WOAH with mapping and accessible information on zoonoses and other disease outbreaks in both domestic and wild animals (Carrion & Madoff, 2017; Caceres et al., 2020).

CURRENT INTERNATIONAL HUMAN HEALTH REGULATIONS

As mentioned above, politically driven quarantine decisions for control of yellow fever, cholera, and plague were increasingly commonplace when the International Sanitary Conventions were convened to coordinate a set of agreed upon international rules and policies to approach infectious disease control. Shortly after the founding of the WHO, member states adopted the International Sanitary Regulations in 1951, which initially covered six specific diseases (cholera, plague, yellow fever, as well as relapsing fever, smallpox, and typhoid). This list was consolidated in 1969 (to only cholera, plague, yellow fever) and created the International Health Regulations (IHR), which had the explicit goal of "maximum security against the international spread of disease with a minimum interference with world travel" (World Health Organization, 1983).

At the turn of the 21st century, WHO member states recognized the need to revise the IHR to better meet the needs and challenges of a globalized world encountering novel infectious disease threats. Key efforts were made in the IHR revisions that went into effect in 2005: widen the scope beyond specific diseases, focus on public health capacity building, develop a framework for accountability, and create a mechanism for coordination when outbreaks of concern do occur. The goal of the 2005 IHR is stated as: "to prevent, protect against, control and provide a public health response to the international spread of disease in ways that are commensurate with and restricted to public health risks." One can see the transition from the narrow scope of a few diseases with "maximal security" to a broader public health approach (World Health Organization, 2016).

The new IHR are potentially more rigorous and enforceable, although critics have suggested that a purely "top-down" approach with global codification of risk ignores, at its peril, what happens on the ground (Calain, 2007). A better multi-level stakeholder engagement strategy would allow for greater adaptability and faster response to observed risk. At the same time, we do observe that the IHR system does evolve. It is not just a legal agreement to prevent or limit the global spread of a select group of infectious disease; WHO member states have committed to map and monitor core capacities for disease prevention and health promotion through the IHR lens. Over the past decade of IHR implementation, there had been steady increases in investments in developing and strengthening core capacities; however, starting in 2018, such focus and investments began to decline (International Health *Regulations Capacities*, 2023).

CURRENT ANIMAL HEALTH REGULATIONS

Animal diseases, particularly those originating in a foreign location or transboundary, are an emerging threat to countries around the world due to their potential for significantly impacting animal health, production, and trade. Immediate control and eradication of any identified foreign animal disease (transboundary animal disease) is necessary to protect the long-term success of a country's animal agriculture sector. In 2022 alone, avian influenza in the United States killed over 50 million domestic birds (CDC, 2022).

With global markets expanding, animal production practices intensifying, and infectious pathogens continuing to evolve, these challenges have made a distinct impact on how we develop and implement national animal health systems, including risk-based animal health legislation to safeguard against animal disease incursions. From the WTO to the development of the Agreement on the Application of Sanitary and Phytosanitary Measures (SPS) Agreement and the role of the WOAH along with the role of U.S. Department of Agriculture (USDA)

Animal Plant Health Inspection Services (APHIS), the United States and the international community have made numerous advances over the past decades in creating legislation and initiating standards that prioritize improving global animal health and safeguarding trade.

Throughout human history, the desire to trade has always existed and can greatly benefit individuals and countries. First, trade gives people access to goods they may not otherwise be able to get themselves. Second, producing a commodity that a country can sell to others provides potential revenue streams. As a result, international trade increases the overall wealth of a country. Although trade is mutually beneficial to both trading partners, it is not without its challenges. Trading partners may be interested in protecting their national markets from outside competition. To protect national markets, countries may impose trade barriers, such as tariffs, or import quotas. These practices make imported goods more expensive or limit their supply, creating an advantage for domestically produced goods. It might appear that import restrictions would promote healthy national economies and internal stability; however, these policies often have a negative effect on consumers by increasing the overall price of goods and reducing choice in the marketplace.

In 1995, the WTO was established to address the needs and challenges associated with international trade and market liberalization in the modern era. One of the first orders of business for the group was to address non-tariff trade barriers and agricultural trade. Their collaborative efforts resulted in the development of the SPS Agreement, which outlines basic rules for measures intended to protect human, animal, and plant health. These measures should be based on internationally agreed guidelines and recommendations, as well as be technically justified and based on an assessment of risk. To summarize, the SPS Agreement allows countries to protect themselves from unwanted pests or disease agents that could harm human, animal, or plant health, but in ways that do not unfairly restrict trade.

The transparency provisions in the SPS Agreement are designed to ensure that standards are made known to both domestic and international stakeholders. Because new requirements must be published promptly, and other members can request an explanation of the reasons for new requirements, trading partners experience much less uncertainty. Lack of clarity arises when requirements are not transparent, are not based on risk or scientific evidence, or are applied in a discriminatory manner. By providing risk assessment guidance and a formal means through which disputes can be resolved, the WTO and SPS Agreement encourages members to base requirements on science and to apply them consistently among all trading partners. Trade relationships are more stable when trade policies are established within international standards to ensure that trade flows as smoothly, predictably, and freely as possible. For this reason, the international agricultural trade community benefits when countries adhere to the WTO and SPS Agreement.

Under the WTO/SPS Agreement, WOAH is recognized as the international standard-setting organization for animal health. One of the primary missions of WOAH is to provide international standards for animal health to safeguard the trade of animals and animal products. The main standard-setting documents are: the Terrestrial Animal Health Code, the Manual of Diagnostic Tests and Vaccines for Terrestrial Animals, the Aquatic Animal Health Code, and the Manual of Diagnostic Tests for Aquatic Animals. The animal health standards and recommendations in the codes and the manuals should be used by the veterinary authorities of member countries to provide for early detection, reporting, and control of agents that are pathogenic to animals or humans, and to prevent their transfer via international trade in animals and animal products, while avoiding unjustified sanitary barriers to trade.

While the SPS Agreement is the foundation upon which all agricultural trade policy with respect to food safety and animal and plant health is based, WOAH specifically interprets the key concepts of the SPS Agreement pertaining to animal health into international guidelines for the safe trade of animals and animal products. Recognizing the interconnectedness of the WTO, the SPS Agreement, and WOAH is vital to understanding the safeguards, policies, and practices put into place to maintain global animal health while supporting international trade efforts.

COMPLIANCE AND AUTONOMY

International regulations and joint global efforts are relevant tools in the global health arsenal to address the control of zoonotic and more-than-human disease, as well as other global challenges (e.g., climate change; the North Pacific Plastic Gyre). But by their very nature regulations and global collaboratives are also political, and subject to higher diplomacy and sometimes unfathomable geopolitical mechanisms. The entities that we have described above often seem to be superseded or overpowered by national trade interests, regulated by the WTO and its trade-related aspects of intellectual property rights (TRIPS) mechanism (Gleeson & Labonté, 2020). A profound illustration to this point is provided by Anne-Emanuelle Birn in her unconventional re-telling of the heroic smallpox eradication tale (Birn, 2011). This was not just the victory of joint efforts of all of humanity over a deadly scourge, as often suggested. More than that, smallpox eradication came about through the fortunes of socio-technical networks and the failings of a bipolar (United States vs USSR) world (Loon, 2005). Rules, regulations, and agreements are wonderful, but only through continuous monitoring, social engagement, awareness, and vigilant preparedness can they be maintained and remain relevant.

The limited political capabilities of the IHR were exposed quite dramatically in the coronavirus disease (COVID)-19 pandemic, which was officially declared a "Public Health Event of International Concern" by the WHO. As Gostin, Halabi, and Klock observe: "(the) pandemic has exposed severe limitations in both the International Health Regulations (IHR) and the WHO's institutional capacities. The IHR, fundamentally revised in the aftermath of SARS-CoV-1, exposed gaps in global governance...." The authors go on to highlight how the Severe Acute

Respiratory Syndrome coronavirus (SARS-CoV)-2 outbreak was inhibited by "the WHO's inability to independently verify state reports...; weak WHO compliance mechanisms to enforce IHR obligations and its own recommendations; WHO's lack of power to monitor, investigate, and remediate harmful actions; insufficient transparency and international exchange of scientific data; and lack of global cooperation, especially in the equitable allocation of vaccines and other medical resources." As a result, global deliberations have commenced to formalize a global pandemic preparedness treaty. However, a limitation to specific disease outbreaks of global concern takes away the potential (and need) to frame a broader global health resilience compact. This would, as Nikogosian and Kickbusch have observed, include a scope for planetary health equity in the face of the still looming climate emergency and its ecological and humanitarian unfolding disasters (Nikogosian & Kickbusch, 2021). The driver of a process toward such a global binding agreement should be the UN and not one of its subsidiaries (i.e., the WHO), though. This may create more degrees of freedom (the UN's bread and butter is treaty development, implementation, and monitoring, whereas the WHO only ever has developed one – the Framework Convention for Tobacco Control (FCTC) which some have claimed is toothless) but also greater political and diplomatic complexities (Hoffman et al., 2022).

ROLE OF FUNDING

As with many modern challenges, there seems to be a paucity of funding for sustained growth in public health preparedness and response. The largest contributions to such efforts seem to come from individual country's investments in their own public health agencies and infrastructure as well as international assistance through development assistance agencies such as the United States Agency for International Development (USAID) and the UK Foreign, Commonwealth & Development Office (FCDO). Further, multilateral funding, generally through time-limited projects and grants, from entities like the Global Fund and World Bank, can also provide countries (particularly those with higher economic need) with opportunities for support. After the 2014 Ebola virus outbreak, the World Bank launched the Pandemic Emergency Financing Facility (PEF) to create a donor-funded pool of monies for the world's poorest countries to quickly draw from in a disease outbreak (Pandemic *Emergency Financing Facility*, 2024). After a slower than expected dissemination of funds in response to the COVID-19 pandemic, the PEF was closed and seems it will not be renewed. Many local and international non-governmental organizations play crucial roles in funding preparedness strategies, laboratory training and capacity, and real-time response. Unfortunately, work to raise funds and rally a response effort after an outbreak has already begun to snowball and over-stretch the local response capacity, proving to be incredibly difficult. Preventive public health work can be expensive and time consuming, and often is not prioritized at the local or international level. Even the robust WHO IHR does not bring forth any suggestion, minimum, or accountability mechanism to ensure adequate prevention work is done.

For animal health in a U.S. context, support has spanned many efforts particularly with the importance of sustaining our food resources and continuing trade between countries. Within the United States, a massive funding package, known as the Farm Bill, exists. The farm bill is an omnibus, multiyear law that governs an array of agricultural and food programs (Johnson & Monke, 2024). It provides an opportunity for policymakers to collectively and periodically address agricultural and food issues. In addition to developing and enacting farm legislation, Congress is involved in overseeing its implementation. The farm bill typically is renewed every five years. Since the 1930s, Congress has enacted 18 farm bills always having three original goals – to keep food prices fair for farmers and consumers, ensure an adequate food supply, and protect and sustain the country's vital natural resources. The goals originate in response to the economic and environmental crises of the Great Depression and the Dust Bowl.

As an example, USDA APHIS is awarding $15.8 million to 60 projects led by 38 states, land-grant universities, and industry organizations to enhance the nation's ability to rapidly respond to and control animal disease outbreaks. The funding is awarded through the 2018 Farm Bill's National Animal Disease Preparedness and Response Program (NADPRP).

NADPRP funding is part of an overall strategy to help prevent animal pests and diseases and reduce the spread and impact of potential disease incursions with the goal of protecting and expanding market opportunities for U.S. agricultural products. Farm bill-funded preparation activities are vital to helping safeguard U.S. animal health, which in turn allows American producers to continue to feed the United States and the world.

Nevertheless, even with all the preparations done, once there is an animal disease outbreak, such as the outbreak of highly pathogenic avian influenza (HPAI) in 2022, additional funding may be needed without guarantee of receipt. Therefore, agriculture and animal producers have to do their part to ensure biosecurity measures are in place at their facilities to continue the production of food needed for U.S. and international trade.

COMPETENCY AND APPLICATION

An understanding of the existing parties and regulations that contribute to the well-being of humans, animals, and the environment is crucial to anyone working on health at any level, from a local, national, or global lens. The One Health approach is much broader in scope than any one practitioner, but public health, human medicine, veterinary medicine, and environmental health all require interdisciplinary understanding and collaboration. The future will likely bring novel challenges, and this will require adaptable practitioners. We have seen that ribonucleic acid (RNA) viruses, because of the biological mechanisms of replication, are at high risk of developing mutations and variants. Similarly, influenza virus coinfection can lead to novel strains and new methods of spread and unique patterns of morbidity and mortality. Further, One Health is an issue that has clear political and social dimensions, beyond the obvious clinical and epidemiological ones. Current and future global health practitioners must take a leading role in guiding local, national, and global policy priorities, and generate support for adequate prevention and response funding. As a global health practitioner, you will likely play more than one role in the One Health ecosystem across your career and using a network of professional relationships, and growing your understanding of the interplay between the numerous moving pieces, will prepare you and our planet for a more successful future.

CASE EXAMPLE: COLLABORATIVE DISEASE MONITORING

The African Regional Disease Surveillance Systems Enhancement (REDISSE) project has been working since 2016 to strengthen the human and animal health systems of 16 West and Central African countries. In the past several years, the World Bank has financed hundreds of millions of dollars toward a tripartite goal of: "addressing weaknesses in human and animal health systems for disease surveillance and response, building capacity for effective cross-sectoral and cross-border collaboration, and providing an immediate and effective response to emergencies" (Epidemic *Preparedness and Response*, 2024). Improved and standardized policies, stockpiling of supplies and vaccines, developing a regional network of reference laboratories, and enhanced laboratory and epidemiological training have improved the quality of care and resilience to evolving health threats.

A similar regional collaboration has taken shape between Thailand, Viet Nam, and Indonesia, via the South East Asia Infectious Disease Clinical Research Network (SEAICRN). With funding from the participating country governments, as well as external funding from the United States and the United Kingdom, the SEAICRN network leverages hospitals and research institutions to improve knowledge and the clinical management of infectious disease, in order to guide evidence-based policymaking (SEAICRN - About SEAICRN, 2024). Ultimately, the group is working toward mitigating and preventing future disease outbreaks.

CASE EXAMPLE: EVOLUTION OF YELLOW FEVER CONTROL

The International Sanitary Conventions initially were focused on cholera and plague, but over the coming decades also came to include a focus on yellow fever, and throughout most of the 20th century, yellow fever was a disease of intense international focus (Vanderslott & Marks, 2020). Yellow fever is a mosquito-borne viral infection that can infect both humans and other primates. The United States and other European geopolitical and commercial interests found that ill soldiers and employees were not good for war or business, respectively. As a result, scientific minds focused research on the transmission and control of the disease, with large credit given to the work of Cuban physician Carlos Findlay, as well as Dr. Walter Reed of the United States (Brink, 2016). Their efforts led to vector control programs and are credited with saving the Panama Canal project. An effective attenuated live vaccine was developed in the 1930s, which has been a cornerstone of prevention, while concurrent efforts around vector control have also proved useful. Numerous countries require a yellow fever vaccination certificate for arriving travelers. To this day, International Certificates of Vaccination are printed on yellow paper stock, given the outsized role yellow fever vaccination has played in international immigration policies. While yellow fever and other specific diseases are no longer the sole focus of the IHR, the historic role of this viral illness has shaped policy and practice around the globe.

The mosquito vector that spreads the yellow fever virus, *Aedes aegypti*, is common in many tropical climates, and after originating in Africa has spread, via human movement, to the Americas, Middle East, and is particularly effective at spreading in urban environments. There is also concern that Asia is at risk of future outbreaks given the presence of the *Aedes aegypti* vector, as well as susceptible primates. The shipping of goods and the slave trade allowed for the historical spread of disease, and there is appropriate concern and monitoring for viral spread into previously unaffected regions. While yellow fever has faded from intensive international focus in the 21st century, an understanding of the history of yellow fever can help prevent and prepare to respond to future outbreaks.

CASE EXAMPLE: HUMANS, ANIMALS, AND THE ENVIRONMENT

In the 1980s, there were approximately 3.5 million cases of dracunculiasis (*Dracunculus medinensis*, or Guinea Worm), predominantly in Africa and South East Asia; in comparison in 2016, there were 25 documented cases worldwide (Molyneux & Sankara, 2017). Humans are infected with the parasite when swallowing fresh water contaminated with microscopic copepods, and the life cycle requires female worms to painfully erupt through the host's skin, which subsequently propagates contamination of fresh water sources. While it was previously recognized that other mammals could be infected with dracunculiasis, it was long believed that such infections were either a distinctive strain of the parasite or perhaps only presented in close proximity to human clusters/epidemics (Callaway, 2016). However, in Chad, as the number of human cases declined, it was found that hundreds of dogs were infected with the same parasite as humans, and infected dogs also were found in several other countries.

As the scientific evidence became clearer, there was urgent recognition that this change could not only negatively impact local communities, but could lead to disease spread to countries which were previously certified as free from human infection. Control and eradication efforts had to adapt to this revised understanding of biology and ecology. Subsequently, extensive work has gone into community education around the need for adequate drying and cooking of fish, safe disposal of fish entrails (which could be infected), and rewards for reporting and tethering infected dogs to break transmission cycles (Molyneux & Sankara, 2017). In 2021, there were 14 human cases of dracunculiasis, and through September of 2024, there were only seven recorded cases (Guinea Worm Case Totals, 2024).

REFERENCES

Birn, A.-E. (2011). Small (pox) success? Ciencia & Saude Coletiva, *16*, 591–597. SciELO Brasil.

Bond, K. C., Macfarlane, S. B., Burke, C., Ungchusak, K., & Wibulpolprasert, S. (2013). The Evolution and Expansion of Regional Disease Surveillance Networks and Their Role in Mitigating the Threat of Infectious Disease Outbreaks. *Emerging Health Threats Journal*, *6*. https://doi.org/10.3402/ehtj.v6i0.19913

Brink, S. (2016, August 28). Yellow fever timeline: The history of a long misunderstood disease. *NPR*. https://www.npr.org/sections/goatsandsoda/2016/08/28/491471697/yellow-fever-timeline-the-history-of-a-long-misunderstood-disease

Caceres, P., Tizzani, P., Ntsama, F., & Mora, R. (2020). The World Organisation for Animal Health: Notification of animal diseases. *Revue Scientifique et Technique (International Office of Epizootics)*, *39*(1), 289–297.

Calain, P. (2007). From the field side of the binoculars: A different view on global public health surveillance. *Health Policy and Planning*, *22*(1), 13–20.

Callaway, E. (2016). Dogs thwart end to Guinea worm. *Nature*, *529*(7584), 10–11.

Carrion, M., & Madoff, L. C. (2017). ProMED-mail: 22 years of digital surveillance of emerging infectious diseases. *International Health*, *9*(3), 177–183.

Centers for Disease Control and Prevention (CDC). (2022, November 3). *U.S. approaches record number of avian influenza outbreaks*. Centers for Disease Control and Prevention. https://www.cdc.gov/bird-flu/spotlights/nearing-record-number-avian-influenza.html

Centers for Disease Control and Prevention (CDC). (2024, May 16). *About Zoonotic diseases*. One Health. https://www.cdc.gov/one-health/about/about-zoonotic-diseases.html

Collaborative Arrangement for the Prevention and Management of Public Health Events in Civil Aviation—CAPSCA. (2024). https://www.icao.int/safety/CAPSCA/Pages/About-CAPSCA.aspx

de Leeuw, E., & Simos, J. (2017). Healthy cities: The theory, policy, and practice of value-based urban planning. In *Healthy Cities: The Theory, Policy, and Practice of Value-Based Urban Planning* (p. 515). https://doi.org/10.1007/978-1-4939-6694-3

Epidemic Preparedness and Response. (2020). [Text/HTML]. World Bank. https://www.worldbank.org/en/results/2020/10/12/epidemic-preparedness-and-response

Gleeson, D., & Labonté, R. (2020). *Trade agreements and public health*. Springer.

Guinea Worm Case Totals. (2024, April 16). The Carter Center. https://www.cartercenter.org/health/guinea_worm/case-totals.html

Hoffman, S. J., Baral, P., Rogers Van Katwyk, S., Sritharan, L., Hughsam, M., Randhawa, H., Lin, G., Campbell, S., Campus, B., & Dantas, M. (2022). International treaties have mostly failed to produce their intended effects. *Proceedings of the National Academy of Sciences, 119*(32), e2122854119.

Home—ProMED. (2024). ProMED-Mail. https://promedmail.org/

International Health Regulations Capacities. (2023). Our world in data. https://ourworldindata.org/grapher/ihr-core-capacity-index-sdgs?time=2010..2020

Johnson, R., & Monke, J. (2024). Farm bill primer: What is the farm bill? *Congressional Research Service*. https://crsreports.congress.gov/product/pdf/IF/IF12047#:~:text=The%20farm%20bill%20is%20an,address%20agricultural%20and%20food%20issues.

Kickbusch, I., & de Leeuw, E. (1999). Global public health: Revisiting healthy public policy at the global level. *Health Promotion International, 14*(4), 285–288. Oxford University Press.

Lee, J. A., Kharel, R., Naganathan, S., Karim, N., Aluisio, A. R., & Levine, A. C. (2022). Frameworks for Global Health Collaboration in Pandemic Disease. *RI Med J*, 2013, 33–37.

Lloyd, B. J. (1930). The Pan American Sanitary Bureau. *American Journal of Public Health and the Nations Health, 20*(9), 925–929.

Loon, J. van. (2005). Epidemic space. *Critical Public Health, 15*(1), 39–52.

Molyneux, D., & Sankara, D. P. (2017). Guinea worm eradication: Progress and challenges—Should we beware of the dog? *PLOS Neglected Tropical Diseases, 11*(4), e0005495. https://doi.org/10.1371/journal.pntd.0005495

Nikogosian, H., & Kickbusch, I. (2021). The case for an international pandemic treaty. *BMJ, 372*, n527. https://doi.org/10.1136/bmj.n527

Pandemic Emergency Financing Facility. (2022). [Text/HTML]. World Bank. https://www.worldbank.org/en/topic/pandemics/brief/pandemic-emergency-financing-facility

SEAICRN - About SEAICRN. (2024). http://www.seaicrn.org/info.aspx?pageID=103

Sealey, P. A. (2011). *The League of Nations health organisation and the evolution of transnational public health*. The Ohio State University.

Vanderslott, S., & Marks, T. (2020). Health diplomacy across borders: The case of yellow fever and COVID-19. *Journal of Travel Medicine, 27*, (5), taaa112. https://doi.org/10.1093/jtm/taaa112. Oxford University Press.

World Health Organization. (1983). International Health Regulations (1969). World Health Organization. https://apps.who.int/iris/bitstream/handle/10665/96616/9241580070.pdf?sequence=1&isAllowed=y

World Health Organization. (2016). International Health Regulations (2005). World Health Organization. https://apps.who.int/iris/rest/bitstreams/1031116/retrieve

4 Major Stakeholders and Navigating Engagement

Sarah E. Baum, Catherine Machalaba, and Shannon Mesenhowski

INTRODUCTION

Successfully operationalizing One Health requires a solid understanding of the relevant stakeholders and effective strategies for engaging them. The One Health approach can help to identify potential trade-offs as well as co-benefits to achieving optimal outcomes for healthy humans, animals, and the environment with sustainable development – but particular objectives and inputs to achieve them may vary (One Health High-Level Expert Panel (OHHLEP) et al. 2022). Understanding the systems and stakeholders that relate to a given issue (and its possible solutions) is key to identifying entry points for a One Health approach where greater collaboration and coordination across actors and disciplines could add value and create synergies. The use and practical benefits of stakeholder mapping extend well beyond One Health; it informs stakeholder engagement, which is a key element of leading change in general (see Box 4.1). Policy agenda setting and implementation are influenced by the relative power of different stakeholders, their mandates, and priorities. Outlining these dynamics can ensure policies are being designed and implemented to meaningfully address existing gaps facing citizens and in a way that maximizes scarce resources.

BOX 4.1 MAJOR ELEMENTS OF LEADING CHANGE

1. Envisioning the future state;
2. Engaging key stakeholders and coalition building;
3. Identifying barriers, and breaking down resistance to change; and
4. Institutionalizing change.

Source: Machalaba and Sleeman (2022); Kotter (1996)

Broadly, stakeholders are defined as anyone with a stake in an issue including whether they benefit or are disadvantaged by it, in terms of impacts or interventions. Take zoonotic tuberculosis (zTB) for example. zTB causes bovine tuberculosis in domestic animals, primarily cattle, and wildlife. It is often transmitted

DOI: 10.1201/9781003232223-4

from cattle to humans and is estimated that 10% of tuberculosis cases in human are a result of *Mycobacterium bovis* (World Organisation for Animal Health 2024). Transmission from cattle to humans primarily occurs through the consumption of contaminated meat and dairy products. Airborne transmission and transmission from handling contaminated tissues affect individuals in close contact with animals, such as farmers, veterinarians, or abattoir workers. zTB has economic consequences for communities whose livelihoods depend on animal husbandry. Animals may be less productive, and control efforts may require the removal of contaminated carcasses and impose restrictions on local and international trade (Olea-Popelka et al. 2017).

Multi-lateral organizations, such as the World Organisation for Animal Health (WOAH), the World Health Organization (WHO), the Food and Agriculture Organization (FAO), and the International Union Against Tuberculosis and Lung Disease, guide global prevention and control efforts informed by donors and academic experts and launched their joint *Roadmap for Zoonotic Tuberculosis* in 2017. Public and private veterinarians and human healthcare providers, laboratory specialists, epidemiologists, food safety and wildlife authorities, and local and international non-governmental organizations (NGOs) carry out prevention, surveillance, and response activities at national and sub-national levels (World Health Organization, Food and Agriculture Organization of the United Nations, and World Organisation for Animal Health 2017). Farmers and protected area managers determine vaccination, surveillance, and reporting strategies at a localized level. Individuals and households experience impacts on their health and potentially on their livelihoods, economic, and social status from infection. These may be mediated by protective measures, safety nets, and adjustments to trade regulation. Each sector and stakeholder group may assign priority and resources differently to address zTB. They may routinely coordinate their activities and share data and knowledge across sectors or may default to coordination only in emergency situations. They may also have no coordination mechanisms, instead taking single-sector approaches. Against this backdrop, donors allocate resources targeting specific aspects of zTB prevention, detection, and control, such as awareness building activities, or broad capacities, such as sub-national laboratory capacity, that may deliver shared benefits for zTB, among other health priorities. However, the entry points and investments selected, and their mode of operations, are not necessarily the most efficient or effective – thus the potential utility of a One Health approach.

As evidenced by zTB, but in general across the human-animal-environment interface, the landscape of relevant actors is often comprised of a complex and fragmented network of multi-lateral organizations, donors, regional networks, national governments, implementers, universities, private sector entities, and communities – thus reinforcing the need, and value, for One Health approaches to orient and align strategies for resource efficiencies and effectiveness. Stakeholders hail from a diverse range of sectors and disciplines. The scale of operations varies from individuals and self-organized community decision-making entities, to government ministries that operate within a country's borders, to regional and global multi-lateral organizations, such as WHO or WOAH, that operate across countries. The mandates dictating each entities' objectives and priorities are rarely the same. It may be disease-specific, in which case they typically operate laterally from disease prevention to response. They may also target a specific competency, such as integrated disease surveillance or veterinary laboratory capacity, or they may pursue a combination of both. Regional and country context, including cultural, economic, and historical factors, influence how organizations view the landscape, their role, and their collaborations with others.

Given the proliferation and diversity of stakeholders and disciplines engaged in operationalizing One Health, the scale at which they operate, and their mandates, the One Health ecosystem is constantly evolving. In Africa alone, it has been estimated that there are over 100 One Health initiatives as of 2021 (Fasina et al. 2021). Understanding the stakeholder landscape may help to avoid duplicating resources and fragmentation and identify gaps that have yet to be addressed where additional stakeholders are important to engage or coordination would be useful. The One Health Joint Plan of Action (2021–2026) provides an international guiding framework for coordination between the four quadripartite organizations,[1] itself identifying key stakeholders to engage in activities under its six main action tracks (FAO et al. 2022). Having a broader sense of stakeholders in the landscape can also help build more resilient public health systems and improve health outcomes beyond zoonotic diseases where efficient coordination and resource allocation yield optimal benefits for biodiversity, climate action, livelihoods, food systems, and other sustainable development objectives.

ONE HEALTH LANDSCAPE

Stakeholder mapping is a practical and highly useful tool for navigating this often complex landscape. Beyond identifying relevant stakeholders, mapping exercises shed light on individual mandates and priorities as well as potential synergies between stakeholders to increase engagement and coordination, particularly those that could benefit from information and resource sharing or determine duplicative activities. In a national setting, while entities may generally be known, their topical (and sometimes geographic) jurisdiction may be incorrectly assumed – such as responsibility for monitoring and managing wildlife trade, which could potentially involve dozens of relevant departments, agencies, budgets, or laws, depending on factors such as setting (e.g., air or sea port, urban markets, protected or conserved areas, private reserves, rescue centers, zoos, farms), legal or illegal status, and concern (e.g., sustainable harvest, invasive species, welfare, health, jobs). Laws may be established that are inadvertently conflicting, requiring harmonization. A lack of equity between sectors – especially when it comes to the environmental agencies and their involvement and empowerment – requires clarifying who is equipped to do what, both in terms of mandates and resources such as budget, personnel, and day-to-day procedures. Through stakeholder mapping, gaps and areas of needed clarification can be clearly elucidated in an objective way across stakeholders, promoting common understanding;

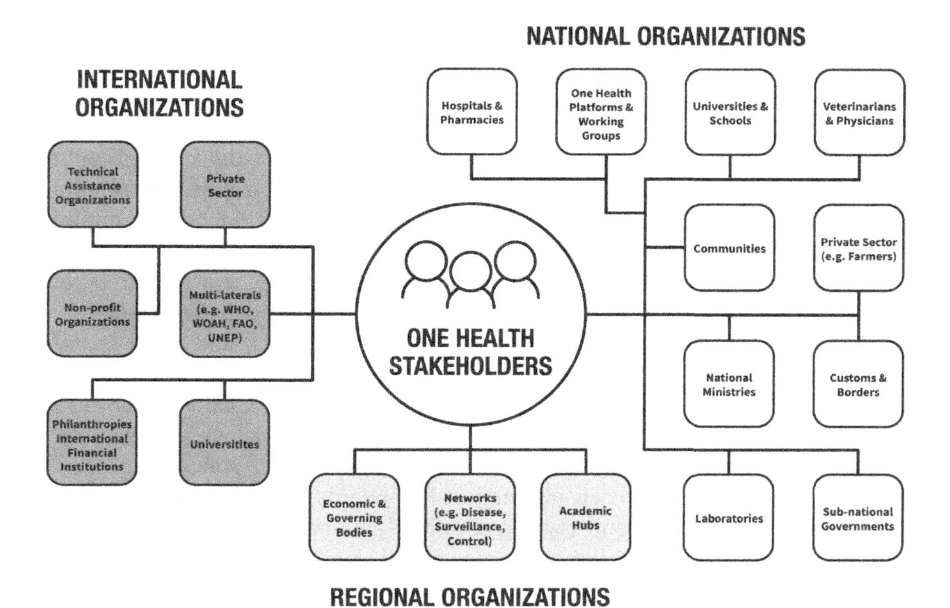

FIGURE 4.1 Examples of potential One Health stakeholders across international, regional, and national levels.

additionally, stated mandates (sometimes on paper only) and assumptions can be ground-truthed further to see where implementation resources and capacity are inadequate to meet expected mandates (Figure 4.1).

The following section provides an overview of key stakeholders engaged in operationalizing One Health at global, regional, and national levels, organized by their roles and responsibilities. Stakeholders may have different roles, such as organizations responsible for standard-setting/regulatory frameworks, assessment, planning, implementation, and expert networks (Berthe et al. 2018). The categories below are used to help show the primary role – and limits – of organizational mandates, though in practice some organizations may serve multiple roles. Similarly, it is important to note that this overview is intended for general orientation; the specific stakeholders and their relevance will always depend on the particular context, which may vary country-country or industry-industry, shaped by factors such as the organization of sectoral departments and epidemiological and ecological factors (e.g., transmission routes and available prevention and control measures).

Standard Setting

International and regional actors develop guidance and set standards for human, animal, and environmental health, contributing to addressing epidemic and pandemic threats and operationalizing One Health at a global level. Frameworks can extend to guiding One Health coordination among multi-lateral partners as evidenced by the 2022 Quadripartite Memorandum of Understanding framework that strengthens cooperation among WHO, WOAH, FAO, and United Nations Environment Programme (UNEP) in tackling One Health challenges. There are also informal frameworks and guidelines, such as the World Bank One Health Operational Framework and Africa Centres for Disease Control and Prevention Framework for One Health Practice, that provide practical guidance for implementing One Health approaches.

In most cases, existing global standards are sector-specific and seek to build country-level capacities (e.g., WHO's International Health Regulations, WOAH's Terrestrial and Aquatic Animal Health Codes and Manuals, and multi-lateral environment agreements, such as the Convention on Biological Diversity and the United Nations Framework Convention on Climate Change). Chapter 3 outlined additional guidance on existing frameworks at the animal-human-environment interface. Implementation is intended at the country level, typically allowing for flexibility in ways to reach international standards. Few standards encompass capacities across the entirety of the animal-human-environment interface. International and regional actors are responsible for harmonizing existing frameworks to ensure guidance is not conflicting and that country obligations for implementation and reporting are not duplicative, especially as sectoral silos break down and coordination increases. However, as seen with the many objectives in the Sustainable Development Goals, pursuing individual goals can inadvertently have consequences for others – thus the need for a coordinated approach.

National policymakers, such as Ministries of Agriculture, Health, and Environment, ensure national policies align with global strategies and oversee the implementation of strategies in their country. Where they exist, national One Health strategies and action plans contextualize global guidance and outline key priorities and responsibilities for each sector. At times, local and national mandates may align or conflict with international frameworks and standards. International and regional actors monitor the country's progress toward meeting their agreed upon obligations. For instance, WHO's Joint External Evaluation is a voluntary assessment of progress toward implementing the International Health Regulations. WOAH's Performance of Veterinary Services Pathway identifies strengths and weaknesses in national veterinary services.

Funding and Resource Allocation

When engaging in international One Health, it is important to consider how funding influences the focus and prioritization of the effort. Funding that supports One Health interventions, program design, and strategic frameworks internationally comes from a variety of sources, each with its own objectives and subsequent approaches. This presents opportunities for coordination and synergistic efforts as well as challenges from competing priorities and silos. There is also an abundance of funding which is not categorized as One Health funding specifically, but is intended to address critical One Health topics such as antimicrobial resistance (AMR), zoonotic diseases, and neglected tropical diseases, among others. When considering this type of funding, there is a substantial increase in the number and type of funders, significantly complicating coordination and prioritization challenges.

Financing is an important component of operationalizing One Health, particularly with regard to equity among sectors and in shifting from primarily relying on a reactive response to improved prevention and preparedness to minimize the frequency and impact of disease events. Key funders of One Health can be grouped into five categories: national governments, bilateral governments, private sector, philanthropy, and international investment banks.

National (and local) governments mobilize resources for their country, partially in the form of domestic financing. In many low and middle-income countries (LMICs), available funding is often limited, given limited tax revenue capture. As a result, resources for government operations may be scarce – though can still go a significant way in building and sustaining capacity. In addition to domestic financing, governments can shape investments from international donors – whether in the form of loans, grants, credits, or in-kind technical assistance. This can take on conventional lending models, such as targeted support to one agency for vertical programs (e.g., on malaria or maternal and childhood health), or increasingly, to strengthen systems as a whole to address multiple hazards. Coordination between agencies in One Health fashion can thus guide a common resource mobilization agenda; as such, national action planning exercises (e.g., National Action Plans for Health Security) are encouraged to include inputs from multiple sectors. In some countries, innovative financing mechanisms are being developed to support One Health coordination itself, including flexible funding pots that can be directed to agencies as needed (e.g., Tripartite Zoonoses Guide).

Bilateral governments provide resources to support One Health programs, such as the efforts of Germany's BMZ (Federal Ministry for Economic Cooperation and Development 2022) and the U.S. Government through CDC (Centers for Disease Control and Prevention 2022) Agency for International Development (USAID 2023), and USDA (U.S. Department of Agriculture 2024). In these cases, a country government provides the resources to support One Health initiatives, platforms, and programs. These resources are often designed to support partner country governments, NGOs, and international organizations (such as WOAH, WHO, FAO) to build One Health capacity, tools, and expertise under the umbrella of international development. Bilateral funding can come in the form of grants, contracts, or loans and much of this funding falls under the framework for Official Development Assistance (ODA) funding (OECD 2024).

The private sector is another significant funder of One Health-relevant activities and agendas, and it presents a distinct potential for sustainability with respect to finance and innovation. This was highlighted during the coronavirus disease (COVID)-19 pandemic where collaboration with the private sector was critical for expediency (Runde, Metzger, and Abdullah 2020). The type of private sector currently engaged in One Health spans from global pharmaceutical companies such as Merck ("One Health" 2024), to Zoetis, to more local private sector industry associations engaging in Public Private Partnerships (PPPs) to advance One Health implementation (World Organisation for Animal health 2024). Financing from the private sector comes through grants, loans, or in-kind support, and is underpinned by the common objective that issues relevant to health drive their engagement across human, animal, and environmental dimensions, and is aligned to their business model. It is important to keep in mind, as stakeholder mapping can inform us, that there may be financing opportunities in other sectors beyond health entities that can help to achieve intended effects. For example, we could imagine how risk reduction investments taken by logging, mining, or tourism companies could help to reduce practices that increase pathogen spillover risk.

Philanthropies are private entities that provide funding to international One Health programs to fill gaps or supplement the resources needed to advance One Health objectives and activities, and include organizations such as the Wellcome Trust (Viergever and Hendriks 2016), the Dorothy A. Johnson Center (Dorothy A. Johnson Center for Philanthropy 2022; DeLeeuw 2022), and the Bill & Melinda Gates Foundation, among many others. Through their resources, a strategic approach to One Health which is tailored to the institution's mission is brought to global, national, and local efforts, often intended to supplement or complement existing funding which may be more constrained.

The fifth category of One Health financers is international financial institutions (IFIs). These are institutions which sponsor loans, grants, equity investments, and blended finance instruments to achieve policy and implementation goals around One Health. These institutions such as the World Bank Group and the International Finance Corporation (IFC) also offer technical assistance and expertise in thematic areas such as One Health, to support the design and implementation of work being financed through their partnerships. However, it should be recognized that the overwhelming amount of financing to countries follows single sector-led programs, and thus is not inherently developed with a One Health mindset during the design, implementation, or evaluation process. As a result, to date there are often missed opportunities for relevant entry points and synergies in investments and their results frameworks (The World Bank 2021).

IMPLEMENTATION

Implementation of One Health initiatives can take many forms, potentially involving and leveraging international, regional, national, and local stakeholders. Again, the specific actors of greatest relevance for implementation will depend on the target objectives and context.

At a national level, implementation (and implementation resourcing) is frequently guided by capacity assessment and action planning – the latter of which is also often aligned to global priorities and commitments (e.g., tackling AMR). Key action planning processes relevant to One Health coordination are shown in Table 4.1.

Toward broad national plans, specific programs and targets (e.g., rabies elimination) are often the focus of day-to-day implementation. In a One Health lens,

TABLE 4.1
Key Action Planning Processes Relevant to One Health Coordination

Plan Name	Scope (Typical Topics and Sectors)
National Action Plan for Health Security	Epidemic prevention, detection, and response
National Action Plan Against AMR	Curbing the emergence and spread of AMR
National Biodiversity Strategy and Action Plan	Conservation and biodiversity protection
National Adaptation Plan	Climate change mitigation and adaptation
Health-National Adaptation Plan	Health-relevant risks and vulnerabilities related to climate change

standard activities across all countries typically include some level of surveillance, laboratory activities, reporting, and disease management strategies. However, these may be unevenly pursued across pathogens, species, settings, or geographic locations. Practical constraints or enablers can play a role in whether implementation occurs; for example, if electricity is not reliably available in rural areas, routine deployment of cold-stored vaccines is unlikely to be viable in the long run (implementation strategies, of course, may find suitable workarounds such as portable coolers that can transport vaccines over large distances, or working with energy providers to reinforce the electrical grid in healthcare facilities). Again, this speaks to the importance of liaising with the right stakeholders to fully understand the issues and refine implementation approaches as needed. The *plan-do-study-act* cycle is often used as a simple framing for continuous improvement of programs (Solaimani, Haghighi Talab, and Van Der Rhee 2019); a stakeholder analysis could be completed at each step of the process.

Several functions are well understood as contributing to disease prevention, detection, and response, among them laboratory screening for disease diagnosis (e.g., in an active outbreak) or general monitoring of trends (e.g., to track antimicrobial or insecticide resistance). In some cases, institutions or countries may not have the capacity or resources to perform necessary functions or may seek external technical or resource support for confirmation or quality assurance services. International human and animal reference laboratories, such as those through WHO, FAO, and WOAH, provide this on a global scale; academic and research entities, in some cases non-governmental, public, or private sector, may also provide key services for detection, diagnosis, or research (e.g., the Institut Pasteur networks).

These functions may not be inherently operating with (nor necessarily require) a One Health scope; however, the interpretation and use of information – and even the selection of screening panels – can be informed by One Health coordination. Additionally, considering the broader role of these functions in overall systems provides agility that can deliver the value addition from One Health approaches. A clear example of this was the surge capacity provided by veterinary laboratories in many parts of the world to support COVID-19 detection and response efforts.

Disease surveillance is a major focus of implementation efforts where stakeholder engagement should be built into the process. Information management and flow across sectors is crucial to harness and use information input in a timely manner. For example, detection of a disease event in wild animals may signal a public and/or domestic animal health threat, a concern for conservation of wildlife and ecosystems, or just be a routine illness in one species that has no significant impact. Without effective and efficient information flows, the pieces of the puzzle may not be connected for full understanding, which may result in inconsistent messaging and a range of detrimental and avoidable impacts (e.g., delayed human or animal health treatment, inappropriate killing of wildlife out of fear, loss in tourism revenue)

Because there are many potential stakeholders of relevance to a topic, mapping may identify different levels of groups – which may offer efficiencies for implementation. For example, while veterinary services and agricultural extension services may target individual farms in the management of specific outbreaks, an industry association (such a wool grower cooperative) may be able to weigh in on and help make decisions on behalf of producers with their best interest in mind like incentives to encourage vaccination and the optimal distribution of safety net resources. Similarly, information dissemination channels to reach target stakeholders should consider how they acquire trusted information and information flow needs – and whether this is best served by radio, television, social media, cartoons, mobile phone messaging, or discussions under the mango tree in a village with local leaders.

One Health Coordination Platforms and other multi-sectoral coordination mechanisms, including working groups on specific diseases or objectives, are thus important for bringing together information. With increased adoption of digital technologies and a goal of interoperable systems, mechanisms are being developed to issue alerts and disseminate information; these should consider both who needs to know as well as how they can use it to promote action where necessary.

Technical Assistance from Expert and Advisory Groups

We can imagine how a wide range of entities could provide technical assistance across the scope of prevention, detection, response, and recovery for a given disease or health threat. At the global level, several expert networks and advisory groups have been formed to provide technical advice that helps guide strategic plans and accordingly technical assistance; their terms of reference vary. A brief description is provided of several groups with One Health relevance supporting prioritization, coordination, and collaboration:

- GOARN: The Global Outbreak Alert and Response Network, formed in 2000, brings together entities and individuals willing to support internationally coordinated responses in the event of infectious disease outbreaks.

- OFFLU: WOAH/FAO of the United Nations expert group on animal influenzas, which tracks viral diversity and supports information exchange with WHO for the annual influenza vaccine virus selection (Pavade 2020).
- OHHLEP: The One Health High-Level Expert Panel serves as a multi-disciplinary advisory group to the Quadripartite and their joint work to address issues at the animal-human-environment interface. Formed in 2021, its first task was to develop a harmonized working definition of One Health; it also assisted in the development of a OH Joint Plan of Action.
- IUCN Specialist Groups: The International Union for the Conservation of Nature maintains expert networks for specific taxonomic and thematic groups, among them a Wildlife Health Specialist Group, overseeing the Red List of Threatened Species assessments and the development of conservation action plans.

It should be noted that, at the intergovernmental level, many expert working groups involve experts in a voluntary capacity, thus facing practical limitations on time commitments and resources for on-the-ground activities.

Initiatives, such as the Global Health Security Agenda and international treaties (e.g., the Paris Agreement), help to prioritize and mobilize technical assistance across countries and, in some cases, accompanying financial assistance around common objectives and action packages, helping to bring together stakeholders from governments, donor entities, the private sector, and civil society toward agreed commitments. Expert networks can play a role in helping to assemble the state of knowledge and issue evidence-based recommendations to help direct resources.

In addition to expert networks, technical assistance is typically provided by international organizations, such as the quadripartite organizations (WHO, WOAH, FAO, UNEP), and the World Bank; regional entities; and NGOs, in addition to donor agencies. Technical assistance can take many forms across the One Health spectrum depending on the needs of the partners requesting and receiving technical assistance, the comparative advantage of technical assistance-providing institutions, such as program implementation or framework development, and where technical assistance is tied to specific funding, the grant, or contract stipulations. Technical assistance may look like strategic planning to identify and prioritize needs and gaps in programming, coordination of stakeholders and programming across sectors, knowledge sharing in the design of programs, training and capacity building, assessment of systems and projects, and supporting the design of and advocacy for One Health policies.

Technical assistance with prioritization and coordination supports identifying and aligning priorities of actors at global, regional, and national levels to mobilize funding and efforts. These are most often facilitated by international and multi-lateral agencies who have experience convening a wide range of stakeholders. Efforts like the WHO's Research and Development (R&D) Blueprint, for instance, steward global strategic coordination for epidemic preparedness and response. This has focused on fast-tracking effective health technologies. In 2016, the WHO convened a forum of technical advisors with expertise from the medical, scientific, and regulatory fields to establish a list of priority diseases with epidemic potential, including Ebola/Marburg, Lassa Fever, and Nipah. Building from this framework, WHO produces R&D Roadmaps for each priority disease that outline priorities for the accelerated development, manufacturing, and distribution of tests, treatments, and vaccines to better guide rapid and effective response measures should a future epidemic occur (World Health Organization 2024). However, without the necessary One Health lens, one can see how these efforts are quickly overwhelmingly focused on detection methods and medical countermeasures in humans. These efforts sometimes come at the cost of missing animal and environmental links to spillover risk and opportunities for human behavior change, effective risk communication, sentinel detection, and prevention at source.

In addition to honing global guidance and frameworks, international actors also support workshops and expert convenings that bring local partners to the table. For example, the U.S. CDC supports and facilitates One Health Zoonotic Disease Prioritization (OHZDP) workshops with country partners and regional governing bodies. An aim of these efforts is to identify national and sub-national One Health stakeholders and understand their roles and responsibilities. Similar to the R&D Blueprint, OHZDP workshops begin by identifying a set of priority zoonotic diseases, which provides the bedrock for developing action plans (Centers for Disease Control and Prevention 2022). International partners may also serve on country One Health platforms and specific expert working groups as advisory members. Technical assistance may also be provided to advance progress on targets set by international actors, such as meeting the International Health Regulations, implementing feedback from Joint External Evaluations, and operationalizing National Action Plans. In addition to the role of technical assistance in creating or optimizing hard infrastructure (e.g., in the construction or use of veterinary laboratories), the value of technical assistance in strengthening "soft skills" (e.g., for effective coordination, communication, and collaboration) is increasingly recognized.

More traditional conceptions of technical assistance as a top-down exchange of expertise from donor or international agency to build the capacity of grantees has shifted in recent years. While not specific to One Health, recent efforts have called into question the effectiveness of technical assistance for global health and called for further evaluation and new paradigms of co-creation between partners (Kanagat et al. 2021). Greater emphasis is being made to align assistance with national policy priorities, to be responsive to local contexts, and ensure the engagement of communities, rather than being driven by the policy agendas of multi-laterals and donors alone. With the aim of building sustainability, this shift has spurred greater focus on ensuring partner buy-in and ownership of decision-making and programs. It has also emphasized the need for technical assistance to contribute to strengthening systems, rather than short-term assistance subject to funding periods and more siloed oversight on specific projects. This reinforces that framework and action plans that are developed through the coordination and priority-setting processes need continual monitoring and evaluation to ensure that they are implemented effectively.

Training and Research

The One Health workforce encompasses a diverse range of disciplines (e.g., animal health, human health, public health, epidemiology, entomology, environmental, wildlife experts, nursing, ecology) that contribute to One Health implementation and research. One Health research can identify sources of risk and effective strategies for disease prevention, control, and response; training ensures that professionals have the skills needed to implement programs effectively.

Pre- and in-service training are the main components of the One Health training pipeline. Pre-service training pipeline includes academic and professional institutions that prepare professionals prior to entering the workforce or a new field. Academic institutions and regional networks around the world offer One Health-specific training opportunities and, in doing so, empower the next generation of One Health champions. These range from One Health modules, such as in epidemiology, risk analysis, or infectious disease management, all the way to full university degree programs. These are often at the master's level. Several regional networks, such as the Africa One Health University Network (AFROHUN) and Southeast Asia One Health University Network (SEAOHUN), bring together regional universities to develop One Health technical competencies, curriculums, and research capacity.

In-service training, or continuous training, ensures that professionals already in the workforce are practicing according to the most up-to-date guidelines and tools. National governments often have their own required in-service education programs, particularly for veterinarians and physicians. These may also be mandated by international authorities, such as WHO and WOAH, to ensure guidelines are harmonious across countries. More and more efforts are being made to integrate One Health topics, such as AMR, into in-service trainings.

One Health competencies are also being actively integrated into key public health capacity-building initiatives. Field Epidemiology Training Programs (FETPs) have played an instrumental role in equipping professionals to engage in disease prevention, surveillance, and response. One Health-focused competencies related to animal surveillance and investigation, zoonoses, and transboundary animal diseases are being integrated into existing FETPs (Seffren et al. 2022). Where formal continuous education may be less accessible, massive open online courses (MOOCs) or continuing education workshops, such as those offered by the World Bank on operationalizing One Health, are available. Additional detail on training and One Health is covered later in this book.

Academic institutions and regional networks have been instrumental in supplying and generating demand for One Health research. Research themes span understanding the dynamics of disease spillover and transmission, identifying sources of risk or risky behavior, all the way to designing cost-effective strategies for disease prevention, detection, and response. This includes a host of scientific experts with backgrounds in ecology, virology, animal health, human health epidemiology, pathology, environmental health, and economics.

Several international and/or regional research consortiums exist. Consortiums will often include universities, NGOs, independent research networks, and government research centers. For instance, the One Health European Joint Programme (OHEJP) collaborates with veterinary, food, and human health laboratories; government agencies; and research institutions across the continent. There are also independent research stakeholders, such as the International Livestock Research Institute, which is a research center within the Consultative Group on International Agricultural Research (CGIAR) with a focus on food security in East and Southern Africa. In addition to carrying out research projects, stakeholders also advance research capacity. The Afrique One-African Science Partnership for Intervention Research Excellence (ASPIRE) was one project that aimed to build research capacity related to endemic zoonotic diseases in partnership with nine academic institutions and four partner organizations in Africa.

Across many fields, donors, guided by their funding priorities, play a key role in deciding what research is conducted, where, and by whom. This influences who sets the research agenda. Research on One Health is no exception where, in many cases, research is led by institutions in the Global North to conduct research in the Global South. Achieving effective collaborations whether in the remit of research, capacity building, or implementation requires that those collaborations are equitable. It also requires efforts to engage stakeholders across the spectrum, such as environmental experts, that at times are left out.

CASE STUDIES

ANTIMICROBIAL RESISTANCE

AMR, or drug-resistant infections, is identified as a serious and growing threat to health and economies (Jonas et al. 2017). There are many possible sources and sectors contributing to the development and spread of AMR; thus, a One Health approach is crucial to identify and prioritize risks and necessary actions. A Global Action Plan was launched in 2015, with five overarching objectives (World Health Organization 2015):

- to improve awareness and understanding of AMR through effective communication, education, and training;
- to strengthen the knowledge and evidence base through surveillance and research;
- to reduce the incidence of infection through effective sanitation, hygiene, and infection prevention measures;
- to optimize the use of antimicrobial medicines in human and animal health; and

- to develop the economic case for sustainable investment that takes into account the needs of all countries and to increase investment in new medicines, diagnostic tools, vaccines, and other interventions.

A One Health Global Leaders Group on AMR has been formed, and at a country-level, multi-sectoral coordination mechanisms are increasingly being developed to implement National Action Plans on AMR. There are many stakeholders of importance for tackling AMR, requiring a whole of society effort: depending on the context, this might include human, animal, and environmental health laboratories to support effective and timely diagnostics, residue detection, and antimicrobial susceptibility testing; law enforcement to tackle the movement, sale, and use of counterfeit and substandard medicines; human and animal medicine prescribers; pharmacists; farmers and human patients administering or taking medicines; environmental and sanitation managers (e.g., at waste and waste water treatment facilities); risk communicators; among others.

RABIES

Rabies causes almost 60,000 human deaths annually. Most deaths occur in Africa and Asia as a result of transmission from dogs (World Health Organization 2023). The tools for preventing and controlling rabies are available. Chronic implementation gaps at all levels demonstrate how engaging stakeholders through a One Health lens can help address existing fragmentation in surveillance, prevention, and control.

In 2012, the WHO, WOAH, FAO, and the Global Alliance for Rabies Control (GARC) launched the United Against Rabies collaboration. The collaboration developed and oversees progress toward implementing the Zero by 30 strategy, which aims to eliminate human deaths from dog-mediated rabies by 2030 via improvements in access to post-exposure prophylaxis, dog vaccination, community awareness of outbreaks, and rabies surveillance and reporting.

National governments may have their own rabies elimination strategy that is aligned with the Zero by 30 strategy. In some cases, these are designed in collaboration with WHO, WOAH, and GARC, among others. Where they exist, national rabies technical working groups reflect stakeholders from the Ministry of Agriculture and Ministry of Health, national universities, academia, NGOs, and WHO, FAO, and WOAH and provide expert insight into programmatic implementation. WHO, FAO, and WOAH support countries in conducting rabies surveillance and national veterinary laboratories in detecting and confirming animal cases.

Regional networks, such as GARC's Pan-African Rabies Control Network (PARACON) in sub-Saharan Africa, support countries in developing national strategies and ensure standardization. GARC developed and supports countries in evaluating their performance via the Stepwise Approach toward Rabies Elimination (SARE) and provides overall advocacy support to keep rabies on national and global human and animal health agendas to overcome its neglected status. National rabies prevention and control strategies are not always coherent or standardized across regions. Vaccination may be required in one country, but not in a neighboring country. This can undermine even the most successful within-country efforts where borders are porous and cross-border animal movement is common.

The Ministries of Agriculture (which often house Veterinary Services) and Health are typically tasked with operationalizing national strategies. Yet, there is often debate on who has jurisdiction over rabies prevention and control (Tiwari, Gogoi-Tiwari, and Robertson 2021). More often than not, associated budgets, sharing of surveillance data, and human and animal vaccination are siloed between Veterinary Services and the Ministry of Health. National budgets allocated for rabies control are often insufficient to cover the necessary activities and countries rely on external donors for access to human and canine vaccines and rabies immunoglobulin for post-exposure prophylaxis. Domestic and international financing via governments and donors, such as Gavi and the private sector, helps to finance the provision of the rabies post-exposure vaccine as well as mass dog vaccination campaigns, often during World Rabies Day celebrations or via private pet owner expenditure in countries where animal rabies vaccinations are required and enforced by law.

At a sub-national level, district veterinary and health services, primary healthcare providers, animal and community health workers, and private veterinarians and physicians engage in rabies prevention and control. Depending on the context, they work closely with members of community, including community leaders, livestock keepers and pastoralists, physicians, and students, on building awareness through risk community, vaccination, and responding to suspected cases. Local and international NGOs, such as Veterinarians Sans Frontiers or the International Livestock Research Institute's One Health Center in Africa, play a key role in supporting animal vaccination and awareness building campaigns; capacity building of health extension workers, veterinarians, and para-veterinarians; and dog population management. These efforts are often stymied by issues related to vaccine procurement and storage, and specimen transport. Stockouts of testing supplies like reagents are common. Healthcare facilities are not adequately staffed, and veterinary services are not available. These issues are most acute in rural areas, in which not only availability but also accessibility of vaccine and testing resources are out of reach. Post-exposure prophylaxis (PEP) can be cost-prohibitive, especially for the most vulnerable, including children.

Depending on context, the landscape of stakeholders engaged in rabies surveillance, prevention, and control may be highly complex and fragmented; in some cases, efforts are ad hoc, and priorities vary across sector and stakeholder. At the local level, stakeholder mapping can target synergies with existing human and animal initiatives that target similar populations or geographies. These could be vaccination campaigns targeted to rural areas (e.g., polio, peste des petits campaigns) or outreach efforts by community health workers in which rabies vaccination or risk communication could be integrated. In addition to human and animal health, we could think about a wider group of stakeholders that could promote success of rabies elimination campaigns, for example, through education initiatives that reduce risks in children, who are especially susceptible to dog bites.

Mapping of domestic stakeholders and inputs is also necessary to understand how existing budget allocations support the spectrum of rabies prevention and control. In addition to clarifying key gaps or duplications in service provision, it can also ensure that domestic funding streams are not inadvertently reinforcing siloed efforts and inequities across sectors. Understanding the mandates and priorities of external donors, on the other hand, may elucidate whether and how integrated services could be provided or how existing resources could support building infrastructure for more sustained, rather than ad hoc efforts. At a regional level, stakeholder mapping can highlight comparative advantages in terms of producing vaccines and reagents for testing that could alleviate chronic shortages.

ENGAGING KEY STAKEHOLDERS IN OPERATIONALIZING ONE HEALTH

Key Opportunities and Challenges to Engaging Stakeholders and Building Collaborations (across Levels/at Each Level)

Clearly, understanding the stakeholder landscape and effective stakeholder engagement are important – and necessary – for operationalizing One Health. That said, there is no one perfect way to conduct stakeholder mapping and foster collaboration; context must always be taken into account. Similarly, it is fair to expect that sectoral biases – as well as biases from past experiences, which may be negative in some cases – will affect which stakeholders are more readily brought to the table. This has been apparent, for example, with the relatively limited inclusion of the environment sector in One Health efforts thus far (in some cases because several agencies have an environment-focused remit, spanning wildlife, climate, soil, water, and more, resulting in fragmented mandates and lack of clarity on who is doing what and key entry points, which stakeholder mapping can help clarify). Being aware of these biases, and providing a neutral, objective space and process can help to build trust over time. National One Health

Coordination Platforms and the Quadripartite agreement are an example of how countries and global institutions are fostering mechanisms for routine coordination, collaboration, communication, and capacity strengthening (referred to as "the 4Cs").

It is important to note that genuine stakeholder engagement will recognize that different entities may have different, and sometimes competing, priorities. Non-traditional partnerships may be formed between stakeholders to achieve a common goal even while their core objectives or incentives are not fundamentally aligning on other issues (e.g., for-profit company that benefits from the sale of a product harmful to public health or the environment).

In the pursuit of specific objectives (e.g., strengthening health security, rabies elimination), stakeholder engagement should consider broader aspects of equity and inclusion. The inclusive definition of One Health, presented in Chapter 2, is accompanied by a series of underlying principles, which together reinforce the importance of minimizing trade-offs and maximizing co-benefits to different stakeholders (see Box 4.2) (One Health High-Level Expert Panel [OHHLEP] et al. 2022). A broad view of stakeholders is thus needed to determine which stakeholders are impacted positively and negatively, including marginalized populations which may not always be afforded the same voice in decisions. The granting of legal status to rivers, as seen in Bangladesh, India, and New Zealand, is also expanding what may be formally designated and represented as a stakeholder (Anderson et al. 2019).

BOX 4.2 KEY UNDERLYING PRINCIPLES ACCOMPANYING THE ONE HEALTH DEFINITION

1. equity between sectors and disciplines;
2. sociopolitical and multicultural parity (the doctrine that all people are equal and deserve equal rights and opportunities) and inclusion and engagement of communities and marginalized voices;
3. socioecological equilibrium that seeks a harmonious balance between human-animal-environment interaction and acknowledging the importance of biodiversity, access to sufficient natural space and resources, and the intrinsic value of all living things within the ecosystem;
4. stewardship and the responsibility of humans to change behavior and adopt sustainable solutions that recognize the importance of animal welfare and the integrity of the whole ecosystem, thus securing the well-being of current and future generations; and
5. transdisciplinarity and multi-sectoral collaboration, which includes all relevant disciplines, both modern and traditional forms of knowledge and a broad, representative array of perspectives.

Source: One Health High-Level Expert Panel (OHHLEP) et al. (2022)

With this in mind, stakeholder engagement may result in a shift from a course of action initially proposed to something that is considered more acceptable and viable among stakeholders. Of course, what works for one population may be entirely unacceptable to another, including due to cultural, economic, food security, or other factors. Identifying barriers and incentives to changes sought is thus a critical component of building sufficient political will in the stakeholder engagement process (whether from government or non-governmental stakeholders who can help push forward the agenda).

Strategies for Identifying Key Stakeholders and Potential Opportunities and Challenges That Can Emerge

At its core, stakeholder engagement is practical, and various tools have been developed to assist countries and others in the process (Table 4.2). For example, WHO's resource mapping exercises have examined where health security resources are distributed in a country, and the IHR-PVS National Bridging Workshops look at specific common priorities and capacity needs to identify pragmatic actions and points of collaboration. Simulation exercises, which may practice plans or use scenarios such as a disease event, may help determine who is missing from the table to fully operationalize programs; after-action reviews can also take stock of key stakeholders that played a valuable role, as well as additional stakeholders where collaboration must be cultivated (e.g., to reduce spillover or tackle misinformation). Monitoring and evaluation frameworks can also help track engagement as well as uptake, with metrics of relevance varying by sector or group.

Building trust may require gaining the buy-in and involvement of local leaders or others who can effectively serve as a champion or be a bridge to a target stakeholder. The importance of this engagement was clear during the Ebola epidemic in West Africa, when religious leaders played an essential role in championing respectful but safe burial processes, thus helping to reduce transmission and ultimately end the epidemic.

Finally, since the root causes of issues are not always apparent, methods such as the "5 Whys" can help to understand the upstream causes and target efforts appropriately to relevant stakeholders to best mitigate or adapt to an issue.[2] Ultimately, a One Health approach reminds us that important stakeholders may not always be the most visible – thus why many initiatives fail to effectively engage those needed for sustained change.

TABLE 4.2
Illustrative Stakeholder Mapping Tools Available

Tool	Description
WHO Health Security Resource Mapping (REMAP) Tool	REMAP examines national and sub-national health security interventions with the aim of identifying gaps in investment and promoting harmonization across projects being carried out under NAPHS.
IHR-PVS National Bridging Workshops	These workshops bring together actors across animal and human health sectors to jointly identify gaps and develop a road map for strengthening compliance with IHR and PVS standards.
WHO and WOAH Simulation Exercises	Both WHO and WOAH offer guidance and toolkits for how to assess country capabilities to respond to a public health emergency or human and animal disease outbreaks.
One Health Zoonotic Disease Prioritization (OHZDP) Workshops	OHZDP workshops bring together stakeholders from human, animal, and environmental health sectors to identify a set of priority zoonotic diseases, clarify stakeholder roles and responsibilities, and develop a set of action plans and recommendations on how to address identified zoonoses.

NOTES

1 The Quadripartite includes the Food and Agriculture Organization of the United Nations (FAO), the United Nations Environment Programme (UNEP), the World Health Organization (WHO), and the World Organisation for Animal Health (WOAH).

2 The "5 Whys" asks a series of "whys" to a problem to understand its complexities and causal factors, often identifying the root cause(s) by the 5th question. They can be helpful when designing as well as evaluating interventions (i.e., in the plan-do-study-act cycle).

REFERENCES

Anderson, E. P., Jackson, S., Tharme, R. E., Douglas, M., Flotemersch, J. E., Zwarteveen, M., Lokgariwar, C., Montoya, M., Wali, A., Tipa, G.T., Jardine, T.D., Olden, J.D., Cheng, L., Conallin, J., Cosens, B., Dickens, C., Garrick, D., Groenfeldt, D., Kabogo, J., Roux, D.J, Ruhi, A., Arthington, A.H. (2019). Understanding rivers and their social relations: A critical step to advance environmental water management. *Wiley Interdisciplinary Reviews: Water, 6*(6), e1381.

Berthe, F. C. J., Bouley, T., Karesh, W. B., Le Gall, F. G., Machalaba, C. C., Plante, C. A., & Seifman, R. M. (2018). *Operational framework for strengthening human, animal and environmental public health systems at their interface (English)*. World Bank Group. https://documents.worldbank.org/curated/en/703711517234402168/Operational-framework-for-strengthening-human-animal-and-environmental-public-health-systems-at-their-interface

Dorothy A. Johnson Center for Philanthropy (2022). "11 Trends in Philanthropy for 2022."

Centers for Disease Control and Prevention. (2022). *One Health Zoonotic disease prioritization (OHZDP)*. One Health. https://www.cdc.gov/one-health/php/prioritization/index.html

DeLeeuw, J. (2022). *One Health & animal protection philanthropy*. Dorothy A. Johnson Center for Philanthropy. August 2, 2022. https://johnsoncenter.org/blog/one-health-and-animal-protection-philanthropy-a-growing-sub-sector/

FAO, UNEP, WHO, and WOAH. (2022). *One Health Joint Plan of Action (2022–2026). Working together for the health of humans, animals, plants and the environment*. FAO; UNEP; WHO; World Organisation for Animal Health (WOAH) (founded as OIE); https://doi.org/10.4060/cc2289en

Fasina, F. O., Fasanmi, O. G., Makonnen, Y. J., Bebay, C., Bett, B., & Roesel, K. (2021). The One Health landscape in sub-Saharan African countries. *One Health, 13*, 100325. https://doi.org/10.1016/j.onehlt.2021.100325

Federal Ministry for Economic Cooperation and Development. (2022). *One Health*. https://www.bmz.de/en/issues/one-health

Jonas, O., Irwin, A., Berthe, F. C. J., Le Gall, F. G., & Marquez, P. V. (2017). *Drug-resistant infections: A threat to our economic future (Vol. 2): Final report (English)*. World Bank Group. https://documents.worldbank.org/curated/en/323311493396993758/final-report

Kanagat, N., Chauffour, J., Ilunga, J.-F., Ramazani, S. Y., Ajiwohwodoma, J. J. P. O., Anas-Kolo, S. I., Maryjane, O., Onuekwusi, N., Ezombe, T., Dominion, J., Sunday, J., Kasongo, J., Ngambwa, G., Asala, C., Nsibu, C., Williams, A., Wendland, M., Klimiuk, E., LaFond, A., Orobaton, N., Kasungami, D. (2021). Country perspectives on improving technical assistance in the Health Sector. *Gates Open Research, 5*(August), 141. https://doi.org/10.12688/gatesopenres.13248.1

Kotter, J. P. (1996). *Leading change*. Harvard Business School Press.

Machalaba, C., & Sleeman, J. M. (2022). Leading change with diverse stakeholders. In C. Stephen (Ed.), *Wildlife population health*. Cham: Springer International Publishing. https://doi.org/10.1007/978-3-030-90510-1_22

OECD. (2024). *Official Development Assistance (ODA)*. https://www.oecd.org/dac/financing-sustainable-development/development-finance-standards/official-development-assistance.htm

Olea-Popelka, F., Muwonge, A., Perera, A., Dean, A. S., Mumford, E., Erlacher-Vindel, E., Forcella, S., Silk, B.J., Ditiu, L., El Idrissi, A., Raviglione, M., Cosivi, O., LoBue, P., Fujiwara, P.I. (2017). Zoonotic Tuberculosis in human beings caused by Mycobacterium Bovis—A call for action. *The Lancet Infectious Diseases, 17*(1), e21–25. https://doi.org/10.1016/S1473-3099(16)30139-6

"One Health." (2024). Merck veterinary manual. https://www.merckvetmanual.com/resourcespages/one-health.

One Health High-Level Expert Panel (OHHLEP), Adisasmito, W. B., Almuhairi, S., Behravesh, C. B., Bilivogui, P., Bukachi, S. A., Casas, N., Becerra, N.C., Charron, D.F., Chaudhary, A., Ciacci Zanella, J.R., Cunningham, A.A., Dar, O., Debnath, N., Dungu, B., Farag, E., Gao, G.F., Hayman, D.T.S., Khaitsa, M., Koopmans, M.P.G., Machalaba, C., Mackenzie, J.S., Markotter, W., Mettenleiter, T.C., Morand, S.,

Smolenskiy, V., Zhou, L. (2022). One Health: A new definition for a sustainable and healthy future. edited by Jeffrey D. Dvorin. *PLOS Pathogens, 18*(6), e1010537. https://doi.org/10.1371/journal.ppat.1010537

Pavade, G. 2020. "OFFLU collaboration in the WHO influenza vaccine virus selection process." *Bulletin de l'OIE, 2020*(2). https://doi.org/10.20506/bull.2020.2.3152.

Runde, D. F., Metzger, C., & Abdullah, H. F. (2020). *Covid-19 demands innovative ideas for financing the SDGs.* https://www.csis.org/analysis/covid-19-demands-innovative-ideas-financing-sdgs.

Seffren, V., Lowther, S., Guerra, M., Kinzer, M. H., Turcios-Ruiz, R., Henderson, A., Shadomy, S., Baggett, H., Harris, J.R., Njoh, E., Salyer, S.J. (2022). Strengthening the Global One Health Workforce: Veterinarians in CDC-supported field epidemiology training programs." *One Health, 14*(June), 100382. https://doi.org/10.1016/j.onehlt.2022.100382.

Solaimani, S., Talab, A. H., & Van Der Rhee, B. (2019). An integrative view on lean innovation management. *Journal of Business Research, 105*(December), 109–120. https://doi.org/10.1016/j.jbusres.2019.07.042.

The World Bank. (2021). *Safeguarding animal, human and ecosystem health: One Health at the World Bank.* World Bank. June 3, 2021. https://www.worldbank.org/en/topic/agriculture/brief/safeguarding-animal-human-and-ecosystem-health-one-health-at-the-world-bank.

Tiwari, H. K., Gogoi-Tiwari, J., & Robertson, I. D. (2021). Eliminating dog-mediated rabies: Challenges and strategies. *Animal Diseases, 1*(1), 19. https://doi.org/10.1186/s44149-021-00023-7.

U.S. Department of Agriculture. (2024). *One Health | USDA.* https://www.usda.gov/topics/animals/one-health.

USAID. (2023). *One Health: Connecting the dots between human health and the environment | Biodiversity | Stories.* U.S. Agency for International Development. August 4, 2023. https://www.usaid.gov/biodiversity/stories/human-health-environment.

Viergever, R. F., & Hendriks, T. C. C. (2016). The 10 largest public and Philanthropic funders of Health Research in the World: What they fund and how they distribute their funds. *Health Research Policy and Systems, 14*(1), 12. https://doi.org/10.1186/s12961-015-0074-z.

World Health Organization. (2015). *Global action plan on antimicrobial resistance.* World Health Organization. https://iris.who.int/handle/10665/193736.

World Health Organization. (2023). *Rabies.* World Health Organization. https://www.who.int/news-room/fact-sheets/detail/rabies.

World Health Organization. (2024). *Research and Development (R&D) Roadmaps.* https://www.who.int/teams/blueprint/r-d-roadmaps.

World Health Organization, Food and Agriculture Organization of the United Nations, and World Organisation for Animal Health. (2017). *Roadmap for Zoonotic Tuberculosis.* World Health Organization. https://iris.who.int/handle/10665/259229.

World Organisation for Animal Health. (2024). *Bovine Tuberculosis.* WOAH - World Organisation for Animal Health. https://www.woah.org/en/disease/bovine-tuberculosis/.

World Organisation for Animal health. (2024). Public-private partnerships in the veterinary domain. WOAH - World Organisation for Animal Health. https:// www.woah.org/en/what-we-offer/improving-veterinary-services/pvs-pathway/public-private-partnerships-in-the-veterinary-domain/

5 Coordinated Surveillance Systems for Infectious Disease at National and International Levels

Lindsey McCrickard Shields, Stephanie J. Salyer, and Katherine A. Feldman

INTRODUCTION

Surveillance systems are critical for tracking events, syndromes, and diseases that impact human, animal, and environmental health. Surveillance can be used to identify outbreaks early, monitor disease trends, and evaluate the impact of control program interventions. For the purposes of this book, we will focus on surveillance systems used to track and monitor infectious diseases and how these systems can be coordinated using a One Health approach to improve overall health. However, this should not preclude the use or coordination with other surveillance systems that track non-infectious diseases (e.g., cancer, chronic disease, reproductive outcomes, nutritional status), climate, or other environmental events which could be used to inform preparedness activities for both infectious and non-infectious diseases or events (e.g., Rift Valley fever outbreaks associated with rainfall, cholera with flooding, toxic gas exposure with volcano eruptions).

The U.S. Centers for Disease Control and Prevention (CDC) defines public health surveillance as "the ongoing, systematic collection, analysis, and interpretation of health-related data essential to the planning, implementation, and evaluation of public health practice, closely integrated with the timely dissemination of these data to those responsible for prevention and control" (Centers for Disease Control and Prevention, 2024). The World Organisation for Animal Health (WOAH) defines surveillance as "the systematic ongoing collection, collation, and analysis of information related to animal health and the timely dissemination of information so that action can be taken" (World Organisation for Animal Health, 2018).

When the need for surveillance is recognized (such as for an emerging infectious disease), the traditional approach globally has been to create a siloed system

DOI: 10.1201/9781003232223-5

that captures health outcomes in a single population or for a single setting. But what happens when an infectious disease is impacted by environmental changes or spreads between human and animal populations? It has long been recognized that diseases in animals can present risk to human populations, but recent disease events have highlighted the critical need for coordinated disease surveillance systems that monitor the health of various populations and the environment. For instance, in 2002, Severe Acute Respiratory Syndrome (SARS) emerged in China and the critical interlinkages between humans, animals, and the environment became apparent to the world at large (Holmes, 2022). To ensure that the world is ready to identify and respond to both known and emerging infectious diseases, it has become crucial to develop One Health surveillance systems that capture data from across multiple sectors to inform action for safeguarding public and animal health. The Tripartite Surveillance and Information Sharing Operational Tool (SISOT) has formally defined a coordinated surveillance system as "the platform or system that allows for the collection, aggregation, and analysis of surveillance elements across multiple sectors collaborating at the human-animal-environmental interface, to enable them to effectively work together toward their aligned objectives and goals" (World Health Organization et al., 2022).

SURVEILLANCE AND ONE HEALTH SURVEILLANCE APPROACHES

Surveillance systems can be broadly classified as active or passive systems. Active surveillance systems rely on actively seeking out potential cases while passive surveillance systems wait for reports to be submitted from healthcare providers or systems (Centers for Disease Control and Prevention, 2024). Surveillance systems can be further classified into event-based or indicator-based systems, depending on the type of data collected. Event-based surveillance systems collect unstructured information about potential threats to human, animal, or environmental health. These data could be from multiple sources such as media reports, rumors within the community, health facilities, or hotlines. In contrast, indicator-based surveillance systems, typically collected at health facilities, capture standardized and structured data, often using standard case definitions to ensure systematic and consistent consideration of what constitutes the disease or condition under surveillance. Case definitions include uniform criteria to define the disease under surveillance and may include laboratory results (Centers for Disease Control and Prevention, 2023). Historically, surveillance systems have been developed in silos, focused on one particular disease or syndrome, species, or geographic region. With the increase in the emergence and spread of zoonotic diseases, there has been more focus placed on developing coordinated surveillance systems that integrate information across sectors (Sudhan & Sharma, 2020). Coordinated surveillance systems can be used in several ways: to monitor trends in endemic infectious diseases, to identify outbreaks of known or novel diseases and trigger a multidisciplinary response, or to inform interventions to protect the health of humans, animals, and/or the environment.

Globally, the One Health approach has been gaining traction. The United Nations Environment Programme (UNEP) recently joined the World Health Organization (WHO), WOAH, and the Food and Agriculture Organization of the United Nations (FAO) to implement the One Health approach, creating the Quadripartite. The addition of the UNEP (to what was referred to as the Tripartite) reflects the critical role that the environment plays in the emergence and spread of infectious diseases that impact human and animal health.

Multisectoral coordinated surveillance systems are challenging to establish as they require clear policies, data systems, data sharing, and case definitions to ensure coordination across diverse organizations and sectors. The lack of a standardized framework for One Health surveillance has been cited as a major challenge to successful implementation of these systems (Bordier et al., 2020). Recently developed guidance documents (developed since 2015) address the management of zoonotic diseases and encourage coordinated surveillance systems; these include the Tripartite Guide to Addressing Zoonotic Diseases, CDC's Generalizable One Health Framework, and the One Health Surveillance Codex (Filter et al., 2021). Additionally, there is the SISOT, which aims to support national governments in developing a coordinated surveillance system for zoonotic diseases that aligns with the Tripartite Guide to Addressing Zoonotic Diseases (World Health Organization et al., 2022). These documents provide an overall framework for the key competencies (further outlined below) that are required to successfully implement a coordinated surveillance system.

Principles of One Health Surveillance

The goal of coordinated surveillance is to ensure the timely, appropriate, and coordinated collection and integration of surveillance data used to protect the health of humans, animals, and the environment that they share. To successfully implement coordinated surveillance, appropriate policies, data agreements, infrastructure, and workforce must be in place.

Policy: Coordinated surveillance systems that bring together data and input from multiple government agencies (such as the Ministries or Departments of Health, Agriculture, and the Environment) require clear policies that are supported by senior government officials who can advocate for their implementation. These policies should guide data sharing, identify priority diseases, and describe roles and responsibilities of relevant agencies and authorities. This critical step can provide the framework for coordinated surveillance systems at a country level.

In some countries, the One Health approach has been institutionalized at the national level, providing a policy framework on which to build coordinated surveillance systems. In 2012, Bangladesh created and approved the Health Strategic Framework and Action Plan for the One Health Approach for Infectious Diseases in Bangladesh, which was approved by three key ministries, the Ministry of Health and Family Welfare, the Ministry of Fisheries and Livestock, and the Ministry of Environment, Forest and Climate Change (Debnath, 2020).

The institutionalization of the One Health concept in government policy can facilitate activities related to data sharing and disease prioritization.

However, many countries have taken a disease-specific approach, identifying specific priority zoonotic diseases where surveillance data are shared across sectors despite the absence of an overarching national policy. Identifying priority diseases for One Health surveillance allows countries to develop a focused set of reportable diseases with standardized case definitions. Often, laboratory diagnostic capacity is required to definitively diagnose the priority diseases. To help guide the decision-making on zoonotic disease prioritization, CDC has developed and implemented One Health Zoonotic Disease Prioritization workshops in over 30 countries and one region (the Economic Community of West African States (ECOWAS)) (as of 2022). These prioritization workshops follow a standard process to review the impact of various zoonotic diseases at the country level using criteria jointly determined by participants, bring representatives together from a wide range of technical and governmental stakeholders to review disease prioritization, and inform planning efforts and strategy development related to One Health.

In addition to national policies, guidelines, and plans developed to inform coordinated surveillance systems, there are several globally developed policies and guidelines that can assist in the development of coordinated surveillance systems. The International Health Regulations 2005 (IHR) provide overall legal guidance on public health threats, including zoonotic diseases. These legally binding regulations have been adopted by all WHO member states and outline specific requirements related to reporting and notification of disease outbreaks or events to WHO, require the development of public health capacity, and provide guidance on health and sanitary measures for international movement of people and goods (World Health Organization, 2016). To monitor the progress of countries to achieve these core IHR capacities, the WHO developed the Joint External Evaluation (JEE) tool in 2016. The third edition of the tool was released in 2022 and includes a section for assessment of zoonotic diseases and evaluate the presence of multisectoral mechanisms, systems, practices, and policies to mitigate the spread of zoonotic diseases from animals to humans. Countries must show the presence of coordinated surveillance systems to achieve the highest levels of performance in this area (World Health Organization, 2022).

There are several other guidance documents that have been developed to assist countries in implementing surveillance activities that advocate for a One Health approach to surveillance. The 2019 revision of WHO's Integrated Disease Surveillance and Response (IDSR) framework provides a standardized approach to setting up national and subnational surveillance systems for government-identified priority diseases, provides standardized approaches on how to track case counts, and has been adopted by nearly all countries in Africa (World Health Organization African Region, 2019). Additionally, the Africa Centres for Disease Control and Prevention (Africa CDC), a specialized agency within the African Union, has revised their Event-Based Surveillance Framework, which provides countries with a standardized approach to implement event-based surveillance, to incorporate aspects of a multisectoral, One Health approach into early warning systems (Africa CDC, 2023).

Data Agreements: For data to be easily and safely shared across sectors, it is critical to have clear definitions of the type of data that will be collected, shared, and analyzed and ensure that data protections can be maintained (e.g., under the Health Insurance Portability and Accountability Act (HIPAA) in the United States). To assist in developing the framework for data sharing and/or integration across ministries, it is often required to develop data sharing memorandums of understanding (MOUs) that provide guidance on what information will be shared, how it will be shared, what the data will be used for, and the frequency of data sharing or data integration. The agreements provide protection for all involved agencies. The agreements also provide information on where the data itself will be integrated within one lead agency, or within a national multisectoral platform developed for that purpose with representation from all agencies involved.

Data sharing and data integration steps can be highly technical, working to link systems with often differing terminology or variables. For example, administrative units (i.e., cities or facilities) within the Ministry of Health may not always match administrative units in the Ministry of Agriculture or Ministry of Environment, making reporting and data integration more complicated. There may also be differences in the reporting mechanisms used by different ministries; for instance, electronic reporting may be used in one sector, but the related disease reporting is paper-based for the other ministries. Data sharing agreements can provide information on how the data will be shared to facilitate the rapid sharing and integration of surveillance data, especially in cases where the data may be used to trigger outbreak response.

Infrastructure: Surveillance systems, regardless of sector, require at least some basic infrastructure to be in place to ensure that cases are identified and reported. Infrastructure can include the community-based reporting tools used to report suspected cases of disease by community or animal health workers such as cell-phone-based reporting tools (e.g., FrontlineSMS (2024), RapidPro (UNICEF, 2024), DHIS-2 android app (University of Oslo, 2023), EMA-I (Food and Agriculture Organization, 2015)). In Senegal, the United Nations Children's Fund (UNICEF) developed an SMS-based reporting tool for community-based surveillance that integrated both human and animal health disease reporting using RapidPro called mInfoSante. mInfoSante has been rolled out to more than 10 districts in Senegal in the Tambacounda and St. Louis regions through support from development projects such as the U.S. Agency for International Development (USAID) Infectious Disease Detection and Surveillance and MEASURE Evaluation. To report with this tool, community and animal health workers must have access to a cell phone, internet, or cellular access for reporting, and health officers receiving the message must be able to access the system through their computers at the district, regional, and/or national level (UNICEF, 2024; Digital Square, 2021).

Ideally, surveillance should be conducted using electronic, internet-connected tools that allow for near real-time reporting of suspected cases. In many countries, reporting for routine surveillance is implemented through electronic or paper-based tools with variable timeliness and accuracy of reported data. For coordinated surveillance systems, there is added complexity as data being collected often comes from multiple systems, which must then be aligned and merged. In many countries, the animal and environmental sectors have less developed surveillance systems than the human sectors and may rely more often on paper-based reporting tools and simple spreadsheets.

In addition to the surveillance system infrastructure, there is also a critical need for infrastructure for laboratory diagnosis and/or specimen transportation, depending on the disease under surveillance. Adequate laboratory capacity is especially important for diseases that have common symptoms or must be differentiated from other critical illnesses. In some areas, samples must travel long distances prior to reaching a diagnostic laboratory and it is important to have infrastructure in place (e.g., standard mechanisms for safe and rapid sample transportation) to ensure that viable samples reach the laboratory in a timely fashion to ensure appropriate diagnosis.

Workforce: As coordinated surveillance requires a wide range of technical capacities and dedicated staff, it is critical to have a trained and committed workforce to support the implementation of coordinated surveillance. Coordinated surveillance often starts at the community level where there is a sick animal or human that is identified by a local health facility, veterinary officer, or community health worker. These local community members are essential for identifying events or cases of priority diseases and must be trained on how to recognize the clinical signs and symptoms of priority diseases, signals, or events that may warrant a response. In addition, they must know how and to whom to report suspect cases and events. While this reporting may be common for persons involved in human or animal clinical care, in remote areas where clinical care is harder to obtain, local community members supporting animal, human, or environmental health may need additional training. In some instances, it is best practice to train community health workers representing both human and animal health sectors together to ensure that they are both able to recognize the clinical signs and symptoms of priority diseases in both human and animal populations.

Once a suspected case has been reported, there is a need for a trained workforce in both the surveillance and laboratory sectors. For some diseases, a laboratory technician is required to perform laboratory tests on the animal or human specimen to confirm the suspected diagnosis. This requires consistent access to a laboratory and a mechanism by which the confirmed laboratory result can be reported both to the surveillance officer and back to the community where the case was identified. Surveillance officers play a critical role in linking the data from the community and laboratory, as well as integrating data across sectors and interpreting the data. Data integration and analysis require technical capacities for epidemiology and a linkage with emergency response systems in case the results trigger a multisectoral investigation or outbreak response. A well-trained workforce is at the center of a successful coordinated surveillance system.

APPLICATION AS A GLOBAL HEALTH PRACTITIONER

Aligned with the Tripartite Guide to Addressing Zoonotic Diseases and the to-be-published SISOT, there are several steps required when establishing a coordinated surveillance system:

1. Pre-planning and Assessment
2. Planning and Buy-in
3. Implementation
4. Monitoring and Evaluation

Pre-planning and Assessment

When establishing a new coordinated surveillance system, it is important to clearly identify the goals of the system, as they will impact the type of information reported, the frequency of reporting, and how generated data can be used. In addition, the goals of the surveillance system will help to identify key stakeholders, sectors to be covered, and geographic areas of focus. In some instances, the pre-planning step may also require identification of priority diseases that could or should be covered by the coordinated surveillance system. This prioritization can be done using a variety of methodologies or tools such as the CDC's One Health Zoonotic Disease Prioritization workshop as described above (Centers for Disease Control and Prevention, 2022). Another standardized tool useful at this step is the WHO Joint Risk Assessment Operational tool which can assist countries to identify, quantify, and prioritize based on the risks posed by various zoonotic diseases (Food and Agriculture Organization of the United Nations et al., 2020).

Potential stakeholders that need to be involved at local, regional, and/or national level will depend on the goals of the surveillance system. Key stakeholders include those persons involved with any aspect of data collection or data use and could include representatives from relevant ministries, health practitioners, veterinarians, national parks, academia, the private sector, and laboratories, among others. Representatives from these groups should be involved in all aspects of the pre-planning and assessment phase.

During the pre-planning phase, it is also critical to identify or develop a standard case definition to classify whether a human or animal is considered a case. Case definitions provide a standardized way to determine whether the human, animal, or population has the disease or condition under surveillance. Depending on the surveillance system and targeted disease or event under surveillance, there are standardized case definitions that can be used that are developed by global or national organizations such as the Council for State and Territorial Epidemiologists (CSTE) in the United States. Using global standards for case definitions is critical especially when attempting to compare disease situations across countries. While case definitions vary, they can include the following items:

- Clinical signs (e.g., vomiting, diarrhea, fever)
- Clinical findings (e.g., abnormal electroencephalogram (EEG), abnormal radiographic findings)
- Laboratory results

Case definitions are updated routinely, and can include subclassifications of suspected, probable, or confirmed, which vary in clinical features, epidemiologic linkages, and requirements for laboratory confirmation. For example, the CSTE defines a confirmed case of rabies in a human as a "clinically compatible case that is laboratory confirmed" and provides specifications on what clinical description and laboratory diagnostics can be used for confirmation (Hampson et al., 2016). When global health practitioners respond to an outbreak, they will often use the international or local standard case definition and add additional requirements for person (or animal), place, and time to ensure cases are grouped appropriately spatially and temporally for epidemiologic studies.

Once goals of the surveillance system have been identified, a case definition has been identified, and the list of key stakeholders has been finalized, assessment activities should begin, including an assessment of existing surveillance systems for the selected disease or event. These surveillance systems may be currently siloed in specific ministries, or part of a larger routine disease reporting tool. The assessment of existing surveillance systems can be guided by the use of standard assessment tools such as the FAO Surveillance Evaluation Tool (SET) or the CDC Guidelines for Evaluating Public Health Surveillance Systems (German et al., 2001; Vasconcelos Gioia et al., 2021). Both of these tools can assist in identifying where gaps are present in existing surveillance systems and identify technical discussion points for key stakeholders to review how to best develop coordinated surveillance systems. Assessment of existing surveillance systems includes review of key technical areas required for a successful surveillance system such as institutional organization, data collection, information systems, data quality, representativeness, completeness, and timeliness. The output of the assessment can serve as the basis for the organizational meetings required for the planning and buy-in steps described below.

Planning and Buy-In

Once the pre-planning and assessment phase has been completed, it is time to focus on the practical aspects of the coordinated surveillance system including the development of policies, guidance, MOUs and data sharing agreements, site selection, data analysis planning, and budgeting.

- Policy/Guidance: For successful implementation, teams should review existing policies or guidance documents that can provide a framework for agreements across sectors. When these policies or guidelines do not exist, it may be necessary to develop a standard guidance document on the approach the coordinated surveillance system will take and ensure cross-sectoral buy-in. This step is crucial for ensuring success of a coordinated surveillance system but can be a lengthy process if the system is large, nationally based, or requires buy-in from a wide range of stakeholders.

- MOU/Data Sharing Agreements: For many of the coordinated surveillance systems, it is essential to put in place MOUs and/or data sharing agreements for the data owners across sectors. For example, a data sharing agreement may outline what data would be shared by the Ministry of Health related to an outbreak of anthrax with the Ministry of Agriculture and vice versa. Having standard data that can be shared across sectors is critical to operationalize a coordinated surveillance system.
- Site Selection: In some cases, coordinated surveillance systems will need to be site-specific due to disease etiology, budgetary constraints, or for other reasons. The process of site selection should be participatory with inputs from key stakeholders and based on mutually agreed criteria such as burden of disease, buy-in from local leaders, availability of resources, and/or stakeholder interest. Site selection may also need to consider requirements for laboratory analysis and an understanding of geographic distance to the nearest laboratory capable of implementing the required testing. Questions to be considered include: What is the additional burden on the laboratory to process these samples? Are there considerations for sample transportation that need to be included? Site selection both for data collection and laboratory analysis can have a profound impact on the type, quality, and quantity of data the system can generate.
- Data Analysis Planning: Once the coordinated surveillance system begins generating data, it is important to have a plan in place for the analysis and use of this data. This should be determined during the planning phase (in advance of data being generated) to show where data will be analyzed, the type of analysis that will be completed, and what the data can be used for. Coordinated surveillance systems must show impact using the outputs for public health action.
- Budgeting: The budget for a coordinated surveillance system can be highly variable depending on the disease of focus and availability of funding to support these activities. In some instances, it may be prudent to keep the scope of the coordinated surveillance system small in the beginning to show proof of concept and impact and ensure sustainability. Budgeting may need to consider requirements of various stakeholders and variation in capacities among those stakeholders to implement the surveillance activities in the targeted locations.

Once these steps have been completed, and all stakeholders are in agreement of the focus and goals for the coordinated surveillance system, it is time to implement.

Implementation

As a global health practitioner, nothing feels better than watching the hard work involved in establishing a surveillance system finally generate data for public health action. Once implementation begins, teams are often busy troubleshooting, conducting data analyses and interpreting the data, and making recommendations for action.

- Troubleshooting: One of the most time-consuming aspects of implementing a coordinated surveillance system is addressing issues as they manifest. When a new system is set up and implementation starts, there may be many operational or technical issues that arise, compromising the success of the overall system. Examples may include issues with the reporting tools, challenges with obtaining or accessing data once reported, and competing priorities (e.g., disease outbreaks) that pull staff from surveillance activities. Global health practitioners must be ready to provide technical and operational support to address these issues as they arise. Addressing these challenges also requires a transparent and participatory approach to avoid conflicts between key stakeholders.
- Data Analysis and Interpretation: The entire purpose of implementing a surveillance system is to generate data for action. This relies on data analysis and interpretation using the analysis plan that was prepared during the planning phase. This may require training key stakeholders on how to conduct the analysis itself, perhaps using analytic software such as R studio, and will almost certainly include a substantial effort cleaning the data. Interpreting these data, including considering what biases might be present, is critical to ensure that data are not taken out of context. For example, a sudden increase in the number of reports of avian influenza among chickens in a rural location may provoke concern but may simply be the result of improved and increased surveillance in the area. Interpreting the data generated through coordinated surveillance systems will also need to consider what is relevant across multiple stakeholders.
- Recommendations for Action: As data are generated, analyzed, and interpreted, this information can be used to inform recommendations for public health action. Data may need to be presented in different ways for various audiences, such as distilling dense scientific results to more straightforward messaging for the general public or policymakers. The results of this coordinated surveillance may inform improved guidance for disease control or vaccination efforts or trigger an outbreak response. Depending on the goals of the surveillance system, these recommendations may be generated immediately when the results are available (such as in the case of a disease outbreak), or on an annual basis (as needed for vaccine-preventable diseases).

Monitoring and Evaluation

The goal of coordinated surveillance is to ensure the timely, appropriate, and coordinated collection and integration of surveillance data used to protect the health of humans, animals, and the environment that they share. Those involved in the implementation of coordinated surveillance need to use surveillance data to rapidly address identified events and priority diseases; assess the impact of interventions; and provide updates to stakeholders on implementation progress. Thus, there is a need for coordinated surveillance implementers to develop a monitoring and evaluation (M&E) plan to routinely review surveillance performance as well as to account for key program activities and needed resources.

An ideal M&E plan should provide timely information on whether a system is functioning properly and meeting targets, while providing data to guide continuous performance improvement. An M&E plan should ideally describe why, how, and when changes toward a desired public health surveillance goal are achieved. In brief:

> **Monitoring** is the process of continuously tracking progress or delay in inputs, activities, outputs, and outcomes. Monitoring helps track implementation processes and provides a basis for re-adjustments based on performance plan metrics.
>
> **Evaluation** is the process of periodically assessing the relevance, effectiveness, and impact of a program or system. Evaluation ensures that the coordinated surveillance system meets the objectives for which it was set by providing evidence-based explanations for achievements and shortcomings and recommending its improvements.

An M&E plan can be based on a logical framework or theory of change (TOC) that creates indicators to track inputs, activities, outputs, outcomes, and impacts of the coordinated surveillance system.

- Inputs: all the resources required for the implementation of the coordinated surveillance program. Inputs include implementation documents (e.g., legal frameworks, policies, guidelines); curriculum and tools; human resource, time, finance, materials, infrastructure, stakeholders (e.g., communities, public health and healthcare workers, national and subnational leadership, multisectoral partners) and other resources.
- Activities: any tasks, actions, processes, or procedures undertaken during the coordinated surveillance program implementation through the use of the inputs. Activities rely on a well-thought-out strategy for the successful surveillance implementation. They include tasks such as planning meetings, procurement of supplies, training and sensitizations, and implementation.
- Outputs: the immediate gains of activities during the coordinated surveillance implementation activities. For example, outputs can include number of people trained, number of signals triaged, or number of events responded to.

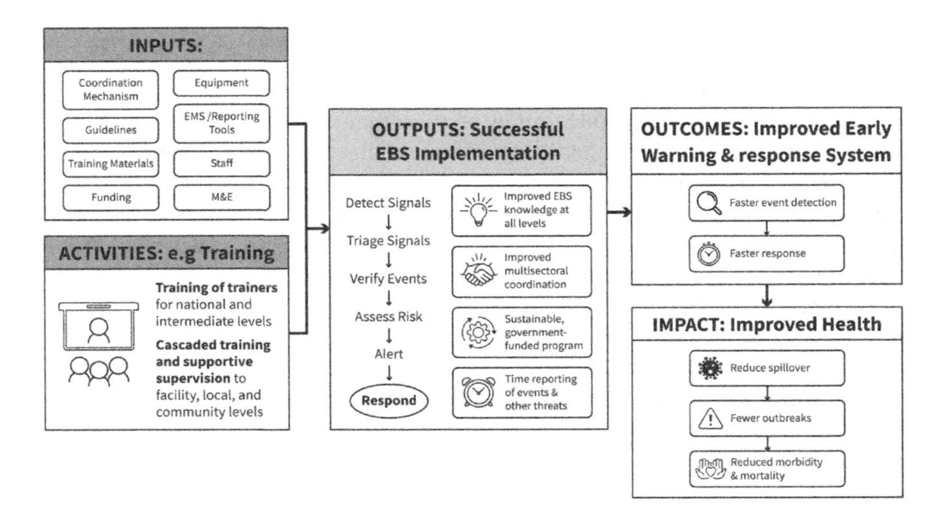

FIGURE 5.1 This figure was taken from Africa CDC's Event-based Surveillance (EBS) Framework and illustrates a TOC that was created to inform the implementation of multisectoral, One Health EBS. This TOC illustrates that specific resources (inputs) are required to undertake program tasks (activities) whose accomplishments (outputs) bring about system changes (outcomes) that eventually lead to overall improved health (impact). Based on the coordinated surveillance goals, this TOC could be adapted to help visualize a coordinated surveillance program's actions to desired results, over a defined period of time.

- Outcomes: short-term and medium-term direct changes resulting from the implementation. These include implementation outcomes that demonstrate changes in the timeliness and accuracy of reporting, success of response activities, or other interventions put in place to prevent and control priority diseases.
- Impacts: the overall long-term improvements in health outcomes attributed to the coordinated surveillance program implementation. The impacts are aligned to the surveillance goals and may be due to the implementation outcomes only or in combination with the outcomes of other health programs. Impacts include reduction in health emergencies and/or reduction in mortalities, disabilities, and morbidities due to priority diseases diseases (Figure 5.1).

When developing an M&E plan, the plan should:

- Make references to existing baseline data or begin with a baseline evaluation
- Be developed in a participatory fashion and involve all program stakeholders, including implementers and beneficiaries
- Respect and protect the rights and confidentiality of all participants

- Be integrated with other surveillance systems for sustainability beyond the life of the program
- Be considered a living document that needs to be reviewed on an annual basis and updated to reflect any changes in referenced technical guidelines or whenever the coordinated surveillance system is modified

Example sector-specific tools that can be adapted to the country context or global health practitioner's needs include: the FAO's SET (Vasconcelos Gioia et al., 2021) or the USAID MEASURE Evaluation Performance of Routine Information System Management (PRISM) tool (MEASURE Evaluation, 2024).

Summary

Coordinated surveillance systems can provide a great opportunity for integrating data across sectors and providing more informed data for decision-making. These systems are complex to set up, but critical in responding to the threats facing society today, ranging from new emerging zoonotic diseases to battling the effects of climate change.

CASE CALLOUTS

EXAMPLE 1: INTEGRATED BITE CASE MANAGEMENT FOR RABIES

To help control and eliminate rabies, the Tripartite (WHO, WOAH, FAO) developed the "Zero by 30" global strategic plan to end human deaths from dog-mediated rabies by 2030 (Hampson et al., 2023; Nyasulu et al., 2021; WHO Rabies Modelling Consortium, 2020). As part of this strategy, Integrated Bite Case Management (IBCM) has been promoted globally as an advanced surveillance method that incorporates a One Health approach and involves the investigation of suspected rabid animals and sharing of information between animal and human health investigators to inform risk assessments and recommendations for appropriate post-exposure prophylaxis (PEP) administration (Wallace et al., 2015; World Health Organization, 2018) (Figure 5.2).

IBCM has been shown to increase rabies case detection (Lushasi et al., 2020; Wallace et al., 2015) and improve the administration and cost-effectiveness of PEP (Undurraga et al., 2017). IBCM has the potential to help monitor the effectiveness of canine vaccination campaigns and verify a country's freedom from rabies (Lushasi et al., 2020; Undurraga et al., 2017).

IBCM can be implemented using existing systems; however, to support the rapid notification and response that successful implementation calls for, several applications and platforms have been built to assist countries with IBCM implementation (i.e., REACT jointly developed by CDC and Mission

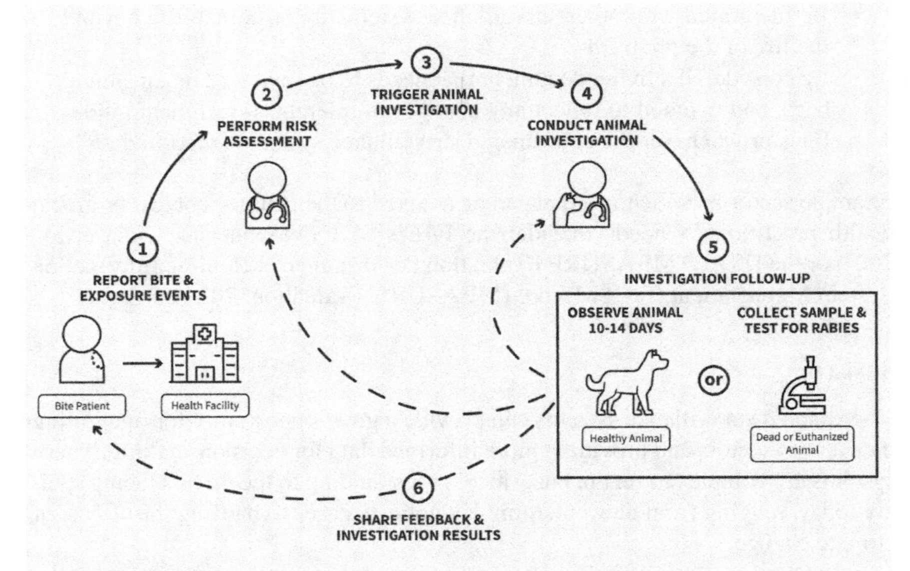

FIGURE 5.2 IBCM explained. The six key activities in IBCM are annotated in red with arrows and numbering indicating the sequential order of these components. (Adapted from Implementing a One Health Approach to Rabies Surveillance: Lessons From Integrated Bite Case Management; Adapted from Swedberg et al., 2022.)

Rabies (Adrien et al., 2019; International Rabies Taskforce, 2024), WVS android app (Mission Rabies, 2024)). IBCM has been implemented and described in over 11 countries: Chad, Cambodia, Haiti, India, Indonesia, Kenya, Namibia, Philippines, Tanzania, Uganda, and Vietnam (Chatterjee et al., 2016; Kisaka et al., 2021; Kitala et al., 2000; Lechenne et al., 2017; Lushasi et al., 2020; Pham et al., 2021; Subrata et al., 2022; Undurraga et al., 2017).

Haiti is one of five remaining countries in the Americas where canine rabies is still a problem. Haiti reports some of the highest rates of human rabies deaths in the Western Hemisphere, with an estimated two deaths each week. In 2013, the Haitian Ministry of Health (MSPP) and Ministry of Agriculture (MARNDR) established the Haiti Animal Rabies Surveillance Program (HARSP), with support from the U.S. CDC, Humane Society International, Christian Veterinary Mission, and the Pan American Health Organization. HARSP incorporates IBCM to provide better treatment for bite victims. As of 2022, HARSP has increased animal rabies detection 18-fold, decreased the risk of dying from rabies by 50%, and increased post-bite rabies vaccination adherence by 230% (CDC, 2024; Undurraga et al., 2017).

The United Against Rabies resource toolbox provides information on what resources are needed to implement this type of surveillance, including the suggested minimal data element set for rabies that should be collected from all sectors (Hampson et al., 2023; United Against Rabies, 2024).

EXAMPLE 2: COVID-19 INTEGRATED SURVEILLANCE

The coronavirus disease (COVID)-19 pandemic, caused by the virus SARS-CoV-2, likely emerged from an animal source (World Health Organization, 2021). It is the exemplification of why a One Health approach to surveillance is so critical, given its likely emergence from animals, rapid human-to-human spread, unknown behavior in different animal species as the virus evolved and spread globally, and the ability to detect SARS-CoV-2 in the environment.

The WHO has provided clear objectives for surveillance of COVID-19 in humans and has recommended that SARS-CoV-2 be integrated into all stages of the surveillance process, from data collection at sentinel sites to sharing of data (Holmes, 2022). Core objectives for COVID-19 surveillance systems include signaling the onset and offset of SARS-CoV-2 activity at defined thresholds; describing the epidemiology and seasonality of SARS-CoV-2; establishing historic surveillance data; and monitoring locally circulating virus lineages. Where capacity, policy needs, and resources allow additional objectives for surveillance can include identifying and monitoring groups at high risk for severe outcomes, understanding clinical manifestations associated with different virus strains, assessing the impact of interventions (including vaccination), and detecting clusters or outbreaks.

While it is believed that SARS-CoV-2 emerged from animals, it is unknown from which species or how the virus behaves as it encounters new species. Hence, it is important to monitor SARS-CoV-2 in animals. As of mid-2022, 35 countries have reported SARS-CoV-2 detection in 24 species to the WOAH, including domestic and large cats, white-tailed deer, mink, hamsters, and gorillas (Tizzani, 2022). Free-ranging white-tailed deer in the United States have been shown to be highly susceptible to infection with SARS-CoV-2 and are capable of sustaining transmission in nature (Hale et al., 2022). Similarly, mink have been shown to be particularly susceptible to SARS-CoV-2 and have suffered widespread outbreaks in the Netherlands, the United States, and elsewhere (Eckstrand et al., 2021; Koopmans, 2021; Moreno et al., 2022; Oude Munnink et al., 2021). Transmission from mink to humans has also been documented (Oude Munnink et al., 2021), and raises the concern that SARS-CoV-2 could become established in certain animal species and serve as reservoirs for and as a potential source of infection to humans.

Because SARS-CoV-2 can be shed in feces, it is possible to monitor for the presence of the virus in wastewater. Wastewater surveillance provides valuable information that can complement other sources of surveillance data and can be particularly useful when diagnostic testing is unavailable or underutilized, or if test results are not reported to public health officials (for instance, with at-home test kits). Wastewater surveillance can serve as an early detection system, providing insights into the emergence of new viral variants and community spread before community members get ill, seek medical care, get tested, and the test results are reported. Wastewater surveillance might be particularly useful in certain settings, such as university campuses (Lee et al., 2022).

Just as with COVID-19, future pandemics are likely to have a zoonotic origin (Judson & Rabinowitz, 2021), which underscores the need for an integrated, One Health approach for early detection and ongoing monitoring.

REFERENCES

Adrien, J., Georges, Y., Augustin, P. D., Monroe, B., Gibson, A. D., Fenelon, N., Fleurinord, L., Crowdis, K., Mandra, A., Joseph, H. C., Etheart, M. D., Wallace, R. M., Haiti-Rabies Field Response Team, Haiti-Rabies Field Response Team, Blanton, J., Cleaton, J., Condori, R. E., Petersen, B., Ross, Y., … King, A. (2019). *Notes from the field*: A multipartner response to prevent a binational rabies outbreak — Anse-à-Pitre, Haiti, 2019. *MMWR. Morbidity and Mortality Weekly Report*, *68*(32), 707–709. https://doi.org/10.15585/mmwr.mm6832a6

Africa CDC. (2023, March 28). Africa CDC Event-based Surveillance Resources. *Africa CDC*. https://africacdc.org/download/africa-cdc-event-based-surveillance-framework-2/

Bordier, M., Uea-Anuwong, T., Binot, A., Hendrikx, P., & Goutard, F. L. (2020). Characteristics of One Health Surveillance Systems: A Systematic Literature Review. *Preventive Veterinary Medicine*, *181*, 104560. https://doi.org/10.1016/j.prevetmed.2018.10.005

CDC. (2024, May 23). *CDC in Haiti*. Global Health. https://www.cdc.gov/global-health/countries/haiti.html

Centers for Disease Control and Prevention. (2022). *One Health Zoonotic Disease Prioritization (OHZDP)*. One Health. https://www.cdc.gov/one-health/php/prioritization/index.html

Centers for Disease Control and Prevention. (2023, April 18). *Event-based Surveillance | Division of Global Health Protection | Global Health | CDC*. https://stacks.cdc.gov/view/cdc/120880/cdc_120880_DS1.pdf

Centers for Disease Control and Prevention. (2024, May 8). *Public Health 101 Series*. https://www.cdc.gov/training-publichealth101/php/index.html

Chatterjee, P., Kakkar, M., & Chaturvedi, S. (2016). Integrating One Health in National Health Policies of Developing Countries: India's Lost Opportunities. *Infectious Diseases of Poverty*, *5*(1), 87. https://doi.org/10.1186/s40249-016-0181-2

Debnath, N. (2020). *Journey of One Health in Bangladesh*. https://rr-asia.woah.org/app/uploads/2020/01/2-3_share-experience-from-regional-champion_bangladesh.pdf

Digital Square. (2021). *Digital Health Systems to Support Pandemic Response in Senegal.* PATH. https://static1.squarespace.com/static/59bc3457ccc5c5890fe7cacd/t/60da42563e56f742edc89629/1624916569830/M%26M-brief-Senegal.pdf

Eckstrand, C. D., Baldwin, T. J., Rood, K. A., Clayton, M. J., Lott, J. K., Wolking, R. M., Bradway, D. S., & Baszler, T. (2021). An Outbreak of SARS-CoV-2 with High Mortality in Mink (Neovison vison) on Multiple Utah Farms. *PLOS Pathogens, 17*(11), e1009952. https://doi.org/10.1371/journal.ppat.1009952

Filter, M., Buschhardt, T., Dórea, F., Lopez De Abechuco, E., Günther, T., Sundermann, E. M., Gethmann, J., Dups-Bergmann, J., Lagesen, K., & Ellis-Iversen, J. (2021). One Health Surveillance Codex: Promoting the adoption of One Health solutions within and across European countries. *One Health, 12*, 100233. https://doi.org/10.1016/j.onehlt.2021.100233

Food and Agriculture Organization. (2015). *EMA-i: A Mobile App for Timely Animal Disease field reporting to enhance surveillance* (FCC-EMPRES Information Sheets). https://openknowledge.fao.org/bitstreams/8fc65826-6153-415a-864a-db6362d835e1/download#:~:text=Using%20Smartphones%2C%20animal%20disease%20information,the%20information%20is%20safely%20stored.

Food and Agriculture Organization of the United Nations, World Organisation for Animal Health, & World Health Organization. (2020). *Joint Risk Assessment Operational Tool (JRA OT); An operational tool of the tripartite Zoonoses guide—Taking a multisectoral, One Health approach: A tripartite guide to addressing Zoonotic diseases in countries.* https://iris.who.int/bitstream/handle/10665/340005/9789240015142-eng.pdf?sequence=1

FrontlineSMS. (2024). https://www.frontlinesms.com/

German, R. R., Lee, L. M., Horan, J. M., Milstein, R. L., Pertowski, C. A., Waller, M. N., & Guidelines Working Group Centers for Disease Control and Prevention (CDC). (2001). Updated guidelines for evaluating public health surveillance systems: Recommendations from the Guidelines Working Group. *MMWR. Recommendations and Reports: Morbidity and Mortality Weekly Report. Recommendations and Reports, 50*(RR-13), 1–35.

Hale, V. L., Dennis, P. M., McBride, D. S., Nolting, J. M., Madden, C., Huey, D., Ehrlich, M., Grieser, J., Winston, J., Lombardi, D., Gibson, S., Saif, L., Killian, M. L., Lantz, K., Tell, R. M., Torchetti, M., Robbe-Austerman, S., Nelson, M. I., Faith, S. A., & Bowman, A. S. (2022). SARS-CoV-2 infection in free-ranging white-tailed deer. *Nature, 602*(7897), 481–486. https://doi.org/10.1038/s41586-021-04353-x

Hampson, K., Abela-Ridder, B., Brunker, K., Bucheli, S. T. M., Carvalho, M., Caldas, E., Changalucha, J., Cleaveland, S., Dushoff, J., Gutierrez, V., Fooks, A. R., Hotopp, K., Haydon, D. T., Lugelo, A., Lushasi, K., Mancy, R., Marston, D. A., Mtema, Z., Rajeev, M., … Del Rio Vilas, V. (2016). *Surveillance to establish elimination of transmission and freedom from dog-mediated rabies.* https://doi.org/10.1101/096883

Hampson, K., Lohr, F., Mwangi, T., Ruman Siddiqi, U., Salahuddin, N., Scott, T., Undurraga, E., & Wallace, R. (2023). *Minimum data elements for monitoring and evaluation of national and international rabies control programs* (Version 4). United Against Rabies. https://www.unitedagainstrabies.org/wp-content/uploads/2023/09/2023_MinimumDataElements_V4_Sep2023_EN.pdf

Holmes, E. C. (2022). COVID-19—Lessons for zoonotic disease. *Science, 375*(6585), 1114–1115. https://doi.org/10.1126/science.abn2222

International Rabies Taskforce. (2024). *REACT App—International rabies taskforce.* https://rabiestaskforce.com/toolkit/react-app

Judson, S. D., & Rabinowitz, P. M. (2021). Zoonoses and global epidemics. *Current Opinion in Infectious Diseases*, *34*(5), 385–392. https://doi.org/10.1097/QCO.0000000000000749

Kisaka, S., Makumbi, F. E., Majalija, S., Kagaha, A., & Thumbi, S. M. (2021). "As long as the patient tells you it was a dog that bit him, why do you need to know more?" A qualitative study of how healthcare workers apply clinical guidelines to treat dog bite injuries in selected hospitals in Uganda. *PLoS One*, *16*(7), e0254650. https://doi.org/10.1371/journal.pone.0254650

Kitala, P. M., McDermott, J. J., Kyule, M. N., & Gathuma, J. M. (2000). Community-based active surveillance for rabies in Machakos District, Kenya. *Preventive Veterinary Medicine*, *44*(1–2), 73–85. https://doi.org/10.1016/S0167-5877(99)00114-2

Koopmans, M. (2021). SARS-CoV-2 and the human-animal interface: Outbreaks on mink farms. *The Lancet Infectious Diseases*, *21*(1), 18–19. https://doi.org/10.1016/S1473-3099(20)30912-9

Lechenne, M., Mindekem, R., Madjadinan, S., Oussiguéré, A., Moto, D. D., Naissengar, K., & Zinsstag, J. (2017). The importance of a participatory and integrated One Health approach for rabies control: The case of N'Djaména, Chad. *Tropical Medicine and Infectious Disease*, *2*(3), 43. https://doi.org/10.3390/tropicalmed2030043

Lee, L., Valmond, L., Thomas, J., Kim, A., Austin, P., Foster, M., Matthews, J., Kim, P., & Newman, J. (2022). Wastewater surveillance in smaller college communities may aid future public health initiatives. *PLoS One*, *17*(9), e0270385. https://doi.org/10.1371/journal.pone.0270385

Lushasi, K., Steenson, R., Bernard, J., Changalucha, J. J., Govella, N. J., Haydon, D. T., Hoffu, H., Lankester, F., Magoti, F., Mpolya, E. A., Mtema, Z., Nonga, H., & Hampson, K. (2020). One Health in practice: Using integrated bite case management to increase detection of rabid animals in Tanzania. *Frontiers in Public Health*, *8*, 13. https://doi.org/10.3389/fpubh.2020.00013

MEASURE Evaluation. (2024). *PRISM: Performance of routine information system management series*. https://www.measureevaluation.org/prism

Mission Rabies. (2024). *Rabies App*. https://missionrabies.com/app/

Moreno, A., Lelli, D., Trogu, T., Lavazza, A., Barbieri, I., Boniotti, M., Pezzoni, G., Salogni, C., Giovannini, S., Alborali, G., Bellini, S., Boldini, M., Farioli, M., Ruocco, L., Bessi, O., Maroni Ponti, A., Di Bartolo, I., De Sabato, L., Vaccari, G., … Giorgi, M. (2022). SARS-CoV-2 in a mink farm in Italy: Case description, molecular and serological diagnosis by comparing different tests. *Viruses*, *14*(8), 1738. https://doi.org/10.3390/v14081738

Nyasulu, P. S., Weyer, J., Tschopp, R., Mihret, A., Aseffa, A., Nuvor, S. V., Tamuzi, J. L., Nyakarahuka, L., Helegbe, G. K., Ntinginya, N. E., Gebreyesus, M. T., Doumbia, S., Busse, R., & Drosten, C. (2021). Rabies mortality and morbidity associated with animal bites in Africa: A case for integrated rabies disease surveillance, prevention and control: A scoping review. *BMJ Open*, *11*(12), e048551. https://doi.org/10.1136/bmjopen-2020-048551

Oude Munnink, B. B., Sikkema, R. S., Nieuwenhuijse, D. F., Molenaar, R. J., Munger, E., Molenkamp, R., Van Der Spek, A., Tolsma, P., Rietveld, A., Brouwer, M., Bouwmeester-Vincken, N., Harders, F., Hakze-van Der Honing, R., Wegdam-Blans, M. C. A., Bouwstra, R. J., GeurtsvanKessel, C., Van Der Eijk, A. A., Velkers, F. C., Smit, L. A. M., … Koopmans, M. P. G. (2021). Transmission of SARS-CoV-2 on mink farms between humans and mink and back to humans. *Science*, *371*(6525), 172–177. https://doi.org/10.1126/science.abe5901

Pham, Q. D., Phan, L. T., Nguyen, T. P. T., Doan, Q. M. N., Nguyen, H. D., Luong, Q. C., & Nguyen, T. V. (2021). An evaluation of the rabies surveillance in Southern Vietnam. *Frontiers in Public Health*, *9*, 610905. https://doi.org/10.3389/fpubh.2021.610905

Subrata, I. M., Harjana, N. P. A., Agustina, K. K., Purnama, S. G., & Kardiwinata, M. P. (2022). Designing a rabies control mobile application for a community-based rabies surveillance system during the COVID-19 pandemic in Bali, Indonesia. *Veterinary World*, 1237–1245. https://doi.org/10.14202/vetworld.2022.1237-1245

Sudhan, S. S., & Sharma, P. (2020). Human viruses: Emergence and evolution. In *Emerging and reemerging viral pathogens* (pp. 53–68). Elsevier. https://doi.org/10.1016/B978-0-12-819400-3.00004-1

Swedberg, C., Mazeri, S., Mellanby, R. J., Hampson, K., & Chng, N. R. (2022). Implementing a one health approach to rabies surveillance: Lessons from integrated bite case management. *Frontiers in Tropical Diseases*, *3*, 829132. https://doi.org/10.3389/fitd.2022.829132

Tizzani, P. (2022). *SARS-CoV-2 in animals – Situation report 15*. World Organization for Animal Health. https://www.woah.org/app/uploads/2022/08/sars-cov-2-situation-report-15.pdf

Undurraga, E. A., Meltzer, M. I., Tran, C. H., Atkins, C. Y., Etheart, M. D., Millien, M. F., Adrien, P., & Wallace, R. M. (2017). Cost-effectiveness evaluation of a novel integrated bite case management program for the control of human rabies, Haiti 2014–2015. *The American Journal of Tropical Medicine and Hygiene*, *96*(6), 1307–1317. https://doi.org/10.4269/ajtmh.16-0785

UNICEF. (2024). *Real time information—RapidPro | UNICEF Office of Innovation*. https://www.unicef.org/innovation/rapidpro

United Against Rabies. (2024). *Resources: Toolbox*. United Against Rabies. https://www.unitedagainstrabies.org/resources-toolbox/

University of Oslo. (2023). *Mobile capture and analysis with DHIS2 android*. DHIS2. https://dhis2.org/android/

Vasconcelos Gioia, G., Lamielle, G., Aguanno, R., ElMasry, I., Mouillé, B., De Battisti, C., Angot, A., Ewann, F., Sivignon, A., Donachie, D., Rozov, O., Bonbon, É., Poudevigne, F., VonDobschuetz, S., Plée, L., Kalpravidh, W., & Sumption, K. (2021). Informing resilience building: FAO's Surveillance Evaluation Tool (SET) Biothreat Detection Module will help assess national capacities to detect agro-terrorism and agro-crime. *One Health Outlook*, *3*(1), 14. https://doi.org/10.1186/s42522-021-00045-8

Wallace, R. M., Reses, H., Franka, R., Dilius, P., Fenelon, N., Orciari, L., Etheart, M., Destine, A., Crowdis, K., Blanton, J. D., Francisco, C., Ludder, F., Del Rio Vilas, V., Haim, J., & Millien, M. (2015). Establishment of a canine rabies burden in Haiti through the implementation of a novel surveillance program. *PLOS Neglected Tropical Diseases*, *9*(11), e0004245. https://doi.org/10.1371/journal.pntd.0004245

WHO Rabies Modelling Consortium. (2020). Zero human deaths from dog-mediated rabies by 2030: Perspectives from quantitative and mathematical modelling. *Gates Open Research*, *3*, 1564. https://doi.org/10.12688/gatesopenres.13074.2

World Health Organization. (2016). *International health regulations (*2005). World Health Organization. https://apps.who.int/iris/rest/bitstreams/1031116/retrieve

World Health Organization. (2018). *WHO expert consultation on rabies: Third report*. World Health Organization. https://iris.who.int/handle/10665/272364

World Health Organization. (2021). *WHO-convened global study of origins of SARS-CoV-2: China Part*. World Health Organization. https://www.who.int/docs/default-source/coronaviruse/final-joint-report_origins-studies-6-april-201.pdf?sfvrsn=4f5e5196_1&download=true

World Health Organization. (2022). *Joint External Evaluations (JEE).* https://www.who.int/emergencies/operations/international-health-regulations-monitoring-evaluation-framework/joint-external-evaluations

World Health Organization African Region. (2019). *Technical guidelines for integrated disease surveillance and response in the African Region* (3rd ed.). WHO | Regional Office for Africa. https://www.afro.who.int/publications/technical-guidelines-integrated-disease-surveillance-and-response-african-region-third

World Health Organization, Food and Agriculture Organization of the United Nations, & World Organisation for Animal Health. (2022). *Surveillance and information sharing operational tool.* https://iris.who.int/bitstream/handle/10665/361443/9789240053250-eng.pdf?sequence=1

World Organisation for Animal Health. (2018). *Glossary.* https://www.woah.org/fileadmin/Home/eng/Health_standards/tahc/2018/en_glossaire.htm#terme_surveillance

6 Antimicrobial Resistance – A One Health Issue

Neil Vezeau and Laura H. Kahn

INTRODUCTION TO AMR GLOBALLY

Antimicrobials (AMs) are substances that kill or prevent the growth of microorganisms. This includes antibiotics and antibacterials for bacteria, as well as antivirals, antifungals, and antiparasitics. AMs, alongside vaccines and basic hygienic principles, are the foundation of modern medicine. AMs are also critical for agriculture. The microbes they target have adapted over time to increased use, leading to escalating levels of antimicrobial resistance (AMR). This has decreased their effectiveness in treating infections in humans and animals.

AMR is recognized as a global health crisis (OPGA et al., 2016). It requires interdisciplinary action across human, animal, plant, and environmental health using a One Health (OH) approach (OPGA et al., 2016; Robinson et al., 2016; Toner et al., 2015). The multidimensional nature of AMR makes it "the quintessential One Health issue" (Robinson et al., 2016). Due to the sheer burden of bacteria, and the commonality of antimicrobial use (AMU) in their treatment, this chapter will largely focus on bacterial antibiotic resistance.

AMR in Human Populations

In 2019, there were almost 5 million infections associated with antimicrobial-resistant bacteria (ARB) worldwide, with 1.3 million deaths directly attributable to them (Murray et al., 2022). The World Bank estimates this will cost 3.4 trillion USD from the yearly global domestic product by 2030 (Jonas et al., 2017). The highest number of ARB deaths was attributed to methicillin-resistant *Staphylococcus aureus* (MRSA) (Murray et al., 2022). Other significant ARBs included multidrug-resistant (MDR) tuberculosis, *Escherichia coli*, *Klebsiella pneumoniae*, and *Acinetobacter baumannii*. AMR tends to be lower in higher-income countries (OECD, 2018). Resistance to antibiotics such as third-generation cephalosporins and fluoroquinolones are among the most concerning developments in AMR. In 2012, *S. aureus* received the most published research of any ARB followed by *E. coli*. Resistance to β-lactamases and methicillin received the highest numbers of publications (Brandt et al., 2014).

DOI: 10.1201/9781003232223-6

These are consistent with the World Health Organization (WHO) global priority pathogens list of antibiotic-resistant bacteria. This list states that carbapenem-resistant *A. baumannii*, *Pseudomonas aeruginosa*, and Enterobacteriaceae are the top-priority ARBs in human health (Tacconelli, Carrara, et al., 2018). Enterobacteriaceae that produce extended-spectrum β-lactamases (ESBLs) are also listed as critical to human health. Important ARBs were largely similar between Europe and the rest of the Global North (European Centre for Disease Prevention and Control & World Health Organization, 2022a). In the United States, critical pathogens include carbapenem-resistant *Acinetobacter* and Enterobacteriaceae, resistant *Clostridioides difficile*, *Neisseria gonorrhoeae*, and the fungus *Candida auris* (Centers for Disease Control and Prevention, 2019).

AMR in Livestock

Globally, AMR has been increasing across terrestrial food animal (livestock) species. A key metric employed in research on this topic is the percentage of AMs tested of which more than 50% of bacteria were resistant (P50). Between 2000 and 2018, P50 more than doubled in pigs and poultry in low and middle-income countries (LMICs), and almost doubled in cattle in LMICs (Van Boeckel et al., 2019). The highest proportions of ARB in animals were clustered around urban centers, particularly in India and China. AMR was highest for AMs used most frequently in animal agriculture: tetracyclines, sulfonamides, and penicillins (Van Boeckel et al., 2019).

From 2019 to 2020, European Union (EU) samples of *Salmonella* spp. had high rates of quinolone resistance (European Food Safety Authority & European Centre for Disease Prevention and Control, 2022). While *E. coli* with β-lactamase resistance decreased in food animals, they remained at significant levels. Ciprofloxacin resistance was high but decreasing in *Campylobacter* spp. isolates in many countries. Resistance to critically important fluoroquinolones and cephalosporins was low in animal *E. coli* isolates. There appeared to be an intermixing of human and livestock-associated strains of MRSA in food animals (European Food Safety Authority & European Centre for Disease Prevention and Control, 2022).

From 2018 to 2019, ciprofloxacin resistance in *Salmonella* spp. isolates in the United States rose in chickens from 26% to 32%. In the same period, ciprofloxacin resistance in *Campylobacter jejuni* rose from 21% to 26%. MDR *Salmonella* rose to 31% in turkeys. Extremely drug-resistant *Salmonella* was also present in cattle, swine, and chickens. From 2013 to 2019, there was a ~2.5 and ~4-fold decrease in gentamicin-resistant *Enterococcus faecalis* in turkeys and chickens, respectively. An increase in erythromycin-resistant *Enterococcus* spp. in beef cattle from 2% to 12% occurred during the same period (Center for Veterinary Medicine, FDA, 2024).

AMR in Aquaculture and Fisheries

Aquaculture (farmed aquatic animals) and fisheries (wild-caught aquatic animals) collectively comprise 20% of human animal protein consumption. The most robust figures for AMR in aquaculture come from an analysis of aquaculture and fisheries in 2018. During that year, Asia accounted for 69% of global aquaculture and fisheries production (Schar et al., 2021). From 2000 to 2018, P50 decreased from 52% to 22%. Aquaculture peaked and leveled off at a P50 of 33%. The most common ARBs were *Vibrio* spp., *Aeromonas* spp., and *E. coli*. Similar to livestock in LMICs, AMR was most common against penicillins, sulfonamides, tetracyclines, and macrolides (Schar et al., 2021).

AMR in Companion Animals

Ownership of pets such as cats and dogs, termed "companion animals," has been on the rise worldwide. It is estimated that over half of the 2.3 billion households worldwide have at least one pet (Architecture News and Editorial Desk, 2021; Growth from Knowledge, 2016). AMR and AMU data from companion animals is provided by some national governments, and by university networks in the United Kingdom (UK) and the United States. Additionally, the World Small Animal Veterinary Association is establishing standards and global oversight initiatives. However, livestock remains the largest focus of animal AMR efforts (World Organisation for Animal Health, 2022).

AMU OVERVIEW

Total global AMU was estimated to be approximately 200,000 tons in 2017 between humans, livestock, and aquaculture. By 2030, this overall use is estimated to rise to over 236,000 tons (Schar et al., 2020). In 2017, global human AMU was approximately 50,000 tons, or a rate of 92.2 mg/kg across all human biomass (Schar et al., 2020). In the same year, over 102,000 tons of AMs were sold for use in land-based food animals (livestock), 140 mg/kg, if all were used (Schar et al., 2020; Tiseo et al., 2020). Use in aquaculture was over 10,000 tons, or 164.8 mg/kg (Schar et al., 2020). The human consumption rate is expected to decrease to 91.7 mg/kg by 2030. Overall human AMU is expected to increase with population growth. Aggregate global AMU in crop agriculture is less well characterized (Miller et al., 2022).

Attention for AMR in animals and the environment is most often based on its impact on public health (Cañada et al., 2022). This is why AMU reduction often focuses on types of AMs used in human medicine, often referred to as "medically important antimicrobials" (MIAs). Accordingly, the WHO maintains a list of important AMs in human medicine, respectively (World Health Organization, 2019). The WHO's list is commonly cited to list AMs most important to conserve. The WHO has also created the AWaRe classification system for AMs. It lists AMs that should be: (i) readily Accessible due to the lower likelihood of AMR; (ii) closely Watched for AMR; and (iii) Reserved for last-resort scenarios. The World Organisation for Animal Health (WOAH, formerly OIE) has created a similar listing of AMs important in veterinary medicine (World Organisation for Animal Health, 2021). AMR is primarily driven by AMU but is heavily exacerbated by societal and economic factors including poverty, low education and infrastructure, and poor governance (Collignon et al., 2018).

AMU in Humans

Large-scale estimates of AM usage are often given in defined daily doses (DDD) due to dosage differences between drugs and patients. In 2018, approximately 40.2 billion DDD were consumed globally, or 14.3 DDD per 1,000 people per day (Browne et al., 2021). The latter represents an increase of 46% since 2000 – from 9.8 DDD per 1,000. Though consumption is higher in high-income countries (HICs), global increases are largely driven by LMICs, correlated with their GDP growth (Klein et al., 2018). It is estimated overall use could exceed 100 billion DDD yearly if controls are not implemented. In HICs, per capita use modestly fell by 4% from 2000 to 2015, though overall use increased by 6% (Klein et al., 2018). The use of AMs of last resort, including glycylcyclines, oxazolidinones, carbapenems, and polymyxins, has increased sharply across both socioeconomic categories. Penicillins constitute the highest overall use.

AMU in Livestock

In food animals, AMs have been used as antimicrobial growth promoters (AGPs) for economic and food security purposes. Due to their effects on the gut microbiome and prevention of subclinical infections, AGPs can increase the rate of growth, which creates an economically important use in food production systems. Reduction of animal AMU tends to focus on restriction of MIAs. This largely excludes AMs widely used in animals, such as coccidiostats and ionophores. Animal-source food consumption rises with increasing wealth globally. This results in increased animal production and AMU. As a result, animal AMU will likely increase in prominence in global health research and policy (Tilman et al., 2011).

In 2010, it was estimated that over 63,000 tons of AM were used in livestock. By 2017, estimates for this figure exceeded 102,000 tons. By 2030, the amount is projected to increase to over 105,500 tons. Humans and animals use similar amounts of AM per unit biomass, but total livestock biomass exceeds that of humans by approximately 70% (Bar-On et al., 2018). Rising livestock AMU is expected to be driven by expanding livestock production in LMICs, especially BRICS countries (Brazil, Russia, India, China, and South Africa) (Van Boeckel et al., 2015). It is also projected that proportional growth in animal AMU will outpace that of humans by over a factor of 7. However, official government reporting via WOAH shows animal use to be decreasing on a per-biomass basis (World Organisation for Animal Health, 2022).

Among jurisdictions that do characterize AMU in animals, many report the amount sold or shipped, not the amount administered. Additionally, data are typically not broken down by species or by companion animal or livestock species. Many jurisdictions may not include companion animal AMU (World Organisation for Animal Health, 2022). There are some efforts to report by the amount administered instead of sold (Statens Serum Institut & National Food Institute, Technical University of Denmark, 2022). Standardizing reporting by DDD, as has been done in humans, has proven difficult (Schar et al., 2018).

AMU in Aquaculture

Aquaculture uses the most AMs of all applications per unit biomass. The most-used AM classes in aquaculture, in descending order of use, are: quinolones, tetracyclines, amphenicols, and sulfonamides (Schar et al., 2020). The majority of these were listed as highly important for human medicine by the WHO. In 2017, over 10,000 tons were used, and by 2030, aquaculture AMU is expected to exceed 13,500 tons, or about 5.7% of global human and food animal use (Schar et al., 2020).

AMU in Plant Agricultural Applications

The role of plant agriculture in the current AMR crisis is less characterized compared to that in animal and human sectors, and few nations monitor AMU in plants (World Health Organization et al., 2021). Recent plant AMU estimates range from approximately 163 to 1200 tons – generally under 1% of total animal use (Taylor & Reeder, 2020). Most crop AMU is thought to be in above-ground fruits. These AMs are spread out more thinly; crop biomass is greater than 60 times that of humans and livestock combined (Bar-On et al., 2018).

Many AM substances common in plant agriculture are not used in human and animal medicine. However, several classes are used in all aforementioned applications, particularly aminoglycosides (e.g., streptomycin), tetracyclines, and quinolones. Azole and polyene macrolide antifungals are also used across the three sectors. Pesticides, metal compounds, and other substances used as AMs in plant agriculture can have adverse consequences for human, animal, and environmental health. The use of copper-containing pesticides and some fungicides is known to select or co-select for ARGs relevant to human and veterinary medicine (Miller et al., 2022).

Lower attention is given to crops because the number of pathogens shared between plants, humans, and animals is lesser than the zoonoses shared only by humans and animals. One zoonotic phytopathogen is the WHO priority ARB *P. aeruginosa*, as well as the non-priority *Burkholderia* spp. (Miller et al., 2022; Tacconelli, Carrara, et al., 2018). Several groups of zoonotic fungi can infect plants, including the genera *Fusarium*, *Aspergillus*, and *Claviceps*. The first two are prominent in the WHO's fungal priority pathogens list (World Health Organization, 2022). AMR in plant pathogens can decrease crop yields, exacerbating animal and human health issues via malnutrition, and downstream effects for other crop uses (Miller et al., 2022).

CURRENT KNOWLEDGE ON MECHANISMS OF SPREAD

AMs, and thus AMR, exist naturally – even without anthropogenic AMU (D'Costa et al., 2011). However, human activity appears to heavily exacerbate the presence of AMR in the environment. Increases in AMR largely originate from three main types of AMU, each correlating with the traditional domains of OH:

(i) use in human populations, (ii) use in animal populations, (iii) use in plant populations. Animal populations may be further subdivided into livestock (terrestrial and aquatic), companion animal, and wildlife populations (Figure 6.1). The exact degree to which AMU or AMR in any population leads to resistance in another remains a matter of debate, though most conclude that significant spread does or at least has the potential to occur between humans, animals, and the environment.

ARBs can persist and grow in a large variety of environments, and the presence of AMs in an environment can increase AMR in microbial populations. Additionally, many genes for AMR (ARG) are carried on mobile genetic elements such as plasmids or transposons that can be passed from cell to cell among microorganisms, especially bacteria (Miller et al., 2022). Even viruses can transfer ARGs (He et al., 2020).

Human-to-Human Spread

Humans are most liable to share their environment, and pathogen-drug combinations, with other humans. Globally, hospital-acquired ARB continues to be a significant driver of antimicrobial-resistant infection (ARI) cases (Murray et al., 2022). The highest correlate of AMR emergence in human populations is human AMU (Mendelsohn et al., 2023). Global travel has also been noted as a vector of important resistant strains inter-regionally (Frost et al., 2019).

Human Sewage and Animal Waste

AMs are eliminated via urination and defecation in humans and animals. A significant portion of AMs are eliminated with little biological alteration. Up to 90% of these wastes occur in animal agriculture (Taylor & Reeder, 2020). Tetracyclines and enrofloxacin occur in the highest amounts in animal manure (Jadeja & Worrich, 2022). Waste and waste systems contain large reservoirs of AM, ARB, and ARGs. In 2014, humans and animals produced an estimated 4 trillion tons of feces, 80% of it from animals. This figure is expected to increase alongside population growth, with higher growth from animals (Berendes et al., 2018). Waste systems generally process human waste, not animal waste. In many developing countries, particularly those in South Asia and sub-Saharan Africa, basic sanitation is not available for humans and animals. Hundreds of millions of people openly defecate, increasing the microbial burden in the environment, and increasing the risk of illness and subsequent antibiotic use (Graham et al., 2019).

Soil

Soil is naturally a reservoir of AMs, ARB, and AMR. It is a major resource from which scientists search for novel AMs. Soil is a recipient of human and animal waste for fertilization and disposal purposes and a significant lynchpin in the chain of transmission of AMR. ARGs in soil are most significantly associated with livestock production, as well as other agricultural AM uses, though AMR presence long predates this anthropogenic activity (D'Costa et al., 2011; Zheng et al., 2022).

Approximately half of the habitable land on Earth is used for agriculture, and much of this is irrigated with wastewater and fertilized with solids derived from wastewater (Ritchie & Roser, 2024; Slobodiuk et al., 2021). Soil and the Earth's subsurface are also where the majority of bacteria live (Bar-On et al., 2018). Many AMR inputs come from animal and human waste. AM, ARB, and ARG can be reduced in waste via composting, heating, or other methods (Jadeja & Worrich, 2022; Slobodiuk et al., 2021).

Water

Human and animal waste systems, agricultural and industrial runoff, and effluents flow into the Earth's water systems. These sources contribute AMs, ARG, and ARBs. Though proportional causality is difficult to attribute, significant contributors are human waste treatment effluents, agricultural runoff stemming from livestock waste, or direct livestock waste effluents (He et al., 2020; Ma et al., 2021).

Waste effluents with the most clinically significant human pathogens and resistances most likely come from hospitals (Alexander et al., 2020). However, ARG concentration in swine and chicken waste is three to five orders of magnitude higher than that found in human waste from hospital and municipal sources (He et al., 2020). Fish and cattle waste has comparable ARG levels to that of human waste.

Wastewater treatment can significantly reduce, but not completely eliminate, a variety of ARGs and ARBs (Miłobedzka et al., 2022). Wastewater treatment has a lesser effect on the presence of AMs (Mackuľak et al., 2021). Depending on the jurisdiction, livestock wastewater may not be treated before being allowed to return to river water (He et al., 2020). River systems downstream from water treatment facilities have been shown to carry AMs common in human use. These AMs are also found in the fish in these environments. However, the fish ARBs did not have resistances proportional to the AM levels in the aquatic environment (Ballash et al., 2022). Anthropogenic AM and ARB runoff also leads to increased AMR in groundwater and the ocean (Singer et al., 2016). Due to the variety of diagnostics for AMs, ARBs, and ARGs in aquatic and waste environments, study-to-study comparisons can be difficult, and further standardization is needed (Liguori et al., 2022).

Arthropod-Borne Spread

Arthropods, particularly the housefly, can transfer ARB from livestock to human residences (Graham et al., 2009). These ARB are genetically linked to livestock operations (Zurek & Ghosh, 2014). Although they are primarily mechanical vectors that spread fecal matter, ARB can multiply on the surface of flies and inside their gastrointestinal tracts (Onwugamba et al., 2018).

Airborne

Carriage of ARB in aerosolized particulate matter is a particular risk from livestock fecal matter (McEachran et al., 2015). Additionally, ARGs have been found diffusely in the air of urban environments in mainland China (Li et al., 2018).

Wildlife Reservoirs

Wild animals also carry ARB. Numerous bird species have been found containing ARB linked to nearby livestock facilities. Particularly well-studied species include wild boar, cervids, migratory waterfowl, and small mammals. Migrating avians are considered to be of particular concern for AMR spread (Dolejska & Literak, 2019). They can circumvent land, water, and artificial structures that block non-flighted animals. Their migratory nature allows them to connect disparate communities with otherwise widely different ARG and ARB profiles. Waterfowl and boars have been found to carry bacteria with ESBLs (Dolejska & Literak, 2019; Plaza-Rodríguez et al., 2021). Extensive AMR has also been identified in wild fish, with ARB showing at least 88% resistance to ampicillin, chloramphenicol, kanamycin, and streptomycin (Ozaktas et al., 2012).

Invasive species also present risks, as they lack natural ecological counterbalances to their activity and spread. Research on AMR in nondomestic species has led to wildlife being highlighted as potential sentinels of ARB and ARG prevalence (Ballash et al., 2022; Plaza-Rodríguez et al., 2021).

ARB prevalence in most wildlife populations studied appears to be lower than in humans and domestic animals, and livestock biomass exceeds that of wild mammals and birds by a factor of 10 (Bar-On et al., 2018). Wildlife populations appear to be a net recipient of AMR, largely originating from prescribing and other anthropogenic activity, though this does not preclude their ability to spread it as reservoirs.

Foodborne

One of the most direct ways ARBs and ARGs can be transferred from animal populations to wider human populations is via the food supply. In the EU from 2019 to 2020, ampicillin, tetracycline, and sulfonamide resistance was found to be at moderate to high levels in foods (European Food Safety Authority & European Centre for Disease Prevention and Control, 2022). AMR is also present in consumed wild animals (Guerrero-Ramos et al., 2016). As with all foodborne pathogens, ARB risks can be reduced via cooking and other food preparation techniques. Stomach acid also greatly decreases the pathogenic load of ingesta.

One study found over 10% of U.S. ground meat products have MRSA and over 20% have *Enterococcus* with β-lactamases (Ballash et al., 2021). Some work found moderately lower AMR in "organic" beef advertised to be raised without AMs (Schmidt et al., 2021). Products "raised without antibiotics" are not necessarily free of ARB for beef and pork products (Vikram et al., 2018, 2019).

AMR is in plant-sourced foods because crop agriculture uses AMs, and waste from humans and animals used to irrigate and fertilize them can contaminate them with AMR (Miller et al., 2022). ARG load in plants is higher if raw, untreated manure is used as fertilizer, and ARG may be proportional to AMU of the livestock source of manure. Similar effects are observed in plant agriculture with untreated human wastewater (Guron et al., 2019). Some work suggests plant source food poses a lower AMR risk than animal source food (Ngaruka et al., 2021).

Food Animal-Human Spread

Close contact with animals also appears to be a route of spread. Midwestern U.S. farm workers have been found to carry the same MRSA strain as their swine (Smith et al., 2013). Mutual carriage of ARB in swine and their human carriers has been shown to correlate with AMU in food animals (Smith et al., 2013). Spread from animals and their carcasses to abattoir workers is also known to occur (Neyra et al., 2014).

Companion Animal-Human Spread

Emerging evidence describes bi-directional AMR spread between humans and companion animals. Animals and humans living in close contact appear to share common bacteria (Song et al., 2013). Humans and their pets may also carry the same MRSA genotypes (Morris et al., 2012; Weese & Van Duijkeren, 2010). They can also share antimicrobial-resistant infections (ARIs)s (Marques et al., 2019). ARB such as MRSA may be further spread by other pets, such as cats and dogs, and by veterinarians. Human-to-equine spread of MRSA has also been found (Weese & Van Duijkeren, 2010).

Animal-community Spread

AMR can spread from swine operations to the surrounding community (Neyra et al., 2014), and potentially cause disease (Feingold et al., 2012). MRSA is particularly problematic. Across the EU, consumption by food animals of 3rd- and 4th-generation cephalosporins and fluoroquinolones was associated with increased resistance in human *E. coli* isolates (European Centre for Disease Prevention and Control [ECDC] et al., 2021). A similar situation was found with *Salmonella* and *C. jejuni* tetracycline resistance.

HISTORY OF AMU AND REGULATION

Although AMs such as arsenicals and sulfonamides were used in patients before penicillin, its discovery in 1928 by Alexander Fleming heralded a new age of modern medicine (Landecker, 2019). In 1938, sulphonamide AMs such as Prontosil were marketed for veterinary applications. AGP use developed afterward, marking the beginning of large-scale agricultural applications (Kirchhelle, 2018).

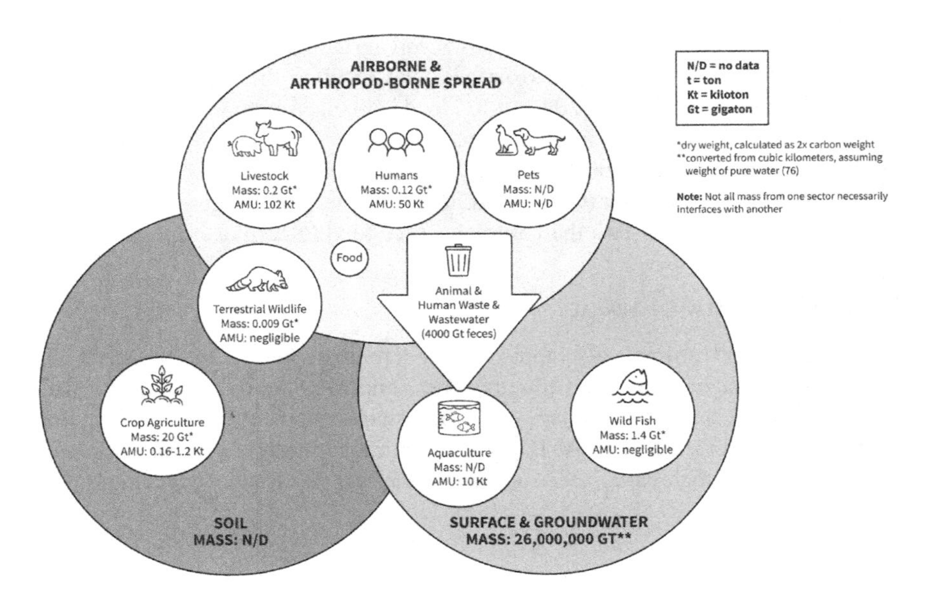

FIGURE 6.1 Conceptual map of AMR reservoirs and routes of spread.
N/D = no data; t = ton; Kt = kiloton; Gt = gigaton
*dry weight, calculated as 2× carbon weight.
**converted from cubic kilometers, assuming weight of pure water (Water Science School, 2019).
Note: Not all mass from one sector necessarily interfaces with another

Use expanded to the prevention of disease and spoilage in animals, plants, and their products. These were used with and without the requirement of prescriptions. Agricultural antibiotic use was led by Europe and spread globally involving regional staple crops such as rice. AMs in crop agriculture grew alongside animal production. Meat and seafood production expanded almost five-fold from 1961 to 2020, and wheat production almost two-fold from 1961 to 2016 (Ritchie et al., 2023; Tadesse et al., 2019). Agricultural uses for AMs propagated worldwide through the 1960s and have continued to this day. AMU increased alongside manufacturing capacity and economic development regionally, and regulations arose unevenly and at different times in different regions (Kirchhelle, 2018).

In 1969, the UK released the Swann report in the wake of several regulatory issues with AM residues (Kahn, 2016). This report is regarded as the first prominent work to raise the issue of AMU in animal agriculture as a public health risk. It eventually led to several restrictions on medically important AMU in the UK in the early 1970s.

In the following decades, various countries began banning the use of medically important AMs as AGPs, some eventually banning all AGP use. From 1971 to 1976, various European countries banned animal uses of various AMs fundamental in human medical care, including tetracycline, penicillin, and streptomycin. Sweden was first to ban all AGPs in 1986 (Kahn, 2016).

Avoparcin is a glycopeptide chemically related to vancomycin, an AM of last resort. Avoparcin's use in animals is thought to have contributed to the rise of vancomycin-resistant enterococci (VRE) in hospitals. AGP use of avoparcin was banned in Denmark in 1995 and Germany in 1996, and the whole EU in 1997. All AGPs were banned in Denmark in 2000 and the whole EU in 2006. The rise of VRE in hospitals was viewed as a substantial instigator for this (Kahn, 2016). This experience from the EU resulted in the WHO recommending ending AGP use (World Health Organization, 2020b). In 2017, the United States issued a voluntary guidance for the food animal industry to stop using MIA as AGPs, resulting in a *de facto* ban on such uses (Kirchhelle, 2018). By 2021, 112 WOAH member countries reportedly banned AGPs, while 46 allowed their use (World Organisation for Animal Health, 2022).

As AGPs were banned in certain jurisdictions, some concomitant increases in livestock AMU occurred for prophylaxis or disease prevention. This resulted in more limited drops in livestock AMU. Successful policies aimed at lowering overall use were developed to target this phenomenon, such as Denmark's "Yellow Card" initiative beginning in 2010 (Food and Agriculture Organization of the United Nations, 2019). In 2022, the EU moved to ban the use of AM prophylaxis for livestock for most use cases (More et al., 2022).

TABLE 6.1
Timeline of Significant Developments and Regulatory Events in AMR through a OH Lens

Year	Event
1935	Sulphonamides first sold for use in humans
1938	Sulphonamides first sold for use in animals
1951– onward	AGP use approved throughout a variety of countries
1961	Blasticidin S. licensed for use against fungal infection in plants
1969	Release of the Swann report in the UK regarding MIA use in agriculture
1971–1976	Scattered penicillin and tetracycline AGP bans across Europe in the wake of Swann report
1986	Swedish ban of all AGPs
1993	VRE detected in Danish pigs, suspected as cross-resistance from avoparcin use
1995	Denmark livestock avoparcin ban
1996	German livestock avoparcin ban
1997	EU livestock avoparcin ban
1997	Japanese ban of avoparcin livestock feed additives
1999	Finalization of Denmark voluntary livestock AGP ban
2017	United States voluntary AGP ban
2022	Significant livestock AM prophylaxis restrictions in EU

Source: Adapted from Kahn (2016); Kirchhelle (2018); More et al. (2022).

FIGURE 6.2 Antimicrobial resistance NAPs.†

Countries in gray have 1 AMR NAP listed on the WHO Library of national action plans, while those in black have 2. Adapted from the WHO Library of national action plans (World Health Organization, 2024).
†The WHO Library also lists the NAP of the partially recognized state Kosovo, not pictured above.

Some jurisdictions disincentivize AMU in animals by limiting profits to veterinarians for prescribing or otherwise dispensing AMs, including Finland, Norway, and Denmark (Levy, 2014; Ollila, 2020). The purchase of AMs without a prescription ("over-the-counter"), even for humans, has comprised a large portion of AMU in LMICs (Morgan et al., 2011). Limited enforcement of laws restricting over-the-counter use and the ability to purchase AMs online inhibits stewardship efforts (OECD, 2018). Regulations also exist for AMU in crop agriculture. These regulations appear to be more common in HICs (Taylor & Reeder, 2020).

AMR GOVERNANCE AND GLOBAL COORDINATION

Global Action Plan on AMR

In 2014, the member states of the WHO voted on the importance of AMR as a global health crisis. The WHO followed by creating a global action plan (GAP) on AMR in 2015. All WHO members signed onto the GAP. It is now among the most important documents targeting global coordination on AMR (Overton et al., 2021). Numerous countries published their own national action plans (NAPs; Figure 6.2), many of which highlighted priorities from the GAP (Cañada et al., 2022;

Van Boeckel et al., 2015). These action plans have become the primary vehicles for publicizing policy efforts on AMR.

Progress on these NAPs is monitored via the Tracking (formerly "Tripartite") AMR Country Self-Assessment Survey (TrACSS). This is a system for countries to centralize evaluations of AMR-mitigating regulations and activities, coordinated by the WHO. The TrACSS database shows self-evaluation data going back to 2016 (World Health Organization et al., 2021).

UN Interagency Coordinating Group on AMR (IACG)

The United Nations (UN) Interagency Coordination Group on AMR (IACG) was a multidisciplinary group formed by the UN to give guidance to streamline global efforts against AMR. In 2016, UN member states adopted the Political Declaration of the High-level Meeting on AMR, advising the IACG's formation. The IACG published the 2019 report, "No Time to Wait," which built upon the GAP and recommended the formation of a variety of international groups to aggregate knowledge and coordinate policy on AMR (Interagency Coordination Group on Antimicrobial Resistance, 2019).

One Health Quadripartite and Related Global Coordination

The One Health Quadripartite is a collaborative framework between the Food and Agriculture Organization of the UN (FAO), the WHO, WOAH, and the UN Environment Programme (UNEP). This framework's purpose is to coordinate the constituent organizations' efforts using a OH approach. They have emphasized the need for a OH approach in global leadership against AMR (World Health Organization et al., 2024).

This relationship began as the "Tripartite" between the FAO, WHO, and WOAH in 2008. They specified AMR as an area of focus in 2018. The UNEP was formally added to create the "Quadripartite" in 2022. This coordination mechanism is further augmented by the AMR Multi-Partner Trust Fund and the Global Leaders Group on AMR, both administered by Quadripartite organizations.

AM Development and Funding Mechanisms

The rate that new AMs and AM classes are being developed has slowed dramatically in the past several decades. This is because the "lowest-hanging fruit" in AM development has already been discovered. This means the cost/benefit ratio of research and development for new AMs has been increasing, making novel drug development for AMs less financially enticing for biopharmaceutical companies. Because of this disincentivization in the private sector, the public sector, intergovernmental organizations, non-governmental organizations (NGOs), and civil society have mobilized resources to both directly contribute and create pipelines of funding for novel AM development.

The Global Antibiotic Research and Development Partnership (GARDP) is one of the most prominent actors in this space. Originally an initiative of the WHO and the Drugs for Neglected Diseases initiative (DNDi), it creates cross-sectoral collaborations to find solutions for AMR. Combating Antibiotic Resistant Bacteria Biopharmaceutical Accelerator (CARB-X) is also a group creating pipelines from NGOs, governments, and other funding sources for AM development. It also funds other AMR solutions such as vaccine initiatives. The Joint Programming Initiative on Antimicrobial Resistance (JPIAMR) is another collaborative platform helping coordinate governmental funding, primarily from the EU. There have even been a variety of civilian science movements to discover naturally produced AMs from soil and other environmental samples.

Despite AM development funding, health system and sanitary interventions other than AM development, such as improving hygiene and prescribing practices, are more cost-effective than AM development (OECD, 2018). Analysis from the World Bank notes that AMR coordination among the WHO, FAO, and WOAH has been underfunded, but this has been improving. Funding for nations developing AMR NAPs has been scarce as well (World Health Organization et al., 2021). It is critical that funding for all AMR fighting initiatives must be global, multi-sectoral, and long term in scope (Jonas et al., 2017).

Other Major Actors

In addition to governmental and intergovernmental actors, there has been no shortage of civil society groups and NGOs attempting to impact positive change on AMR using a OH approach. Some serve to educate, generate analysis on AMR, and offer guidance on AM stewardship. Some of these include the Global AMR R&D Hub and ReAct. The International Centre for Antimicrobial Resistance Solutions (ICARS) helps to develop AMR solutions for LMICs, including policy-focused ones. Some intergovernmental and interorganizational groups exist to help coordinate policy between countries, including the Transatlantic Taskforce on Antimicrobial Resistance (TATFAR) and Joint Action on Antimicrobial Resistance and Healthcare-Associated Infections (JAMRAI). The largest NGOs with a global health focus, including the Bill and Melinda Gates Foundation and the Wellcome Trust, also fund and conduct significant amounts of AMR-related programming. Many organizations related to the UN and the World Bank conduct AMR activities (Wernli et al., 2022).

AMR SURVEILLANCE NETWORKS

Global Antimicrobial Resistance and Use Surveillance System

The most comprehensive country-level data on AMU in human populations has been collected in the WHO's Global Antimicrobial Resistance and Use Surveillance System (GLASS). First implemented in 2015, GLASS aggregates and harmonizes AMR and AMU data from most countries globally and reports on this data annually.

The ARBs targeted for surveillance are *Acinetobacter* spp., *E. coli*, *K. pneumoniae*, *N. gonorrhoeae*, *Salmonella* spp., *Shigella* spp., *S. aureus*, and *S. pneumoniae* (World Health Organization, 2021). Gaps in GLASS data may be covered by scholarly studies using statistical models (Murray et al., 2022). As of writing, there is no equivalently robust global monitoring system for AMR and AMU in animals or the environment. An International FAO Antimicrobial Resistance Monitoring (InFARM) System is being developed for AMR surveillance in food and agricultural settings (Food and Agriculture Organization of the United Nations, 2024).

Regional AMR Surveillance Networks

Many regions have AMR surveillance systems that are roughly contiguous with the WHO regions. Many directly coordinate with WHO regional offices. These include Red Latinoamericana de Vigilancia de la Resistencia a los Antimicrobianos (ReLAVRA), the European Antimicrobial Resistance Surveillance Network (EARS-Net), its newly emerging veterinary counterpart EARS-Vet, the Central Asian European Surveillance of Antimicrobial Resistance (CAESAR), and the Western Pacific Regional Antimicrobial Consumption Surveillance System (WPRACSS) (Mader et al., 2022; World Health Organization, 2021). They represent global coordination to recognize priority pathogens and resistances, and the standardization of laboratory methods and infrastructure.

National Surveillance Networks

There is an expanding number of nationally administered AMR surveillance systems, though their methods and areas of focus vary widely. One study identified 71 country-level surveillance systems worldwide. Thirty-seven were in Europe and 14 were in the Americas, with some nations having more than one. Twelve of these monitored both humans and animals, and six focused on zoonotic pathogens. None monitored environmental samples (Diallo et al., 2020). However, TrACSS data noted in the 2019–2020 period that more countries reported having AMU and sale surveillance systems for animals (Food and Agriculture Organization of the United Nations, 2019) than for humans (Water Science School, 2019). For this same time period, 40% of TrACSS respondents reported monitoring AMU in crop agriculture (World Health Organization et al., 2021). Surveillance systems, especially animal-focused ones, were concentrated in HICs. ARBs of focus were *S. aureus*, *E. coli*, MRSA, and VRE (Diallo et al., 2020). Importantly, the definition of clinical breakpoints in minimal inhibitory concentrations (MICs), among other important reporting parameters, is not standardized between systems. This is especially true between the U.S. Clinical and Laboratory Standards Institute (CLSI) and the European Committee on Antimicrobial Susceptibility Testing (EUCAST), which constitute the globally dominant standards.

Most EU member countries have their own AMR surveillance system, each contributing to EARS-Net (Tacconelli, Sifakis, et al., 2018). The United States has the National Antimicrobial Resistance Monitoring System (NARMS), coordinated between multiple agencies to consolidate data from humans, animals, and animal-source foods (Center for Veterinary Medicine, FDA, 2024).

The differences between these surveillance networks are even more disparate among LMICs (Iskandar et al., 2021). In these areas, basic diagnostic facilities, personnel, and supplies can be particularly lacking. However, the Wellcome Trust, the UK Fleming Fund, and the Bill and Melinda Gates Foundation offer them programmatic support, and GLASS helps them standardize and consolidate data.

TABLE 6.2
Select National AMR and AMU Surveillance Systems

Country	Surveillance Network Name(s)	Animal Reporting Included?
Argentina	WHONET-Argentina	No
Australia	AURA, CARAlert	No
Austria	NRZ	No
Canada	CARSS, CNISP, CIPARS	Yes
China	CHINET, MONHARIN	No
Croatia	ISKRA	No
Denmark	DANMAP	Yes
Finland	FIRE, FINRES-Vet	Yes
France	ONERBA, RESAPATH	Yes
Germany	KISS, ARS, SARI, GERME-Vet	Yes
Greece	GSSAR	No
India	I-AMRSS	No
Italy	MIB, ITAVARM, AR-ISS, ENTERNET	Yes
Japan	JANIS, JVARM	Yes
The Netherlands	ISIS-AR, MARAN	Yes
Norway	NORM, MSIS	No
Philippines	ARSP	No
Romania	n/a	No
Slovakia	SNARS	No
Spain	VIV	Yes (no human component)
South Africa	GERMS-SA	No
South Korea	Kor-GLASS, KONIS	No
Sweden	SVEBAR, SVARM	Yes
Switzerland	CA-MRSA, ANRESIS	Yes
Thailand	NARST	No
Turkey	NNIS	No
United Kingdom	VARSS	Yes (no human component)
United States	ARMOR, NTSS, NARMS, GISP	Yes

Source: Adapted from Diallo et al. (2020).
Note that no national AMR and AMU surveillance systems were found to report environmental data.

Subnational Governmental Efforts

Some subnational entities have also started initiatives in Australia, China, India, the UK, and the United States in implementing a OH approach. They have created AMR action plans, advisory councils, and integrated monitoring structures. These efforts tend to vary in focus between collecting data on AMU, resistance, and education on stewardship.

Whole Genome Sequencing in Surveillance

Whole genome sequencing (WGS) is a technique that allows sequencing an entire pathogen's genome, allowing for a more complete picture of its resistance-encoding genes. WGS also aids in tracking lineages of bacteria and genes over space and time. It allows for more characterization of these and other important metrics compared to older technologies like pulsed-field gel electrophoresis (PFGE). WGS's increasing use worldwide is of critical importance in advancing AMR surveillance infrastructure.

However, detection of the plasmid origin of resistances in the genome, or of plasmids altogether, can be a challenge with WGS (Roosaare et al., 2018). It should be used in combination with phenotypic characterization methods such as MIC and does not replace them (World Health Organization, 2020a). The proliferation of WGS-capable lab equipment is an ongoing and substantial effort across public health laboratories worldwide. These technologies are more common in HICs than in LMICs (World Health Organization, 2020a).

INTERVENTIONS FOR DECREASING AMU AND RESISTANCE

AM stewardship is paramount to mitigating AMR. In addition to preventing increased AMR incidence, increasing rational and judicious use can decrease the selective pressure for resistance. This can in turn decrease AMR.

WASH and Basic Interventions

Improving and maintaining water, sanitation, and hygiene (WASH) behaviors and infrastructure is paramount in mitigating AMR (OECD, 2018). This is true across all domains of OH. Proper WASH can prevent contamination that risks ARI or the need for AM therapy in the first place. This also includes food safety measures and proper cooking. Improved animal manure management is essential. Sanitation systems typically do not process animal waste, only human waste. Human and animal fecal contamination can cause foodborne and waterborne illness leading to AMU and AMR.

Clinician-Level Methods

Vaccination is a first-line method of reducing AM use in human and veterinary medicine. Even in cases where vaccines don't inoculate against ARIs of concern, they can decrease ARI co-infections and reduce differential diagnoses when considering ARIs (OECD, 2018).

Education of prescribing health professionals can be an effective tool, particularly used in combination with hospital-level AM stewardship programs. One important focus of education in a hospital setting includes training on determining which, if any, empiric therapy is appropriate. Another is training on when to de-escalate therapy to less critically important drugs or to discontinue entirely. Following clinical guidelines for selecting differential diagnoses, empiric treatment, and proper diagnostics is paramount to choosing the most prudent course of therapy (OECD, 2018).

Health Systems

Stewardship policies at the hospital/clinic and governmental level can also decrease AMU. Antibiograms help prescribers know which infections may be susceptible to which AMs. Integrating antibiograms across OH domains can ensure further coordination, especially at the local and regional level. At the governmental and intergovernmental levels, improving coordination mechanisms between OH sectors is critical to facilitate collaboration and data-sharing. In livestock populations, better herd management practices, biosecurity, and decreased stocking density decrease the need for AMU (Food and Agriculture Organization of the United Nations, 2019).

Research and Development of Alternate Therapies

Bacteriophages remain an avenue to treat ARI, but the need for a serotype-specific virus and a relative scarcity of research are barriers to implementation. The use of naturally occurring AM peptides for therapy has also been investigated.

CASE STUDIES

OUTBREAK RESPONSE AND PATIENT CARE REGARDING *SALMONELLA* HEIDELBERG ASSOCIATED WITH DAIRY CATTLE IN THE UNITED STATES

From 2015 to 2018, there was an outbreak of *S. enterica* serovar Heidelberg across a variety of states in the United States. There were 64 known human cases associated with the outbreak in total, which peaked in July 2016 (Nichols et al., 2022). Forty of the 64 human patients had contact with cattle, and 11 specifically had contact with young dairy cattle (calves) (Nichols et al., 2022). Evidence, such as shared PFGE patterns, suggested that transmission had occurred to family members from those who had previous contact with cattle, although WGS might have provided stronger evidence if performed. The dairy calves predominantly originated from the U.S. state of Wisconsin. Human cases were also centered around this region. The U.S. CDC handled the association of human cases to track the epidemic. The

USDA's Animal and Plant Health Inspection Service (APHIS) performed animal purchasing traceback and bacterial genomic sequencing. They also lent guidance to dairy producers to prevent disease spread (Department of Agriculture, Trade and Consumer Protection; State of Wisconsin, 2016). Disease control efforts were especially important for calves, due to the lack of AMs to which the strain was susceptible.

Wisconsin's state agencies and university extension program were crucial collaborators. Their veterinary diagnostic laboratory, located within the state's university system, collaborated to provide bacterial genome sequences (Wisconsin Veterinary Diagnostic Laboratory, 2021). This case displays coordination between national and subnational governmental entities, as well as across multiple government agencies at the national level. It highlights the importance of intersectoral coordination and inter-organizational data-sharing, as well as the benefits of WGS. It also highlights environmental samples as an indicator of zoonotic spread. It also shows epidemiologic patterns that ARB can take from livestock to wider human communities. This case additionally highlights the utility of WGS over PFGE in confirming the common descent of disease strains.

INTEGRATION OF SURVEILLANCE AND REPORTING MECHANISMS IN DENMARK

First implemented in 1995, the Danish Integrated Antimicrobial Resistance Monitoring and Research Program (DANMAP) is the oldest national system for monitoring human and animal AMR. Its initiation followed a 1993 report highlighting animal food production as a potential reservoir of VRE. Jointly brought about by Danish Ministries of Health and Food, Agriculture, and Fisheries, DANMAP monitors AMR from humans, animals, and animal-source food. Numerous national AMR monitoring programs have been created since, following this general model. In DANMAP, samples are taken from live animals, animal food products, and humans for ARB. Augmenting AMR reporting, DANMAPs' MedStat and VetStat components report human and animal AMU, respectively. Academia also plays an integral role – the Technical University of Denmark combines and analyzes AMR and AMU data into publicly available annual reports.

DANMAP would soon play an important role in observing and combating AMU for disease prevention purposes in swine. AGPs were banned in adult swine in Denmark in 1998, and in young swine the next year. However, swine AMU continued rising throughout the next decade. This phenomenon of an AGP ban followed by significant increases in another use category had previously been observed in Sweden (Kirchhelle, 2018). It was theorized that different use categories were used to "compensate" for lack of AGPs. To combat this, Denmark implemented its "Yellow Card" system in 2010 (Dupont et al., 2017). This system requires corrective action from producers failing to bring down AMU in livestock below a certain level, regardless of rationale for use. Since the implementation of this program, AMU in livestock has decreased yearly (Food and Agriculture Organization of the United Nations, 2019).

In the absence of AGPs, the industry had to compensate with a variety of other production and health practices, such as decreasing animal stocking density in facilities. This caused pork prices to rise. Since the decrease of macrolide AMU in livestock, macrolide-resistance has also decreased (Food and Agriculture Organization of the United Nations, 2019). The case of Denmark is the model upon which many nations have structured their actions against AMR.

DETECTION AND CONTROL OF COLISTIN-RESISTANT *E. COLI* IN CHINESE SWINE

China is the world's largest consumer of AMs and has seen rapid economic growth in the past several decades (Lim & Grohn, 2021). It is one of the most populous countries in the world and is the largest exporter of food outside of the United States and Europe. AMU is significant in the region. In 2017, China consumed 45% of all AMs used in animals globally (Tiseo et al., 2020).

Colistin has increasingly become an AM of last resort due to rising resistance to other drugs. Until recent years, it was used extensively as an AGP in Chinese swine. In 2015, a novel plasmid ARG was found in *E. coli* associated with these pigs, *mcr-1*. This gene propagated globally through human and animal populations (Wang et al., 2020).

In 2016, the Chinese government banned colistin as an AGP. From 2015 to 2018, colistin premix sales for swine plummeted almost 90% in China. From 2017 to 2018, colistin fecal residues in livestock decreased by a factor of 20. *mcr-1* presence in regional human *E. coli* samples fell from 14.3% in 2016 to 6.3% in 2019 (Wang et al., 2020). Official government figures show this is the largest proportional decrease in livestock AMU ever on a per-year basis. However, more detailed data have not been made available (Zhao et al., 2021). Despite data transparency concerns, China has evidenced how significantly political will and policy action can reverse the tide of AMR.

CONCLUSION

AMR continues to be one of the most complex, "wicked problems" modern society faces, especially when viewed through a OH lens. A single chapter cannot cover every dimension of the global AMR crisis and should be taken as a brief overview. Moving forward, it is critical that those working in and proximal to

AM stewardship efforts have an understanding of the complex and outwardly radiating effects of their work.

Mitigating the prevalence of AMR is possible, but it will require the expansion of stewardship and surveillance efforts across all OH domains. Stewardship and resistance monitoring efforts will be essential to strengthen in LMICs, where AMU will increase the most – in both humans and animals. A variety of HICs have created examples of gathering high-resolution data. Several large LMICs have the opportunity to set global standards for regions without extensive funding and infrastructural resources. Methods for AMU and AMR reduction, such as sanitation standards and vaccines, are becoming increasingly available for these purposes.

For over a century, AMs have served as a foundation of modern medicine. We need to work with nature and equitably across disciplines and borders to ensure they stay effective.

REFERENCES

Alexander, J., Hembach, N., & Schwartz, T. (2020). Evaluation of antibiotic resistance dissemination by wastewater treatment plant effluents with different catchment areas in Germany. *Scientific Reports, 10*(1), 8952. https://doi.org/10.1038/s41598-020-65635-4

Architecture News and Editorial Desk. (2021, November 11). *How many houses are in the world?* Architecture & Design. https://www.architectureanddesign.com.au/features/list/how-many-houses-are-in-the-world

Ballash, G. A., Albers, A. L., Mollenkopf, D. F., Sechrist, E., Adams, R. J., & Wittum, T. E. (2021). Antimicrobial resistant bacteria recovered from retail ground meat products in the US include a Raoultella ornithinolytica co-harboring blaKPC-2 and blaNDM-5. *Scientific Reports, 11*(1), 14041. https://doi.org/10.1038/s41598-021-93362-x

Ballash, G. A., Baesu, A., Lee, S., Mills, M. C., Mollenkopf, D. F., Sullivan, S. M. P., Lee, J., Bayen, S., & Wittum, T. E. (2022). Fish as sentinels of antimicrobial resistant bacteria, epidemic carbapenemase genes, and antibiotics in surface water. *PLoS One, 17*(9), e0272806. https://doi.org/10.1371/journal.pone.0272806

Bar-On, Y. M., Phillips, R., & Milo, R. (2018). The biomass distribution on Earth. *Proceedings of the National Academy of Sciences, 115*(25), 6506–6511. https://doi.org/10.1073/pnas.1711842115

Berendes, D. M., Yang, P. J., Lai, A., Hu, D., & Brown, J. (2018). Estimation of global recoverable human and animal faecal biomass. *Nature Sustainability, 1*(11), 679–685. https://doi.org/10.1038/s41893-018-0167-0

Brandt, C., Makarewicz, O., Fischer, T., Stein, C., Pfeifer, Y., Werner, G., & Pletz, M. W. (2014). The bigger picture: The history of antibiotics and antimicrobial resistance displayed by scientometric data. *International Journal of Antimicrobial Agents, 44*(5), 424–430. https://doi.org/10.1016/j.ijantimicag.2014.08.001

Browne, A. J., Chipeta, M. G., Haines-Woodhouse, G., Kumaran, E. P. A., Hamadani, B. H. K., Zaraa, S., Henry, N. J., Deshpande, A., Reiner, R. C., Day, N. P. J., Lopez, A. D., Dunachie, S., Moore, C. E., Stergachis, A., Hay, S. I., & Dolecek, C. (2021). Global antibiotic consumption and usage in humans, 2000–18: A spatial modelling study. *The Lancet Planetary Health, 5*(12), e893–e904. https://doi.org/10.1016/S2542-5196(21)00280-1

Cañada, J. A., Sariola, S., & Butcher, A. (2022). In critique of anthropocentrism: A more-than-human ethical framework for antimicrobial resistance. *Medical Humanities, 48*(4), e16–e16. https://doi.org/10.1136/medhum-2021-012309

Center for Veterinary Medicine, FDA. (2024, May 1). *2019 NARMS update: Integrated report summary*. FDA. https://www.fda.gov/animal-veterinary/national-antimicrobial-resistance-monitoring-system/2019-narms-update-integrated-report-summary

Centers for Disease Control and Prevention. (2019). *Antibiotic resistance threats in the United States, 2019*. Centers for Disease Control and Prevention (U.S.). https://doi.org/10.15620/cdc:82532

Collignon, P., Beggs, J. J., Walsh, T. R., Gandra, S., & Laxminarayan, R. (2018). Anthropological and socioeconomic factors contributing to global antimicrobial resistance: A univariate and multivariable analysis. *The Lancet Planetary Health, 2*(9), e398–e405. https://doi.org/10.1016/S2542-5196(18)30186-4

D'Costa, V. M., King, C. E., Kalan, L., Morar, M., Sung, W. W. L., Schwarz, C., Froese, D., Zazula, G., Calmels, F., Debruyne, R., Golding, G. B., Poinar, H. N., & Wright, G. D. (2011). Antibiotic resistance is ancient. *Nature, 477*(7365), 457–461. https://doi.org/10.1038/nature10388

Department of Agriculture, Trade and Consumer Protection; State of Wisconsin. (2016, November 29). *Salmonella infections linked to dairy bull calves*. https://datcp.wi.gov/Pages/News_Media/2016.11.29_SalmonellaHeidelberg.aspx

Diallo, O. O., Baron, S. A., Abat, C., Colson, P., Chaudet, H., & Rolain, J.-M. (2020). Antibiotic resistance surveillance systems: A review. *Journal of Global Antimicrobial Resistance, 23*, 430–438. https://doi.org/10.1016/j.jgar.2020.10.009

Dolejska, M., & Literak, I. (2019). Wildlife is overlooked in the epidemiology of medically important antibiotic-resistant bacteria. *Antimicrobial Agents and Chemotherapy, 63*(8), e01167–19. https://doi.org/10.1128/AAC.01167-19

Dupont, N., Diness, L. H., Fertner, M., Kristensen, C. S., & Stege, H. (2017). Antimicrobial reduction measures applied in Danish pig herds following the introduction of the "Yellow Card" antimicrobial scheme. *Preventive Veterinary Medicine, 138*, 9–16. https://doi.org/10.1016/j.prevetmed.2016.12.019

European Centre for Disease Prevention and Control & World Health Organization. (2022). *Antimicrobial resistance surveillance in Europe 2022: 2020 data*. World Health Organization. Regional Office for Europe. https://iris.who.int/handle/10665/351141

European Centre for Disease Prevention and Control (ECDC), European Food Safety Authority (EFSA), & European Medicines Agency (EMA). (2021). *Third joint inter-agency report on integrated analysis of consumption of antimicrobial agents and occurrence of antimicrobial resistance in bacteria from humans and food-producing animals in the EU/EEA, JIACRA III. 2016–2018*. ECDC, EFSA, EMA. https://www.ecdc.europa.eu/en/publications-data/third-joint-interagency-antimicrobial-consumption-and-resistance-analysis-report

European Food Safety Authority & European Centre for Disease Prevention and Control. (2022). The European Union Summary report on antimicrobial resistance in zoonotic and indicator bacteria from humans, animals and food in 2019–2020. *EFSA Journal, 20*(3). https://doi.org/10.2903/j.efsa.2022.7209

Feingold, B. J., Silbergeld, E. K., Curriero, F. C., Van Cleef, B. A. G. L., Heck, M. E. O. C., & Kluytmans, J. A. J. W. (2012). Livestock-associated Methicillin-Resistant *Staphylococcus aureus* in Humans, the Netherlands. *Emerging Infectious Diseases, 18*(11), 1841–1849. https://doi.org/10.3201/eid1811.111850

Food and Agriculture Organization of the United Nations. (2019). *Tackling antimicrobial use and resistance in pig production: Lessons learned in Denmark*. UN. https://doi.org/10.18356/9d63b715-en

Food and Agriculture Organization of the United Nations. (2024). *The International FAO Antimicrobial Resistance Monitoring (InFARM) system*. FAO. https://doi.org/10.4060/cd0805en

Frost, I., Van Boeckel, T. P., Pires, J., Craig, J., & Laxminarayan, R. (2019). Global geographic trends in antimicrobial resistance: The role of international travel. *Journal of Travel Medicine*, *26*(8), taz036. https://doi.org/10.1093/jtm/taz036

Graham, D. W., Bergeron, G., Bourassa, M. W., Dickson, J., Gomes, F., Howe, A., Kahn, L. H., Morley, P. S., Scott, H. M., Simjee, S., Singer, R. S., Smith, T. C., Storrs, C., & Wittum, T. E. (2019). Complexities in understanding antimicrobial resistance across domesticated animal, human, and environmental systems. *Annals of the New York Academy of Sciences*, *1441*(1), 17–30. https://doi.org/10.1111/nyas.14036

Graham, J. P., Price, L. B., Evans, S. L., Graczyk, T. K., & Silbergeld, E. K. (2009). Antibiotic resistant enterococci and staphylococci isolated from flies collected near confined poultry feeding operations. *Science of the Total Environment*, *407*(8), 2701–2710. https://doi.org/10.1016/j.scitotenv.2008.11.056

Growth from Knowledge. (2016, November 22). *Man's best friend: Global pet ownership and feeding trends*. https://www.gfk.com/insights/mans-best-friend-global-pet-ownership-and-feeding-trends

Guerrero-Ramos, E., Cordero, J., Molina-González, D., Poeta, P., Igrejas, G., Alonso-Calleja, C., & Capita, R. (2016). Antimicrobial resistance and virulence genes in enterococci from wild game meat in Spain. *Food Microbiology*, *53*, 156–164. https://doi.org/10.1016/j.fm.2015.09.007

Guron, G. K. P., Arango-Argoty, G., Zhang, L., Pruden, A., & Ponder, M. A. (2019). Effects of dairy manure-based amendments and soil texture on lettuce- and radish-associated microbiota and resistomes. *mSphere*, *4*(3), e00239–19. https://doi.org/10.1128/mSphere.00239-19

He, Y., Yuan, Q., Mathieu, J., Stadler, L., Senehi, N., Sun, R., & Alvarez, P. J. J. (2020). Antibiotic resistance genes from livestock waste: Occurrence, dissemination, and treatment. *NPJ Clean Water*, *3*(1), 4. https://doi.org/10.1038/s41545-020-0051-0

Interagency Coordination Group on Antimicrobial Resistance. (2019). *No time to wait: Securing the future from drug-resistant infections. Report to the Secretary-General of the United Nations* (p. 28). https://www.who.int/publications/i/item/no-time-to-wait-securing-the-future-from-drug-resistant-infections

Iskandar, K., Molinier, L., Hallit, S., Sartelli, M., Hardcastle, T. C., Haque, M., Lugova, H., Dhingra, S., Sharma, P., Islam, S., Mohammed, I., Naina Mohamed, I., Hanna, P. A., Hajj, S. E., Jamaluddin, N. A. H., Salameh, P., & Roques, C. (2021). Surveillance of antimicrobial resistance in low- and middle-income countries: A scattered picture. *Antimicrobial Resistance & Infection Control*, *10*(1), 63. https://doi.org/10.1186/s13756-021-00931-w

Jadeja, N. B., & Worrich, A. (2022). From gut to mud: Dissemination of antimicrobial resistance between animal and agricultural niches. *Environmental Microbiology*, *24*(8), 3290–3306. https://doi.org/10.1111/1462-2920.15927

Jonas, O. B., Irwin, A., Berthe, F. C. J., Le Gall, F. G., & Marquez, P. V. (2017). *Drug-resistant infections: A threat to our economic future* (Vol. 2). The World Bank. https://documents.worldbank.org/en/publication/documents-reports/documentdetail/323311493396993758/final-report

Kahn, L. H. (2016). *One Health and the politics of antimicrobial resistance*. Johns Hopkins University Press.

Kirchhelle, C. (2018). Pharming animals: A global history of antibiotics in food production (1935–2017). *Palgrave Communications*, *4*(1), 96. https://doi.org/10.1057/s41599-018-0152-2

Klein, E. Y., Van Boeckel, T. P., Martinez, E. M., Pant, S., Gandra, S., Levin, S. A., Goossens, H., & Laxminarayan, R. (2018). Global increase and geographic convergence in antibiotic consumption between 2000 and 2015. *Proceedings of the National Academy of Sciences*, *115*(15). https://doi.org/10.1073/pnas.1717295115

Landecker, H. (2019). Antimicrobials before antibiotics: War, peace, and disinfectants. *Palgrave Communications*, *5*(1), 45. https://doi.org/10.1057/s41599-019-0251-8

Levy, S. (2014). Reduced antibiotic use in livestock: How Denmark tackled resistance. *Environmental Health Perspectives*, *122*(6). https://doi.org/10.1289/ehp.122-A160

Li, J., Cao, J., Zhu, Y., Chen, Q., Shen, F., Wu, Y., Xu, S., Fan, H., Da, G., Huang, R., Wang, J., De Jesus, A. L., Morawska, L., Chan, C. K., Peccia, J., & Yao, M. (2018). Global survey of antibiotic resistance genes in air. *Environmental Science & Technology*, *52*(19), 10975–10984. https://doi.org/10.1021/acs.est.8b02204

Liguori, K., Keenum, I., Davis, B. C., Calarco, J., Milligan, E., Harwood, V. J., & Pruden, A. (2022). Antimicrobial resistance monitoring of water environments: A framework for standardized methods and quality control. *Environmental Science & Technology*, *56*(13), 9149–9160. https://doi.org/10.1021/acs.est.1c08918

Lim, M. S. M., & Grohn, Y. T. (2021). Comparison of China's and the European Union's approaches to antimicrobial stewardship in the pork industry. *Foodborne Pathogens and Disease*, *18*(8), 567–573. https://doi.org/10.1089/fpd.2020.2887

Ma, Y., Chen, J., Fong, K., Nadya, S., Allen, K., Laing, C., Ziebell, K., Topp, E., Carroll, L. M., Wiedmann, M., Delaquis, P., & Wang, S. (2021). Antibiotic resistance in Shiga Toxigenic Escherichia coli isolates from surface waters and sediments in a mixed use urban agricultural landscape. *Antibiotics*, *10*(3), 237. https://doi.org/10.3390/antibiotics10030237

Mackuľak, T., Cverenkárová, K., Vojs Staňová, A., Fehér, M., Tamáš, M., Škulcová, A. B., Gál, M., Naumowicz, M., Špalková, V., & Bírošová, L. (2021). Hospital wastewater—Source of specific micropollutants, antibiotic-resistant microorganisms, viruses, and their elimination. *Antibiotics*, *10*(9), 1070. https://doi.org/10.3390/antibiotics10091070

Mader, R., EU- JAMRAI, Demay, C., Jouvin-Marche, E., Ploy, M.-C., Barraud, O., Bernard, S., Lacotte, Y., Pulcini, C., Weinbach, J., Berling, C., Bouqueau, M., Hlava, A., Habl, C., Kernstock, E., Strauss, R., Muchl, R., Buhmann, V., Versporten, A., … Madec, J.-Y. (2022). Defining the scope of the European Antimicrobial Resistance Surveillance network in Veterinary medicine (EARS-Vet): A bottom-up and One Health approach. *Journal of Antimicrobial Chemotherapy*, *77*(3), 816–826. https://doi.org/10.1093/jac/dkab462

Marques, C., Belas, A., Aboim, C., Trigueiro, G., Cavaco-Silva, P., Gama, L. T., & Pomba, C. (2019). Clonal relatedness of Proteus mirabilis strains causing urinary tract infections in companion animals and humans. *Veterinary Microbiology*, *228*, 77–82. https://doi.org/10.1016/j.vetmic.2018.10.015

McEachran, A. D., Blackwell, B. R., Hanson, J. D., Wooten, K. J., Mayer, G. D., Cox, S. B., & Smith, P. N. (2015). Antibiotics, bacteria, and antibiotic resistance genes: Aerial transport from cattle feed yards via particulate matter. *Environmental Health Perspectives*, *123*(4), 337–343. https://doi.org/10.1289/ehp.1408555

Mendelsohn, E., Ross, N., Zambrana-Torrelio, C., Van Boeckel, T. P., Laxminarayan, R., & Daszak, P. (2023). Global patterns and correlates in the emergence of antimicrobial resistance in humans. *Proceedings of the Royal Society B: Biological Sciences*, *290*(2007), 20231085. https://doi.org/10.1098/rspb.2023.1085

Miller, S. A., Ferreira, J. P., & LeJeune, J. T. (2022). Antimicrobial use and resistance in plant agriculture: A One Health perspective. *Agriculture*, *12*(2), 289. https://doi.org/10.3390/agriculture12020289

Miłobedzka, A., Ferreira, C., Vaz-Moreira, I., Calderón-Franco, D., Gorecki, A., Purkrtova, S., Jan Bartacek, Dziewit, L., Singleton, C. M., Nielsen, P. H., Weissbrodt, D. G., & Manaia, C. M. (2022). Monitoring antibiotic resistance genes in wastewater environments: The challenges of filling a gap in the One-Health cycle. *Journal of Hazardous Materials*, *424*, 127407. https://doi.org/10.1016/j.jhazmat.2021.127407

More, S. J., McCoy, F., & McAloon, C. I. (2022). The new veterinary medicines regulation: Rising to the challenge. *Irish Veterinary Journal*, *75*(1), 2. https://doi.org/10.1186/s13620-022-00209-6

Morgan, D. J., Okeke, I. N., Laxminarayan, R., Perencevich, E. N., & Weisenberg, S. (2011). Non-prescription antimicrobial use worldwide: A systematic review. *The Lancet Infectious Diseases*, *11*(9), 692–701. https://doi.org/10.1016/S1473-3099(11)70054-8

Morris, D. O., Lautenbach, E., Zaoutis, T., Leckerman, K., Edelstein, P. H., & Rankin, S. C. (2012). Potential for pet animals to harbour Methicillin-resistant *Staphylococcus aureus* when residing with human MRSA patients. *Zoonoses and Public Health*, *59*(4), 286–293. https://doi.org/10.1111/j.1863-2378.2011.01448.x

Murray, C. J., Ikuta, K. S., Sharara, F., Swetschinski, L., Robles Aguilar, G., Gray, A., Han, C., Bisignano, C., Rao, P., Wool, E., Johnson, S. C., Browne, A. J., Chipeta, M. G., Fell, F., Hackett, S., Haines-Woodhouse, G., Kashef Hamadani, B. H., Kumaran, E. A. P., McManigal, B., … Naghavi, M. (2022). Global burden of bacterial antimicrobial resistance in 2019: A systematic analysis. *The Lancet*, S0140673621027240. https://doi.org/10.1016/S0140-6736(21)02724-0

Neyra, R. C., Frisancho, J. A., Rinsky, J. L., Resnick, C., Carroll, K. C., Rule, A. M., Ross, T., You, Y., Price, L. B., & Silbergeld, E. K. (2014). Multidrug-resistant and Methicillin-resistant *Staphylococcus aureus* (MRSA) In Hog Slaughter and processing plant workers and their community in North Carolina (USA). *Environmental Health Perspectives*, *122*(5), 471–477. https://doi.org/10.1289/ehp.1306741

Ngaruka, G. B., Neema, B. B., Mitima, T. K., Kishabongo, A. S., & Kashongwe, O. B. (2021). Animal source food eating habits of outpatients with antimicrobial resistance in Bukavu, D.R. Congo. *Antimicrobial Resistance & Infection Control*, *10*(1), 124. https://doi.org/10.1186/s13756-021-00991-y

Nichols, M., Gollarza, L., Sockett, D., Aulik, N., Patton, E., Francois Watkins, L. K., Gambino-Shirley, K. J., Folster, J. P., Chen, J. C., Tagg, K. A., Stapleton, G. S., Trees, E., Ellison, Z., Lombard, J., Morningstar-Shaw, B., Schlater, L., Elbadawi, L., & Klos, R. (2022). Outbreak of multidrug-resistant *Salmonella* Heidelberg infections linked to dairy calf exposure, United States, 2015–2018. *Foodborne Pathogens and Disease*, *19*(3), 199–208. https://doi.org/10.1089/fpd.2021.0077

OECD. (2018). *Stemming the superbug tide: Just a few dollars more*. OECD. https://doi.org/10.1787/9789264307599-en

Ollila, K. (2020). *The use of veterinary antibiotics in the Nordic countries and its impact on antibiotic resistance* [University of Eastern Finland]. https://erepo.uef.fi/bitstream/handle/123456789/22271/urn_nbn_fi_uef-20200531.pdf?sequence=1&isAllowed=y

Onwugamba, F. C., Fitzgerald, J. R., Rochon, K., Guardabassi, L., Alabi, A., Kühne, S., Grobusch, M. P., & Schaumburg, F. (2018). The role of 'filth flies' in the spread of antimicrobial resistance. *Travel Medicine and Infectious Disease*, *22*, 8–17. https://doi.org/10.1016/j.tmaid.2018.02.007

OPGA, WHO, FAO, & OIE. (2016, September 21). Press release: High-level meeting on antimicrobial resistance. *At UN, Global Leaders Commit to Act on Antimicrobial Resistance*. https://www.un.org/pga/71/2016/09/21/press-release-hl-meeting-on-antimicrobial-resistance/

Overton, K., Fortané, N., Broom, A., Raymond, S., Gradmann, C., Orubu, E. S. F., Podolsky, S. H., Rogers Van Katwyk, S., Zaman, M. H., & Kirchhelle, C. (2021). Waves of attention: Patterns and themes of international antimicrobial resistance reports, 1945–2020. *BMJ Global Health*, *6*(11), e006909. https://doi.org/10.1136/bmjgh-2021-006909

Ozaktas, T., Taskin, B., & Gozen, A. G. (2012). High level multiple antibiotic resistance among fish surface associated bacterial populations in non-aquaculture freshwater environment. *Water Research*, *46*(19), 6382–6390. https://doi.org/10.1016/j.watres.2012.09.010

Plaza-Rodríguez, C., Alt, K., Grobbel, M., Hammerl, J. A., Irrgang, A., Szabo, I., Stingl, K., Schuh, E., Wiehle, L., Pfefferkorn, B., Naumann, S., Kaesbohrer, A., & Tenhagen, B.-A. (2021). Wildlife as sentinels of antimicrobial resistance in Germany? *Frontiers in Veterinary Science*, *7*, 627821. https://doi.org/10.3389/fvets.2020.627821

Ritchie, H., Rosado, P., & Roser, M. (2023). Meat and dairy production. *Our World in Data*. https://ourworldindata.org/meat-production

Ritchie, H., & Roser, M. (2024). Land use. *Our World in Data*. https://ourworldindata.org/land-use

Robinson, T. P., Bu, D. P., Carrique-Mas, J., Fèvre, E. M., Gilbert, M., Grace, D., Hay, S. I., Jiwakanon, J., Kakkar, M., Kariuki, S., Laxminarayan, R., Lubroth, J., Magnusson, U., Thi Ngoc, P., Van Boeckel, T. P., & Woolhouse, M. E. J. (2016). Antibiotic resistance is the quintessential One Health issue. *Transactions of the Royal Society of Tropical Medicine and Hygiene*, *110*(7), 377–380. https://doi.org/10.1093/trstmh/trw048

Roosaare, M., Puustusmaa, M., Möls, M., Vaher, M., & Remm, M. (2018). PlasmidSeeker: Identification of known plasmids from bacterial whole genome sequencing reads. *Peer J*, *6*, e4588. https://doi.org/10.7717/peerj.4588

Schar, D., Klein, E. Y., Laxminarayan, R., Gilbert, M., & Van Boeckel, T. P. (2020). Global trends in antimicrobial use in aquaculture. *Scientific Reports*, *10*(1), 21878. https://doi.org/10.1038/s41598-020-78849-3

Schar, D., Sommanustweechai, A., Laxminarayan, R., & Tangcharoensathien, V. (2018). Surveillance of antimicrobial consumption in animal production sectors of low- and middle-income countries: Optimizing use and addressing antimicrobial resistance. *PLOS Medicine*, *15*(3), e1002521. https://doi.org/10.1371/journal.pmed.1002521

Schar, D., Zhao, C., Wang, Y., Larsson, D. G. J., Gilbert, M., & Van Boeckel, T. P. (2021). Twenty-year trends in antimicrobial resistance from aquaculture and fisheries in Asia. *Nature Communications*, *12*(1), 5384. https://doi.org/10.1038/s41467-021-25655-8

Schmidt, J. W., Vikram, A., Doster, E., Thomas, K., Weinroth, M. D., Parker, J., Hanes, A., Geornaras, I., Morley, P. S., Belk, K. E., Wheeler, T. L., & Arthur, T. M. (2021). Antimicrobial resistance in U.S. retail ground beef with and without label claims regarding antibiotic use. *Journal of Food Protection*, *84*(5), 827–842. https://doi.org/10.4315/JFP-20-376

Singer, A. C., Shaw, H., Rhodes, V., & Hart, A. (2016). Review of antimicrobial resistance in the environment and its relevance to environmental regulators. *Frontiers in Microbiology*, *7*. https://doi.org/10.3389/fmicb.2016.01728

Slobodiuk, S., Niven, C., Arthur, G., Thakur, S., & Ercumen, A. (2021). Does irrigation with treated and untreated wastewater increase antimicrobial resistance in soil and water: A systematic review. *International Journal of Environmental Research and Public Health*, *18*(21), 11046. https://doi.org/10.3390/ijerph182111046

Smith, T. C., Gebreyes, W. A., Abley, M. J., Harper, A. L., Forshey, B. M., Male, M. J., Martin, H. W., Molla, B. Z., Sreevatsan, S., Thakur, S., Thiruvengadam, M., & Davies, P. R. (2013). Methicillin-resistant Staphylococcus aureus in pigs and farm workers on conventional and antibiotic-free Swine farms in the USA. *PLoS One*, *8*(5), e63704. https://doi.org/10.1371/journal.pone.0063704

Song, S. J., Lauber, C., Costello, E. K., Lozupone, C. A., Humphrey, G., Berg-Lyons, D., Caporaso, J. G., Knights, D., Clemente, J. C., Nakielny, S., Gordon, J. I., Fierer, N., & Knight, R. (2013). Cohabiting family members share microbiota with one another and with their dogs. *eLife*, *2*, e00458. https://doi.org/10.7554/eLife.00458

Statens Serum Institut & National Food Institute, Technical University of Denmark. (2022). *DANMAP 2021: Use of antimicrobial agents and occurrence of antimicrobial resistance in bacteria from food animals, food and humans in Denmark*. Statens Serum Institut. https://www.danmap.org/-/media/sites/danmap/downloads/reports/2021/danmap_2021_version-1.pdf

Tacconelli, E., Carrara, E., Savoldi, A., Harbarth, S., Mendelson, M., Monnet, D. L., Pulcini, C., Kahlmeter, G., Kluytmans, J., Carmeli, Y., Ouellette, M., Outterson, K., Patel, J., Cavaleri, M., Cox, E. M., Houchens, C. R., Grayson, M. L., Hansen, P., Singh, N., … Zorzet, A. (2018). Discovery, research, and development of new antibiotics: The WHO priority list of antibiotic-resistant bacteria and tuberculosis. *The Lancet Infectious Diseases*, *18*(3), 318–327. https://doi.org/10.1016/S1473-3099(17)30753-3

Tacconelli, E., Sifakis, F., Harbarth, S., Schrijver, R., Van Mourik, M., Voss, A., Sharland, M., Rajendran, N. B., Rodríguez-Baño, J., Bielicki, J., De Kraker, M., Gandra, S., Gastmeier, P., Gilchrist, K., Gikas, A., Gladstone, B. P., Goossens, H., Jafri, H., Kahlmeter, G., … Wolkewitz, M. (2018). Surveillance for control of antimicrobial resistance. *The Lancet Infectious Diseases*, *18*(3), e99–e106. https://doi.org/10.1016/S1473-3099(17)30485-1

Tadesse, W., Sanchez-Garcia, M., Tawkaz, S. A., El-Hanafi, S., Skaf, P., El-Baouchi, A., Eddakir, K., El-Shamaa, K., Thabet, S., Assefa, S. G., & Baum, M. (2019). *Wheat breeding handbook at ICARDA*. ICARDA. https://hdl.handle.net/20.500.11766/10723

Taylor, P., & Reeder, R. (2020). Antibiotic use on crops in low and middle-income countries based on recommendations made by agricultural advisors. *CABI Agriculture and Bioscience*, *1*(1), 1. https://doi.org/10.1186/s43170-020-00001-y

Tilman, D., Balzer, C., Hill, J., & Befort, B. L. (2011). Global food demand and the sustainable intensification of agriculture. *Proceedings of the National Academy of Sciences*, *108*(50), 20260–20264. https://doi.org/10.1073/pnas.1116437108

Tiseo, K., Huber, L., Gilbert, M., Robinson, T. P., & Van Boeckel, T. P. (2020). Global trends in antimicrobial use in food animals from 2017 to 2030. *Antibiotics*, *9*(12), 918. https://doi.org/10.3390/antibiotics9120918

Toner, E., Adalja, A., Gronvall, G. K., Cicero, A., & Inglesby, T. V. (2015). Antimicrobial resistance is a global health emergency. *Health Security*, *13*(3), 153–155. https://doi.org/10.1089/hs.2014.0088

Van Boeckel, T. P., Brower, C., Gilbert, M., Grenfell, B. T., Levin, S. A., Robinson, T. P., Teillant, A., & Laxminarayan, R. (2015). Global trends in antimicrobial use in food animals. *Proceedings of the National Academy of Sciences*, *112*(18), 5649–5654. https://doi.org/10.1073/pnas.1503141112

Van Boeckel, T. P., Pires, J., Silvester, R., Zhao, C., Song, J., Criscuolo, N. G., Gilbert, M., Bonhoeffer, S., & Laxminarayan, R. (2019). Global trends in antimicrobial resistance in animals in low- and middle-income countries. *Science*, *365*(6459), eaaw1944. https://doi.org/10.1126/science.aaw1944

Vikram, A., Miller, E., Arthur, T. M., Bosilevac, J. M., Wheeler, T. L., & Schmidt, J. W. (2018). Similar levels of antimicrobial resistance in U.S. food service ground beef products with and without a "Raised without Antibiotics" Claim. *Journal of Food Protection, 81*(12), 2007–2018. https://doi.org/10.4315/0362-028X.JFP-18-299

Vikram, A., Miller, E., Arthur, T. M., Bosilevac, J. M., Wheeler, T. L., & Schmidt, J. W. (2019). Food service pork chops from three U.S. regions harbor similar levels of antimicrobial resistance regardless of antibiotic use claims. *Journal of Food Protection, 82*(10), 1667–1676. https://doi.org/10.4315/0362-028X.JFP-19-139

Wang, Y., Xu, C., Zhang, R., Chen, Y., Shen, Y., Hu, F., Liu, D., Lu, J., Guo, Y., Xia, X., Jiang, J., Wang, X., Fu, Y., Yang, L., Wang, J., Li, J., Cai, C., Yin, D., Che, J., … Shen, J. (2020). Changes in colistin resistance and mcr-1 abundance in Escherichia coli of animal and human origins following the ban of colistin-positive additives in China: An epidemiological comparative study. *The Lancet Infectious Diseases, 20*(10), 1161–1171. https://doi.org/10.1016/S1473-3099(20)30149-3

Water Science School. (2019, November 13). *How much water is there on earth?* USGS Water Science School. https://www.usgs.gov/special-topics/water-science-school/science/how-much-water-there-earth

Weese, J. S., & Van Duijkeren, E. (2010). Methicillin-resistant Staphylococcus aureus and Staphylococcus pseudintermedius in veterinary medicine. *Veterinary Microbiology, 140*(3–4), 418–429. https://doi.org/10.1016/j.vetmic.2009.01.039

Wernli, D., Harbarth, S., Levrat, N., & Pittet, D. (2022). A 'whole of United Nations approach' to tackle antimicrobial resistance? A mapping of the mandate and activities of international organisations. *BMJ Global Health, 7*(5), e008181. https://doi.org/10.1136/bmjgh-2021-008181

Wisconsin Veterinary Diagnostic Laboratory. (2021). *Salmonella*. Wisconsin Veterinary Diagnostic Laboratory. https://www.wvdl.wisc.edu/salmonella/

World Health Organization. (2019). *Critically important antimicrobials for human medicine* (6th rev.). World Health Organization. https://iris.who.int/handle/10665/312266

World Health Organization. (2020a). *GLASS whole-genome sequencing for surveillance of antimicrobial resistance* (p. 59). World Health Organization. https://iris.who.int/handle/10665/334354

World Health Organization. (2020b). WHO guidelines on use of medically important antimicrobials in food-producing animals. [Internet]. World Health Organization; 2017 [cited 2020 Dec 4]. Available from: https://www.who.int/publications/i/item/9789241550130

World Health Organization. (2021). *Global antimicrobial resistance and use surveillance systems (GLASS) report: 2021*. World Health Organization. https://iris.who.int/handle/10665/341666

World Health Organization. (2022). *WHO fungal priority pathogens list to guide research, development and public health action*. World Health Organization. https://www.who.int/publications/i/item/9789240060241

World Health Organization. (2024). *Library of national action plans*. https://www.who.int/teams/surveillance-prevention-control-AMR/national-action-plan-monitoring-evaluation/library-of-national-action-plans

World Health Organization, Food and Agriculture Organization of the United Nations, & World Organisation for Animal Health. (2021). *Monitoring global progress on antimicrobial resistance: Tripartite AMR country self-assessment survey (TrACSS) 2019–2020. Global analysis report*. World Health Organization. https://www.who.int/publications/i/item/monitoring-global-progress-on-antimicrobial-resistance-tripartite-amr-country-self-assessment-survey-(tracss)-2019-2020

World Health Organization, Food and Agriculture Organization of the United Nations, World Organisation for Animal Health, & United Nations Environment Programme. (2024). *Quadripartite Joint Secretariat on Antimicrobial Resistance*. https://www.qjsamr.org

World Organisation for Animal Health. (2021). *OIE list of antimicrobial agents of veterinary importance*. OIE. https://www.woah.org/app/uploads/2021/06/a-oie-list-antimicrobials-june2021.pdf

World Organisation for Animal Health. (2022). *Annual report on antimicrobial agents intended for use in animals* (6th ed.). World Organization for Animal Health. https://www.woah.org/en/document/annual-report-on-antimicrobial-agents-intended-for-use-in-animals/

Zhao, C., Wang, Y., Tiseo, K., Pires, J., Criscuolo, N. G., & Van Boeckel, T. P. (2021). Geographically targeted surveillance of livestock could help prioritize intervention against antimicrobial resistance in China. *Nature Food*, *2*(8), 596–602. https://doi.org/10.1038/s43016-021-00320-x

Zheng, D., Yin, G., Liu, M., Hou, L., Yang, Y., Van Boeckel, T. P., Zheng, Y., & Li, Y. (2022). Global biogeography and projection of soil antibiotic resistance genes. *Science Advances*, *8*(46), eabq8015. https://doi.org/10.1126/sciadv.abq8015

Zurek, L., & Ghosh, A. (2014). Insects represent a link between food animal farms and the urban environment for antibiotic resistance traits. *Applied and Environmental Microbiology*, *80*(12), 3562–3567. https://doi.org/10.1128/AEM.00600-14

7 Global Nutrition and the Intersection of Human and Animal Health

Johanna T. Wong, Robyn G. Alders, Julia de Bruyn, Heather Grieve, Jennifer K. Lane, Gareth Salmon, and Frances Siobhán Ryan

INTRODUCTION

For optimal health and well-being, the body needs to be nourished with a complement of essential macro and micronutrients obtained from the habitual consumption of appropriate quantities of safe, affordable, and culturally appropriate diverse foods. Nutrition security is dependent on food security, and together, they are defined as existing when "all people at all times have physical, social and economic access to food, which is safe and consumed in sufficient quantity and quality to meet their dietary needs and food preferences, and is supported by an environment of adequate sanitation, health services and care, allowing for a healthy and active life" (Committee on World Food Security, 2012).

A healthy diet must also be supported by a healthy environment, including clean water, sanitation, and hygiene (WASH), appropriate care and feeding practices for infants and young children (IYC), and access to education and health services, all set within a supportive sociopolitical context (Figure 7.1, UNICEF, 2021). Evidently, the balance required to achieve optimal food and nutrition security is difficult to achieve, with every country in the world affected by at least one form of malnutrition (Development Initiatives, 2018).

This chapter introduces the different forms of malnutrition relevant to the global One Health practitioner, their prevalence, and immediate and underlying causes (Figure 7.1). Given that the overwhelming burden of malnutrition in women and children in low- and middle-income countries (LMICs) is undernutrition (including underweight in women, acute (wasting) and chronic (stunting) undernutrition in children, and micronutrient deficiencies, e.g., anemia), special consideration is given to the contributions of nutritionally dense animal-source

 DOI: 10.1201/9781003232223-7

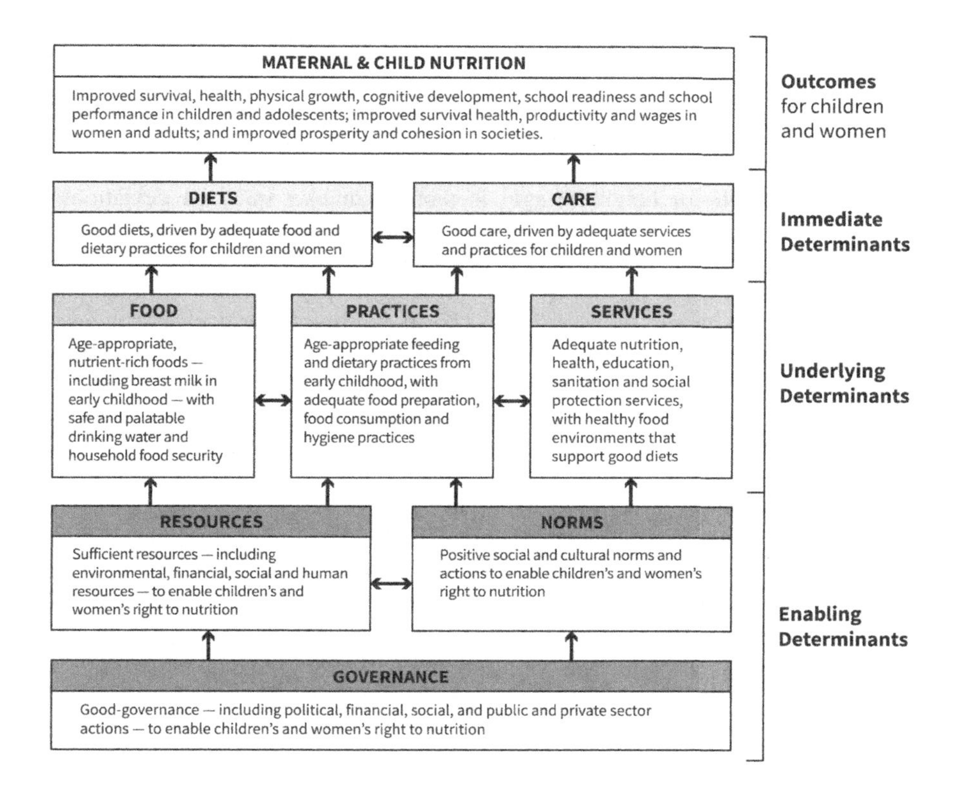

FIGURE 7.1 UNICEF conceptual framework on the determinants of maternal and child nutrition, 2020. (UNICEF, 2021.)

foods (ASF). Historically and today, ASF provide vital micronutrients – some of which cannot be acquired from plant-source foods. The production and consumption of ASF, however, is not without risk – this chapter situates livestock production in the context of One Health, exploring the positive and negative relationships between livestock, humans, and the environment.

BURDEN OF MALNUTRITION

Terminology and Prevalence

The term "malnutrition" encompasses the entire spectrum of poor nutrition, including undernutrition (including micronutrient deficiencies) and overnutrition (including overweight and obesity). Common terms that are used to describe the different forms of malnutrition are defined below.

Hunger, or Undernourishment: This refers to the inadequate intake of dietary energy,[1] or the distress associated with this. In 2021, it was estimated that between 8.9% and 10.5% of the world's population (702–828 million people) were affected by hunger. Since 2019, this number has increased by almost 150 million, with the coronavirus disease (COVID-19) pandemic and its consequences cited as the cause of this rise. More than half of the people affected by hunger reside in Asia, while more than one-third reside in Africa (FAO et al., 2022).

Stunting: Impaired linear growth in children under five years of age, to the degree that length- or height-for-age[2] is more than two standard deviations (z-scores) below the WHO Child Growth Standards' median (World Health Organization, 2024a). Stunting results from chronically poor nutrition, often coupled with repeated infections and inadequate psychosocial stimulation (Development Initiatives, 2021). In 2020, 22% of children under five years of age (149 million children) were stunted. The majority of affected children live in rural areas in LMICs. Stunting is associated with long-term health impacts, including decreased ability to achieve physical and cognitive potential, reduced economic productivity, and increased likelihood of developing non-communicable diseases in adulthood (FAO et al., 2022).

Wasting: Thinness in children under five years of age, to the degree that weight-for-length or weight-for-height is more than two z-scores below the median (Development Initiatives, 2021; World Health Organization, 2024a). Wasting is an acute form of undernutrition, and caused by inadequate food intake, poor nutrient absorption and/or illness, particularly infectious diseases, and can be life-threatening. In 2020, 6.7% of children under five years of age (45.4 million children) suffered from wasting. This situation has worsened with COVID-19, and from the country-specific losses in gross national income (GNI) per capita, it is estimated that there could be at least a 14.3% increase in the prevalence of moderate or severe wasting among children younger than five years (Headey et al., 2020). Wasted children are more likely to live in poorer households in LMICs (FAO et al., 2022).

Underweight: Children with low weight for their age are underweight. Underweight children may be a result of stunting, wasting, or both (World Health Organization, 2024c).

Overweight and Obesity: The prevalence of overweight and obesity has been rising in all age groups over recent decades, fundamentally as a result of decreasing physical activity combined with excessive intake of energy-dense foods high in fat and/or sugar (Development Initiatives, 2021). Classification of overweight and obese differs by age – definitions are presented in Table 7.1. In 2020, 5.7% (38.9 million) of children under five years of age were overweight (FAO et al., 2022). In adults, 40.8% of women and 40.4% of men – 2.2 billion people in total – are overweight or obese, and no country has been able to halt rises in adult obesity (Development Initiatives, 2021). Overweight children are more likely to live in wealthier households in middle-income countries (FAO et al., 2022).

Hidden Hunger: A term used to describe micronutrient deficiencies in general; mild deficiencies are often subclinical, and therefore "hidden." Although availability of more recent population-based data on the prevalence of micronutrient deficiencies is limited (FAO et al., 2022), globally, iron, vitamin A, and iodine deficiencies are the most common, with these micronutrients being major

TABLE 7.1
Definitions of Overweight and Obesity by Age Range

Age Range	Definitions	
	Overweight	Obese
Children <5 years of age	Weight-for-height >2 z-scores above median (WHO, 2022)	Weight-for-height >3 z-scores above median (WHO, 2022)
Children 5–19 years of age	BMI-for-age >1 z-score above median	BMI-for-age >2 z-scores above median
Adults	BMI greater than or equal to 25	BMI greater than or equal to 30

Source: World Health Organization (2024d).

contributors to child morbidity and mortality (Bhutta & Salam, 2012; Muthayya et al., 2013; World Health Organization, 2024c). Iron deficiency affects between 10% and 20% of children under five in every region of the world and has a significant detrimental effect on child growth and development (R. E. Black et al., 2013).

Anemia: Anemia is defined as a low hemoglobin level (<110 g/L for children under five years and pregnant women, <120 g/L for non-pregnant women of reproductive age (WRA), and <130 g/L for men), either through insufficient levels of hemoglobin itself or due to a low red blood cell count (World Health Organization, 2011). In 2019, it was estimated that 31.2% (571 million) of WRA (15–49 years) were anemic, which not only affects their own health, but can also lead to poor pregnancy and neonatal outcomes (R. E. Black et al., 2013; FAO et al., 2022). Anemia is a global issue prevalent across 161 countries (Development Initiatives, 2021). Although multiple micronutrient deficiencies can cause anemia, iron-deficiency anemia is thought to be the most common underlying cause, with a prevalence of approximately 25% in many LMICs (M. M. Black et al., 2011).

Key Concepts

Although these terms are defined separately, multiple forms of malnutrition can be found within a country, community, household, or even individual across the life course. In many LMICs, it is increasingly common to see the coexistence of undernutrition with overweight, obesity, and associated non-communicable diseases – commonly termed the "double burden" of malnutrition (World Health Organization, 2024b). Combined with micronutrient deficiencies, this becomes the "triple burden" (Pinstrup-Andersen, 2007).

Around the globe, LMICs are grappling with these increasing burdens as they go through the "nutrition transition," whereby economic growth, urbanization, globalization, and technology including food processing capabilities are changing dietary patterns (Popkin & Ng, 2022). Generally, the associated dietary changes are linked to increases in overweight and obesity related to decreasing levels of physical activity combined with increased consumption of energy-dense, but nutrient-poor, ultra-processed foods. These foods are often less costly than nutritious options, meaning that overweight and obesity are increasingly seen in lower wealth populations (Popkin et al., 2020).

These problems are exacerbated by undernutrition early in life, particularly stunting, as this increases the risk of overweight, obesity, and non-communicable diseases in adulthood. Ensuring adequate nutrition for WRA and IYC – particularly during the "first 1000 days" of life, from conception to two years of age – is crucial to breaking the cycle of the double and triple burdens of malnutrition. The first 1000 days is a period of high nutrient requirement for the health and well-being of pregnant and breastfeeding women and to support children's rapid growth and development *in utero* and during early childhood (Black et al., 2013). To combat undernutrition and micronutrient deficiencies, there is a significant role for culturally acceptable, nutrient-dense foods. For WRA, recent efforts to rank foods according to their nutrient density revealed that 16 of the 17 foods with high or very high concentrations of commonly deficient micronutrients are ASF (Beal & Ortenzi, 2022).

PATHWAYS FROM LIVESTOCK AND FISH TO HUMAN NUTRITION

The potential of ASF to improve the dietary quality of diets around the world cannot be understated. Worldwide, livestock and fisheries contribute to the livelihoods and food and nutrition security of more than 1.3 billion people, while being invaluable for draft power, transport (including to haul water), manure, social capital, and cultural roles in LMICs (Herrero et al., 2013). They offer an avenue to support women and youth empowerment, while also acting as financial assets and savings accounts (Food and Agriculture Organization, 2022a). The roles of livestock vary greatly by region, livestock species, livelihood system, and season. In many settings around the world, livestock assets are often sold, bartered, saved, or used for other income-generating activities, and are less likely to be directly consumed by individual households. Livestock sales may be a primary contributor to household income and play important roles in household resilience to financial and climatic shocks (Frankenberger et al., 2012; Ramilan et al., 2022). Importantly, healthy livestock are a necessary prerequisite to a healthy ASF supply.

BOX 7.1 A NOTE ABOUT FISH AND INSECTS

Aquatic ASF (including fish from freshwater and marine environments, wild caught and farmed; also referred to as seafood) encompass diverse, highly nutritious foods that are often not included in the same conversations as terrestrial livestock. Globally, more than a billion people depend on aquatic foods to meet their nutritional needs, with seafood representing nearly 20% of animal protein consumed in diets around the world (Golden, Allison, et al., 2016; Food and Agriculture Organization, 2022c). Seafood is an important source of essential vitamins, minerals, and amino acids, and as demand for seafood globally has increased, aquaculture (farmed seafood production) practices are growing rapidly (Food and Agriculture Organization, 2022c). It is estimated that aquaculture now produces just over half of all seafood consumed (Food and Agriculture Organization, 2022b); recognizing the vast diversity of fishery and aquaculture practices, their impacts on livelihoods and on the environment, and ensuring all seafood harvesting methods are environmentally sound, sustainable, and nutrition sensitive is critical. A global movement to catalyze this transition is referred to as the Blue Transformation (Food and Agriculture Organization, 2022c). It has been suggested that aquaculture can support food security with a lower environmental footprint than other ASF (Garlock et al., 2022) but there are challenges (Belton et al., 2020). More recently, the development of land-based fish farming using a recirculating aquaculture system poses direct concerns for land, water, and energy use. Important unanswered questions remain; for instance, from a One Health perspective, do these systems present disease risks to humans and other livestock species (Berggren et al., 2019)?

Increasingly, fish and insect farming are referenced in the food security literature. For insect farming, the scale of the sector is growing (Tanga et al., 2021). While livestock typically refers to vertebrate, land-based livestock systems, fish and insect farming share similarities with livestock systems in terms of requiring inputs and an animal husbandry approach in order to provide food in a managed system (Wilkinson, 2006; Gahukar, 2016). Insect farming has been dubbed "mini-livestock" production in some areas of the literature (Dickie et al., 2020).

Primary Pathways between Livestock Production and Nutrition Outcomes

Livestock have the potential to help alleviate malnutrition through several primary pathways. A common but inaccurate assumption is that increased livestock production will automatically result in improved nutritional status of those in livestock-keeping households (Carletto et al., 2015). This is not always the case but other beneficial pathways have been documented. While these may vary, there is a general consensus on four main ones including:

1. Direct consumption of products;
2. Income from the sale of livestock products;
3. Factors related to gender; and
4. Food prices (Ruel & Alderman, 2013).

Livestock ownership and ASF production have the potential to lead to improved nutrition through all four identified pathways; however, this is highly context specific, and depends on access to and nature of markets, seasonal variations, and decision-making at the household level. At different times of the year, any combination of pathways may be in play in a particular household. Furthermore, the pathways between livestock production and nutrition outcomes can be further complicated by the potential negative impacts of increased exposure to zoonotic diseases, foodborne illnesses, and/or contamination of water supplies by livestock (Randolph et al., 2007). Ultimately, the pathways by which livestock can impact household nutrition are diverse, highly contextual, variable, and influenced by many factors.

Impact Measurement of Nutrition-Sensitive Livestock Programs

Approaches to understand the nutritional impact of agricultural projects have been the focus of substantial research and discussion over the last decade. Given the complex pathways between agricultural production and nutrition outcomes, there has been growing recognition that measuring the impact of agricultural programs on diets and nutritional status of the families and communities growing and raising agricultural commodities are important, yet complicated.

Multiple reviews have highlighted the lack of empirical evidence to support the nutritional impact of agricultural interventions, mostly due to inappropriate study designs, sample size, and indicators (Webb & Kennedy, 2014; Herforth & Ballard, 2016). Currently, it is recommended to focus on a target population and measure individual dietary intake and nutritional adequacy. Many tools exist for dietary assessment and capturing dietary diversity, with common tools and methods presented in Table 7.2. Each varies in time needed, skills required to complete the assessment, validation, and cost, and all have limitations and strengths.

Recognizing the challenges and costs associated with accurately assessing dietary quality, global efforts to improve this include the International Dietary Data Expansion Project (INDDEX) and the efforts on metric and tool development by the Intake Center for Dietary Assessment (Herforth & Ballard, 2016; J. L. Leroy et al., 2020).

Coupled with dietary indicators, measures of nutritional status are also used but only when programs include specific interventions designed to address these and other indicators including micronutrient deficiencies, overweight, and obesity, etc. It is not recommended to rely on anthropometric outcomes (i.e., child stunting or wasting, maternal weight/body mass index (BMI)) to measure the nutritional impact of agricultural programs (J. L. Leroy et al., 2020), because nutrition-sensitive agricultural projects are intended to address the underlying causes of nutrition, rather than the immediate causes, which are targeted by nutrition-specific interventions (see Box 7.2). Agricultural interventions alone are unlikely to produce a measurable change in nutritional status in the timeframes and sample sizes of most projects; therefore, indicators of dietary quality and food intake are considered more suitable (J. L. Leroy et al., 2020).

TABLE 7.2
Common Dietary Assessment Methods and Tools

	Methods of Assessing of Dietary Intake
Weighed food records	Gold standard. Prospective, direct measurement and recording of quantities of foods consumed over a period of 1–7 days. This is labor intensive.
24-hour recall	Retrospective, quantitative recall of foods consumed in the previous 24-hour period.
Food frequency questionnaire	Retrospective, semi-quantitative assessment of frequency of food or food group consumption, often over an extended period of time.
Common Diet Quality Indicator Types	
Dietary diversity indicators	Retrospective, qualitative tools giving a simple score to represent the variety of food groups consumed over a given period as a proxy for nutritional adequacy. Different tools available for different population groups.
Combination indices	Indices are usually constructed from measures of dietary variety, portion sizes, and macro- and micronutrient adequacy or dietary risk factors. Examples include the Diet Quality Index, Healthy Eating Index/Alternate Healthy Eating Index, Healthy Diet Indicator, and the Mediterranean-Diet Quality Index.

Sources: (Gil et al., 2015; Food and Agriculture Organization, 2018).

BOX 7.2 NUTRITION-SENSITIVE AND NUTRITION-SPECIFIC INTERVENTIONS – DEFINITIONS

Nutrition sensitive – Interventions or programs that address the **underlying determinants** of nutrition and development and incorporate specific nutrition goals (i.e., agriculture and food security, women's empowerment, and WASH).

Nutrition specific – Interventions or programs that address the **immediate determinants** of nutrition and development (i.e., adequate food and nutrient intake, feeding, caregiving and parenting practices, and low burden of infectious diseases) (Herforth & Harris, 2014).

CONTRIBUTION OF ASF TO HUMAN NUTRITION AND HEALTH

The role of ASF, including meat, eggs, milk and dairy products, fish and seafood, and insects, in human diets is complex and has been the focus of polarized debate in recent decades. The imperative to transform our current food systems to deliver sustainable, healthy diets (High Level Panel of Experts on Food Security and Nutrition of the Committee on World Food Security, 2020) raises questions about how ASF can continue to play an important role in addressing undernutrition balanced with promoting health and protecting the environment (L. Iannotti, 2021). The EAT-Lancet Commission's call for a "Great Food Transformation" proposed a reference diet for human and planetary health, primarily made up of plant-source foods with little or no ASF (Willett et al., 2019). This high-profile report and the underlying data on which it is based has attracted criticism for the extent of its opposition to red meat (F. Leroy et al., 2022), its limited acknowledgment of global variations in physical and economic access to proposed diets (Hirvonen et al., 2020), and the lack of clarity behind dietary risk factor modeling (Stanton et al., 2022).

Looking Back at ASF Intake

ASF have played an important role in hominid evolution. Key changes in anatomy and physiology, including increased stature, body mass, and brain size, have been attributed to the greater consumption of nutrient-dense ASF by our omnivorous ancestors in the Paleolithic period (Kuipers et al., 2012). An evolutionary perspective of nutrition, as presented in the seminal work of Eaton and Konner (1985), contends that the human genome adapted over 2.3 million years ago to diverse diets, high in animal protein and low in carbohydrates. The concept of "genome-nutrition divergence" suggests that many adverse health outcomes – linked to both under- and overnutrition – may be attributed to the misalignment between our genome and modern diets (Eaton & Iannotti, 2017).

Globally, ASF intake varies widely. National food supply data reveal an upward trend in the per capita supply of most ASF since records became available in 1960 (Grünberger, 2014). Pork meat supply has grown most substantially at a steady rate, while increases in egg and milk supply have been notable since 1990 (Food and Agriculture Organization, 2023). One exception has been bovine meat, for which per capita supply has declined in recent years. High global averages are not shared across regions, however, and obscure the persistent low ASF supply in many low-resource settings. While population and income growth are expected to drive strong future demand for ASF in Asia, Latin America, and the Middle East (Henchion et al., 2017), consumption by many vulnerable populations continues to be constrained by cost, availability (due to disease, productivity, and supply chain factors), nutritional awareness, cultural values assigned to livestock, and the importance of livestock to smallholder households as financial assets (Smith et al., 2013; De Bruyn et al., 2017).

Positive Effects on Nutrition and Health

As a group, ASF are a key source of high-quality protein, essential fatty acids, and a wide array of micronutrients, including iron, calcium, vitamin A, vitamin B12, zinc, selenium, and choline (Allen, 2008; Food and Agriculture

Organization, 2022a). Multiple metrics rank animal proteins, which are highly digestible and have a distribution of essential amino acids aligned with human requirements, above plant proteins (Katz et al., 2019). Several micronutrients of public health significance are more biologically available in ASF than plant-source foods, where their absorption may be limited by the chemical form of nutrients, surrounding food matrix, and presence of antinutrients (Murphy & Allen, 2003; Gibson et al., 2006; Schönfeldt et al., 2014). Vitamin A, essential for growth, vision, reproduction, and immune function, is substantially more bioavailable as retinol in ASF than in its precursor form as carotenoids in plant-source foods (Tang, 2010). *Heme* iron in meat is better absorbed, and less affected by inhibitory compounds, than non-*heme* iron in vegetables and legumes (Gupta, 2016).

This high density and bioavailability of nutrients makes ASF valuable in settings where balanced plant-based diets are difficult to achieve, particularly for nutritionally vulnerable groups. There is high potential for ASF to meet nutrient demands during the critical first 1,000 days (from conception to 2 years), when micronutrient deficiencies are common. Linear programming analyses, used to guide food-based recommendations, often support increasing ASF intake to address micronutrient gaps in existing diets (Hlaing et al., 2016; Raymond et al., 2017; Mejos et al., 2021).

Despite the nutritional potential of ASF, studies which demonstrate their impact on nutrition and health outcomes remain limited, particularly in low-resource settings. Systematic reviews have attributed this evidence gap to poor study design, inadequately powered trials, the challenges of measuring diets, and disentangling effects in observational studies (Grace et al., 2018; Eaton et al., 2019). One well-cited study which compared the growth and development of Kenyan schoolchildren receiving meat, milk, and plant-based snacks showed improved cognitive function and increased arm muscle area in children receiving meat, and a greater height gain in younger and stunted children receiving milk (Neumann et al., 2007). Effects of ASF intake seem to vary between settings, according to the nutrition situation and existing diets. For example, a randomized controlled trial which provided one egg daily to young children over a six-month period demonstrated improved height-for-age in Ecuador (L. L. Iannotti et al., 2017), but no effect when replicated in Malawi (Stewart et al., 2019).

Overall, evidence of health impacts of ASF is strongest for milk and dairy products, followed by beef and eggs. A recent umbrella review of 41 meta-analyses reported one cup of milk per day reduced the risk of cardiovascular disease, hypertension, colorectal cancer, metabolic syndrome, obesity, and osteoporosis among adults (Zhang et al., 2021). For older adults, a growing body of epidemiological evidence is available from high-income countries, including studies linking high milk and dairy consumption with lower frailty and muscle loss (Cuesta-Triana et al., 2019), and lower risks of dementia and Alzheimer's disease (Bermejo-Pareja et al., 2021).

Potential for Negative Health Impacts

Overconsumption of certain ASF can be associated with obesity and non-communicable diseases. Health impacts of red meat consumption have been extensively evaluated in recent years, largely through prospective cohort studies in high-income countries. There is strong evidence linking high intakes of red meat and processed meat (which have been transformed by salting, curing, smoking, or fermentation) with an increased risk of chronic diseases, including cardiovascular disease, stroke, and colorectal cancer (Bouvard et al., 2015; Cui et al., 2019). Historically, health concerns have surrounded the fat content of ASF, including saturated fats, trans fats, and cholesterol. Dietary guidelines recommend limiting saturated fat intake, but several systematic reviews and meta-analyses have failed to identify a link between saturated fats and the risk of disease or death (Chowdhury et al., 2014; Ramsden et al., 2016; Hamley, 2017). Trans fats have been associated with an increased risk of cardiovascular disease and death and all-cause mortality; however, this is suggested to be linked to industrially produced, rather than ruminant-derived, trans fats (Bendsen et al., 2011; De Souza et al., 2015). While there is strong evidence linking blood cholesterol with heart disease, eggs (a rich source of dietary cholesterol) have been shown not to elevate blood cholesterol, or the risk of stroke or coronary heart disease (Rong et al., 2013; Blesso & Fernandez, 2018; Mah et al., 2020).

ASF also present risks to human health as a potential source of foodborne disease, caused by microbial pathogens, parasites, chemical contaminants, and biotoxins (World Health Organization, 2014). This is discussed further in Sections "Role of Wild Meat in Diets and Livelihoods" and "Health Risks for Humans and Domesticated Livestock."

ROLE OF WILD MEAT IN DIETS AND LIVELIHOODS

Wildlife has supported the diets, livelihoods, and customs of people across the globe for millennia (Brashares et al., 2014). Wild meat, also known as bushmeat, includes muscle and organ meat derived from mammals, birds, reptiles, amphibians, and invertebrates. Agricultural development has greatly reduced reliance on wild animals in temperate regions; however, in tropical and subtropical regions of Africa, Asia, and Latin America, close to 2,000 species continue to be hunted for their meat (Redmond et al., 2006). Hunting of primates is common in Central and South America and has been the focus of media attention in Central and West Africa, yet hunting and market studies indicate that ungulates and large-bodied rodents make up the majority of the wild meat consumed globally (Coad et al., 2019). Wild meat provides a core source of quality protein, fat, and micronutrients for many rural people (Bogin, 2011), but is also linked to zoonotic disease outbreaks with far-reaching consequences (Jones et al., 2008) and significant declines in many animal species (Young et al., 2016).

Benefits for Human Nutrition and Health

The magnitude of wild meat consumption and trade varies between countries and regions, according to availability, governmental controls, socioeconomic status, and cultural values (Fa et al., 2015). The existence of laws and cultural prohibitions (Golden & Comaroff, 2015; Spira et al., 2021) makes this a sensitive area of research, and presents challenges in accurately quantifying wild meat intake. A survey of almost 8,000 rural households in 24 countries in Africa, Asia, and Latin America identified wild meat to be hunted by 39% of households, and consumed by almost all (Nielsen et al., 2018). Analysis of data from multiple studies in the Amazon Basin and Congo Basin suggests average consumption of 63 kg and 51 kg per capita per year, respectively (Nasi et al., 2011), but estimates vary widely.

For many rural people, wild meat is a vital nutritional resource. Wildlife provide access to nutrient-dense foods in areas where other forms of ASF are physically or economically inaccessible (Fa et al., 2016). Wild meat, and the income generated from its sale, can provide a "safety net" in times of economic hardship, agricultural food shortages, or civil unrest (Golden, Gupta, et al., 2016; Coad L. et al., 2019). Consumption of wild meat in urban areas is generally much lower on a per capita basis, where it is typically more expensive and less available than livestock-derived foods. Urban consumers may eat wild meat for its cultural significance (Schenck et al., 2006), positive perceptions of taste, quality, and health attributes, and as a symbol of social status (Chausson et al., 2019). Few studies have directly investigated the effect of wild meat consumption on nutritional status. One longitudinal study in rural north-eastern Madagascar found consuming more wildlife was significantly associated with higher hemoglobin levels in children under 12 years of age (Golden et al., 2011). Empirical modeling suggested that removing access to wildlife would lead to a 29% increase in the number of anemic children, and an alarming three-fold increase among those in the poorest households, where reliance on wild meat is greatest.

Risks to Human Health, Wildlife Populations, and Ecosystems

Alongside nutritional contributions, wild meat consumption and associated contact with wildlife also present the risk of zoonotic spillover events. Zoonoses are estimated to cause around 60% of emerging infectious diseases, with most originating in wildlife (Jones et al., 2008). The greatest risk is associated with handling and butchering wildlife, including within live animal markets (Paige et al., 2014). Such contact has been linked to devastating outbreaks of Ebola in West and Central Africa (E. M. Leroy et al., 2004) and the transfer of Human Immunodeficiency Virus (HIV) from primates to humans (Van Heuverswyn & Peeters, 2007), and implicated in the emergence of severe acute respiratory syndrome (SARS) and SARS coronavirus 2 (SARS-CoV-2) (L.-F. Wang & Eaton, 2007; Holmes et al., 2021). Increasing population density and the lengthening of wild meat value chains, including the sale of live animals in urban markets, have amplified the risk of zoonotic disease emergence and transmission (C. K. Johnson et al., 2020). Efforts to mitigate these risks by enforcing total bans on wild meat consumption and trade would have profound implications for wildlife-dependent people. Effective actions rely on building public trust and supporting improvements in sanitation, hygiene, and cooking practices (Onyekuru et al., 2020).

Unsustainable hunting is a major threat to wildlife and ecosystems globally. Evidence supports the sustainability of hunting when human population density is low, targeted wildlife have high reproductive rates, and hunting is almost entirely for household consumption (Wilkie et al., 2016); however, population growth, demand for wild meat by wealthy urban consumers, and multi-species hunting using indiscriminate methods have led to overexploitation of wildlife in many areas. Hunting is estimated to affect around one-fifth of threatened species on the International Union for Conservation of Nature's Red List (Maxwell et al., 2016). Species with high body mass, longer lifespans, and low reproductive rates, including primates and large carnivores, are particularly vulnerable (Nasi et al., 2011; Ripple et al., 2016).

To support sustainable wild meat use, experts highlight the need to address key knowledge gaps: in the scale and multi-level drivers of wild meat trade, the varying levels of nutritional and economic dependence on wild meat, and effective intervention design (Ingram et al., 2021). Continued investment in livestock extension services and agricultural research and development is essential, as part of a multisectoral response to improving and sustaining food security, public health, and conservation simultaneously.

HEALTH RISKS FOR HUMANS AND DOMESTICATED LIVESTOCK

Risks to human health are not limited to the direct consumption or handling of wild meat – as the human population grows and land is cleared for dwellings and agriculture, habitat loss concentrates wildlife in remaining pockets where there is greater likelihood of contact with livestock. One example is where mixed farms producing mangoes and intensively raised pigs elicited spillover of Nipah virus to humans from infected bats feeding above pig housing (Daszak et al., 2013). Direct or indirect transmission of highly pathogenic avian influenza from wild water birds to domestic poultry, compounded by the mixing of poultry at live bird markets, is another pathway where domesticated animals act as intermediate interfaces between wildlife and humans (Fournié et al., 2013; Tian et al., 2015). Human health risks also occur through interactions between livestock, the environment, and humans, such as soil and water contamination with pathogens like anthrax and leptospirosis. Livestock can act as reservoirs for zoonotic diseases, such as bovine spongiform encephalopathy, brucellosis, and tuberculosis, or for vector-borne parasites like trypanosomiasis (Grace et al., 2012).

Furthermore, ASF from both formal and informal livestock markets and value chains have been singled out as the most important source of foodborne disease and an important factor in the rise of antimicrobial resistance (Grace, 2015a; Grace, 2015b; Hoffmann et al., 2017). Microbial pathogens including salmonellosis, toxigenic *Escherichia coli*, campylobacter, and foodborne parasites such as cryptosporidiosis and tapeworm species are estimated to cause a burden of 36 million disability-adjusted life years in LMICs (Grace, 2015a), while toxins such as aflatoxin and dioxin can accumulate in livestock products before being consumed by humans (Dolan et al., 2010).

Particularly with enteric pathogens in LMICs, the combination of livestock raising and inadequate WASH practices is often implicated in childhood diarrhea and stunting. However, recent evidence finds no linkages between improved WASH and child stunting (Cumming et al., 2019). It is possible that livestock, through contamination of the environment with enteric pathogens, contribute to more insidious conditions like subclinical environmental enteropathy disorder (EED), where intestinal villi are shortened, crypts are hypertrophied, and there is increased gut permeability and inflammation. These have the combined effect of impairing nutrient absorption, increasing susceptibility to infection, and sustaining local or systemic inflammation, and is thought to be the link between recurrent infections and the development of stunting (Budge et al., 2019).

The above examples show the importance of livestock to human health; however, more emphasis is needed on the role of humans in livestock health. There is increasing evidence of reverse zoonoses, where humans transmit pathogens to animals, including livestock, wildlife, and companion animals (Messenger et al., 2014; Fagre et al., 2022). Notable examples of this include the recent transmission of SARS-CoV-2 from humans to 23 animal species but particularly farmed mink (World Organisation for Animal Health, 2020; World Organisation for Animal Health, 2022) and the transmission of influenza viruses, particularly H1N1, from humans to pigs (Ma et al., 2008; Song et al., 2010; Forgie et al., 2011).

The conditions under which humans farm livestock can also influence their susceptibility to disease, as, like humans, livestock require adequate housing, nutrition, WASH, and medical attention to maximize health and welfare (Wong et al., 2022). In a longitudinal One Health study in Kenya, increased incidence of illness in livestock was associated with increased illness in humans, which the authors considered to be due to zoonotic pathogens or factors relating to the environment in which they both live (Thumbi et al., 2015).

Lastly, the role livestock play in immune function and improving the social and economic determinants of health should be considered in parallel with the risks. Good evidence exists showing that childhood contact with livestock can be associated with improved human immune system function, particularly around allergic responses (Von Mutius & Vercelli, 2010; Ege et al., 2011; Stein et al., 2016; Dhakal et al., 2019). Additionally, the study by Thumbi et al. (2015) found that livestock ownership was associated with ASF consumption and increased expenditure on human healthcare, leading the authors to conclude that interventions that improve general animal health may have the greatest impact on human health through increasing household wealth rather than directly reducing pathogen exposure.

ENVIRONMENTAL IMPACTS OF LIVESTOCK PRODUCTION SYSTEMS

LIVESTOCK AND THE ENVIRONMENT

Since the 2006 publication of *Livestock's Long Shadow*, the negative environmental impact of livestock has been brought to the fore (Steinfeld & FAO, 2006). For more than 10,000 years, livestock production has contributed to the development of humanity (Hartung, 2013). Yet in that time, associated production systems have accelerated environmental change, negatively impacting the atmosphere, air quality, biodiversity, water, and land. A changing environment can also be detrimental to livestock, contributing to heat stress, disease, and disease vector dynamics, as well as altering feed and water availability (Rojas-Downing et al., 2017; Bernabucci, 2019).

Currently, demand for ASF is stagnating in high-income countries, but is predicted to continue growing in LMICs (Thornton, 2010; Rojas-Downing et al., 2017). In addition to their multiple other roles, livestock, fish, and insects provide nutrient-dense food and can facilitate nutrient cycling, an important factor when considering the health of the environment in which we produce the food we eat (Adesogan et al., 2020). The United Nations' 17 Sustainable Development Goals (SDGs) provide a blueprint for our future, highlighting the importance of the environment, food security, and equity (United Nations Department of Economic and Social Affairs, 2023). Livestock, fish, and insect production has the potential to contribute negatively and positively to the SDGs and trade-offs need to be considered (Food and Agriculture Organization, 2018). This should be acknowledged and understood for successful One Health outcomes.

NEGATIVE IMPACTS OF LIVESTOCK ON ENVIRONMENT

Modeling suggests 14.5% of global anthropogenic greenhouse gas (GHG) emissions come from livestock production systems, with cattle production (beef and dairy) accounting for 65% of this contribution. Methane from enteric fermentation is the biggest source of the sector's emissions (Gerber et al., 2013). Given the significant challenge climate change poses, it is understandable that a major focus has been on emissions (World Health Organization, 2023); however, other impacts should not be ignored. Livestock production contributes to air pollution through the release of particulate matter when forests are burnt to make space for pasture or when manure is stored or applied to land (Lelieveld et al., 2015; Balasubramanian et al., 2021; Roman et al., 2021). Particulate matter can also be derived from the use of plastics in livestock production (Lim, 2021). Concerning health implications are further exacerbated for farmers, their animals, and local residents when animals and their waste are confined within housing (Cambra-López et al., 2010; De Rooij et al., 2019).

Despite productivity improvements and a decoupling of livestock and land having reduced global pastureland area since 2000 (Blaustein-Rejto et al., 2019), production continues to dramatically and extensively change landscapes. Approximately 83% of farmland worldwide is used by livestock, which provide 37% of human protein (Poore & Nemecek, 2018). This polarity draws attention toward the potential inefficiencies in the rearing of livestock to source protein for human consumption (Westhoek et al., 2011; Berners-Lee et al., 2018). An example snapshot of data suggests that globally livestock production uses almost 40% of habitable land area (Ritchie & Roser, 2024). Land use change often results in a loss of biodiversity and ecosystems services (particularly when forests are cleared) (Benton et al., 2021). Soils can become degraded if not well managed, releasing more carbon into the atmosphere (Abdalla et al., 2018).

In the same way that carbon footprinting can capture the livestock sector's contribution toward climate change, water footprints demonstrate the impact on increasingly scarce water resources. Water footprints include both volumes consumed and polluted. A global assessment suggests that, on average, it takes 4,300 and 15,400L of water to produce a single kilogram of chicken or beef, respectively. On a per gram of protein basis, the water footprint of milk, eggs, and chicken meat is roughly 1.5 times that of pulses, while for beef, it is roughly 6 times (Mekonnen & Hoekstra, 2012).

Evidently, the livestock production systems commonly in use today can have varying impacts, from negative (Nguyen et al., 2010; Gerber et al., 2013; Herrero et al., 2016) to positive (Weber & Horst, 2011) and all levels in between (Poore & Nemecek, 2018). Yet, the reality is that livestock populations will not suddenly disappear; in fact, they will likely grow in some regions as incomes rise. Consequently, their role in providing farmers and the food system resilience in the face of growing climate change challenges cannot be ignored (Joy et al., 2020; Komarek et al., 2021).

What Can Be Done?

Certain populations could reduce consumption of livestock products, benefiting their health and reducing absolute environmental impacts (Willett et al., 2019). For others, where ASF provide a vital source of nutrients, reduced consumption is not an option (Adesogan et al., 2020). In both these scenarios, opportunities to reduce the environmental impact must be considered. The choice of source livestock species can make a difference to emissions. For instance, poultry and pork have a lower emission intensity than beef (less than 100kg CO_2 equivalent per kg of protein compared to more than 300kg CO_2 equivalent per kg of protein) (Gerber et al., 2013).

For centuries, entomophagy (eating insects) has been common practice in some cultures, though such insects were not farmed or not farmed at a large scale (Cadinu et al., 2020). Scaling up insect farming has been seen as an opportunity to reduce environmental impact. Per nutritional unit, both GHG emissions and water footprints have been shown to be lower for insect sourced foods than traditional livestock species (Oonincx & De Boer, 2012). Additionally, insects have the highest land use efficiency and can support wider nutrient upcycling (Alexander et al., 2017; Parodi et al., 2021). However, the evidence base around biosecurity of larger scale insect farms and nutrient bioavailability is currently limited, and it is unclear whether their nutrient profiles will diminish under intensive production as seen in poultry and fish (Y. Wang et al., 2010; De Roos et al., 2017; Bertola & Mutinelli, 2021; Ojha et al., 2021).

As well as choosing different sources of food, there is evidence that improvements in production efficiency could reduce carbon footprints of existing and future patterns of consumption – much of this mitigation potential is in LMIC systems (Gerber et al., 2013; Herrero et al., 2016; Salmon et al., 2018). There is increasing interest in Regenerative Agriculture (RA) as a method of production. Despite some ambiguity around a definition, there is convergence on soil health being the entry point for RA to benefit wider ecosystem services and result in a net positive environmental impact (Newton et al., 2020; Schreefel et al., 2020). For livestock, this could mean greater integration of livestock and crops in mixed farm systems, with animals valued to support soil quality as well as produce food. In addition, evidence strongly suggests that well-managed pasture can offer carbon sequestration potential (De Oliveira Silva et al., 2016; Mottet, Henderson, et al., 2017). Under RA, livestock breeds are chosen for environmental compatibility, rather than to maximize yields (Schreefel et al., 2020). In high-income countries, RA should not be regarded as a step backward, but rather an opportunity for systems to shift from an extractive, high-output focus to a more holistic approach using environmentally harmonious farming methods. For most production systems in LMICs, RA could be a lesson in not making the same mistakes of prioritizing weight and/or volume over environmental health and nutrient density in food produced (see Section "Declining Profiles of Animal-Source Food and Nutrient Loss from the Human Food Chain"), while developing the livestock sector. Omitting consideration of the wider environment impacts and the consequences for human nutrition impacts the development of the sector.

IMPACT OF TRADE, VALUE CHAINS, AND VULNERABILITY TO CRISES

Livestock have been traded for thousands of years, especially in association with their valuable roles in transport and food security (R. G. Alders et al., 2021). However, growth of livestock production, increasing intensification, and globalization of livestock trade since the mid-20th century have contributed to the rise of global crises, for example:

- Climate change through increased GHG emissions associated with vast increases in the numbers of ruminants produced and the transport of livestock and livestock products (IPES-Food, 2017; Li et al., 2022);

- Global health threats due to zoonotic diseases and antimicrobial resistance (R. Alders et al., 2013; Pokharel et al., 2020); and
- Biodiversity loss associated with the production of animal feed such as soy and maize (Ceballos et al., 2020).

It has also contributed to the impact of crises on livestock production, for instance:

- Increased livestock morbidity and mortality associated with the increasing frequency of natural disasters such as droughts, floods, and cyclones (Godde et al., 2021); and
- Decreased consumer demand for livestock products, including chicken meat birds (broilers) following pandemics of avian influenza H5N1 and COVID-19 (Sattar et al., 2021).

These changes in livestock production and trade are also impacting food and nutrition security through two major pathways, i.e., food-feed competition and declining nutrient profiles of ASF.

Food-Feed Competition

The expansion of intensive, large-scale livestock production has increased food-feed competition between humans and animals. The responsible use of available nutrient resources has become a key issue over the past decade as nutrient resources have to meet the needs of increasing numbers of humans, livestock, and companion animals (R. G. Alders et al., 2021). With the intensification of livestock production and the globalization of food systems, competition between humans and animals for human-edible crops has increased. Berners-Lee et al. (2018) have argued that there is no nutritional case for feeding human-edible crops to animals, and that a failure to heed this message will require a 119% increase in edible crops grown by 2050. It is estimated that livestock consume one-third of cereal produced globally, requiring approximately 40% of arable land (Mottet, De Haan, et al., 2017). It is also reported that globally, 1 kg of boneless meat in ruminant systems requires 2.8 kg of feed that would be edible by humans and 3.2 kg of edible human feed is required in pig and poultry systems (Mottet, De Haan, et al., 2017).

Livestock production uses around 2 billion hectares (ha) of grasslands with only 700 million ha of this area considered suitable for crop production. With modest improvements in feed conversion ratios, further expansion of arable land dedicated to feed production could be prevented (Mottet, De Haan, et al., 2017). By employing sustainable and circular bioeconomy principles, Van Zanten et al. (2019) have proposed that livestock production systems that convert by-products from the food system and grass resources into valuable food and manure can contribute significantly to human food supply, while simultaneously reducing the environmental impact of the entire food system (Van Zanten et al., 2019).

Declining Profiles of ASF and Nutrient Loss from the Human Food Chain

Over the past 50 years, a marked increase in the fat content of standard broiler chickens has been reported, with the modern broiler providing more energy from fat than protein (Y. Wang et al., 2010). The intensification of poultry production systems, with a lack of exercise, *ad libitum* access to high-energy food, and selection for rapid weight gain, is considered the key driver of these changes to carcass nutrient profiles.

Minimizing pre-consumer losses in agri-food value chains has been identified as an important contribution to reaching the SDGs. Additionally, it can benefit animal, human, and environmental health. In high-income countries, the consumption of offal has diminished significantly over the past five decades. This practice is contributing to the need to produce more animals to meet the same nutrient yield, increased GHG emissions (Wingett et al., 2019), and the loss of highly nutritious bioavailable nutrients such as iron, vitamins A and B12, and zinc from products such as liver (Morrison, 2020). Utilization of livestock carcasses is generally much more efficient in LMICs. Multiple benefits of consuming as much of the carcass as possible, e.g., organ meats, fat, unpopular and uncommon cuts, include the increased entry of nutrient-dense food into household diets, improving the micronutrient status of the population, decrease GHG emissions, and conservation of vulnerable rangeland and other resources due to increased yield per head of livestock (Wingett et al., 2019; Xue et al., 2019).

The recent growth in the international frozen chicken value chain has not only negatively impacted local producers in importing countries (M. C. Johnson, 2011), this trade has also reduced the range and quantity of nutrients available to consumers. A whole chicken carcass containing the heart, liver, and gizzard contains seven times the amount of vitamin A and folate and triple the amount of vitamin B12 than only the muscle meat provided in a frozen carcass, as well as higher levels of iron, zinc, and protein (De Bruyn et al., 2017).

LINKING IT ALL TOGETHER

The interlinkages between human, animal, and environmental health and malnutrition are many. On one hand, terrestrial and aquatic animals support livelihoods and provide people with valuable nutrients, yet their intensive production and global transport, wild consumption, or interactions with humans can also increase risk to human and environmental health (Figure 7.2).

It is imperative that these interconnections are recognized, as the solutions will require collaboration between sectors and disciplines, and adaptation to the local context. Initiatives like the UN's Decade of Action on Nutrition (2016–2025), SDG2 (Zero Hunger), the Scaling Up Nutrition (SUN) Movement, and U.S. Agency for International Development (USAID) Advancing Nutrition project align global and national priorities and promote multisectoral and multistakeholder actions, while systems-level research programs such as CGIAR's

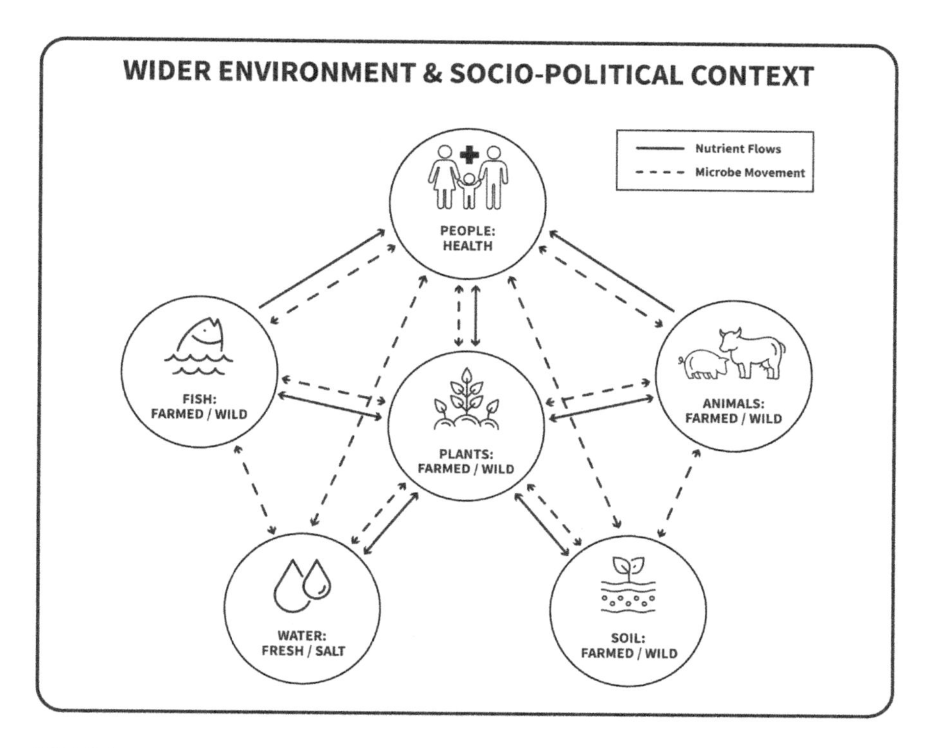

FIGURE 7.2 Nutrient and microbial flows in a One Health system. (Adapted from Alders, 2021)

Agriculture for Nutrition and Health (A4NH) help to realize the potential for agricultural development to contribute to improving global food and nutrition security.

To conclude, it is important to recognize that there will not be a "one approach fits all" solution to the future of livestock production and ASF consumption around the world. Context-specific, culturally appropriate solutions that are balanced with projected environmental and biodiversity challenges are essential. The anticipated changes that climate change and extreme weather events will bring and the resulting impacts on agricultural production, rangelands, fisheries, and disease ecology necessitate flexibility in exploring and implementing a range of solutions. These may include revisiting traditional and indigenous approaches, scaling RA practices, and developing novel options for livestock and fisheries to be sustainable, regenerative, and contribute to climate solutions, rather than exacerbating environmental and climatic challenges. Collectively, solutions must be prioritized that optimize the health of humans, livestock, wildlife and the environment, while simultaneously reducing the persistent risk of known and emerging zoonotic diseases.

BOX 7.3 CASE STUDY – A ONE HEALTH APPROACH TO RESEARCH IN SUPPORT OF RURAL RESILIENCE IN TANZANIA

Organizations: Tanzania Veterinary Laboratory Agency, Tanzania Food and Nutrition Centre, Sokoine University of Agriculture, University of Sydney, Kyeema Foundation, Royal Veterinary College

Location: Twelve rural communities in Singida Region, Tanzania

Project Duration: 5 years

Interdisciplinary research in central Tanzania adopted a One Health approach to design, implement, and evaluate activities in support of human nutrition, health and livelihoods, livestock health and production, and local farming and ecological systems. An overarching project sought to reduce childhood undernutrition, by analyzing and testing opportunities to enhance women's key role in improving poultry and crop integration and efficiency to strengthen household nutrition (R. Alders et al., 2014). With funding from the Australian Centre for International Agricultural Research, the project team worked to understand existing food and nutrition systems and implement community-based interventions. Project activities supported family poultry and crop production, improved nutritional awareness, and built capacity of research students. Quantitative and qualitative results documented the complexity of household food and nutrition systems in semi-arid areas during a time of considerable weather variability. Following the project, the need for government food distribution programs during the hunger periods was reduced in the project area and chicken numbers increased in households that followed a four-monthly schedule of vaccination against Newcastle disease (De Bruyn et al., 2017).

Complementary smaller research projects led by graduate students from Tanzania and Australia contributed to building a broader understanding of the systems underpinning food security and infectious disease challenges. These projects included exploring public health risks of infection associated with bacterial pathogens of food safety importance in village chickens; assessing susceptibility of village chickens to aflatoxin exposure and contamination of grains and chicken products (Magoke, 2021); assessing the antimicrobial efficacy of wood ash lye as a low-cost disinfection option in resource-poor settings; conducting a preliminary investigation into the contribution of field rodents to food security and potential infectious disease risks (Ackland, 2014); using participatory epidemiology to identify the optimal timing of village chicken vaccination campaigns against Newcastle disease in the project area (Kiswaga, 2019); and analyzing village chicken feeding practices across seasons, which revealed limited to no competition for household resources to be used as human food and chicken feed (Katabazi, 2018).

NOTES

1 Note that Sustainable Development Goal 2, "Zero hunger," includes ending all forms of malnutrition, not just undernourishment.
2 Length is measured for children under 24 months of age, while height is measured for children 24 months and older.

REFERENCES

Abdalla, M., Hastings, A., Chadwick, D. R., Jones, D. L., Evans, C. D., Jones, M. B., Rees, R. M., & Smith, P. (2018). Critical review of the impacts of grazing intensity on soil organic carbon storage and other soil quality indicators in extensively managed grasslands. *Agriculture, Ecosystems & Environment, 253*, 62–81. https://doi.org/10.1016/j.agee.2017.10.023

Ackland, L. A. (2014). *Rodents, food security and infectious disease in Tanzania* [BSc(Vet) thesis]. University of Sydney.

Adesogan, A. T., Havelaar, A. H., McKune, S. L., Eilittä, M., & Dahl, G. E. (2020). Animal source foods: Sustainability problem or malnutrition and sustainability solution? Perspective matters. *Global Food Security, 25*, 100325. https://doi.org/10.1016/j.gfs.2019.100325

Alders, R. (2021). One Health in the time of COVID-19 and climate change. *One Health Poultry Hub Blog*, 3 November 2021. Retrieved from: https://www.onehealthpoultry.org/blog-posts/one-health-in-the-time-of-covid-19-and-climate-change/

Alders, R., Aongolo, A., Bagnol, B., De Bruyn, J., Kimboka, S., Kocj, R., Li, M., Maulaga, W., McConchie, R., Mor, S., Msami, H., Mulenga, F., Mwala, M., Mwale, S., Pengelly, B., Rushton, J., Simpson, J., Victor, R., Yongolo, C., & Young, M. (2014). Using a One Health approach to promote food and nutrition security in Tanzania and Zambia. *GRF Davos Planet@Risk, 2*(3), 187–190.

Alders, R., Awuni, J. A., Bagnol, B., Farrell, P., & De Haan, N. (2013). Impact of Avian Influenza on village poultry production globally. *EcoHealth*. https://doi.org/10.1007/s10393-013-0867-x

Alders, R. G., Campbell, A., Costa, R., Guèye, E. F., Ahasanul Hoque, M., Perezgrovas-Garza, R., Rota, A., & Wingett, K. (2021). Livestock across the world: Diverse animal species with complex roles in human societies and ecosystem services. *Animal Frontiers, 11*(5), 20–29. https://doi.org/10.1093/af/vfab047

Alexander, P., Brown, C., Arneth, A., Finnigan, J., Moran, D., & Rounsevell, M. D. A. (2017). Losses, inefficiencies and waste in the global food system. *Agricultural Systems, 153*, 190–200. https://doi.org/10.1016/j.agsy.2017.01.014

Allen, L. H. (2008). To what extent can food-based approaches improve micronutrient status? *Asia Pacific Journal of Clinical Nutrition, 17*(Supp 1), 103–105.

Balasubramanian, S., Domingo, N. G. G., Hunt, N. D., Gittlin, M., Colgan, K. K., Marshall, J. D., Robinson, A. L., Azevedo, I. M. L., Thakrar, S. K., Clark, M. A., Tessum, C. W., Adams, P. J., Pandis, S. N., & Hill, J. D. (2021). The food we eat, the air we breathe: A review of the fine particulate matter-induced air quality health impacts of the global food system. *Environmental Research Letters, 16*(10), 103004. https://doi.org/10.1088/1748-9326/ac065f

Beal, T., & Ortenzi, F. (2022). Priority micronutrient density in foods. *Frontiers in Nutrition, 9*, 806566. https://doi.org/10.3389/fnut.2022.806566

Belton, B., Little, D. C., Zhang, W., Edwards, P., Skladany, M., & Thilsted, S. H. (2020). Farming fish in the sea will not nourish the world. *Nature Communications, 11*(1), 5804. https://doi.org/10.1038/s41467-020-19679-9

Bendsen, N. T., Christensen, R., Bartels, E. M., & Astrup, A. (2011). Consumption of industrial and ruminant trans fatty acids and risk of coronary heart disease: A systematic review and meta-analysis of cohort studies. *European Journal of Clinical Nutrition*, *65*(7), 773–783. https://doi.org/10.1038/ejcn.2011.34

Benton, T. G., Bieg, C., Harwatt, H., Pudasaini, R., & Wellesley, L. (2021). *Food system impacts on biodiversity loss*. Chatham House. https://www.chathamhouse.org/2021/02/food-system-impacts-biodiversity-loss

Berggren, Å., Jansson, A., & Low, M. (2019). Approaching ecological sustainability in the emerging insects-as-food industry. *Trends in Ecology & Evolution*, *34*(2), 132–138. https://doi.org/10.1016/j.tree.2018.11.005

Bermejo-Pareja, F., Ciudad-Cabañas, M. J., Llamas-Velasco, S., Tapias-Merino, E., Hernández Gallego, J., Hernández-Cabria, M., Collado-Yurrita, L., & López-Arrieta, J. M. (2021). Is milk and dairy intake a preventive factor for elderly cognition (dementia and Alzheimer's)? A quality review of cohort surveys. *Nutrition Reviews*, *79*(7), 743–757. https://doi.org/10.1093/nutrit/nuaa045

Bernabucci, U. (2019). Climate change: Impact on livestock and how can we adapt. *Animal Frontiers*, *9*(1), 3–5. https://doi.org/10.1093/af/vfy039

Berners-Lee, M., Kennelly, C., Watson, R., & Hewitt, C. N. (2018). Current global food production is sufficient to meet human nutritional needs in 2050 provided there is radical societal adaptation. *Elementa: Science of the Anthropocene*, *6*, 52. https://doi.org/10.1525/elementa.310

Bertola, M., & Mutinelli, F. (2021). A systematic review on viruses in mass-reared edible insect species. *Viruses*, *13*(11), 2280. https://doi.org/10.3390/v13112280

Bhutta, Z. A., & Salam, R. A. (2012). Global nutrition epidemiology and trends. *Annals of Nutrition and Metabolism*, *61*(Suppl. 1), 19–27. https://doi.org/10.1159/000345167

Black, M. M., Quigg, A. M., Hurley, K. M., & Pepper, M. R. (2011). Iron deficiency and iron-deficiency anemia in the first two years of life: Strategies to prevent loss of developmental potential. *Nutrition Reviews*, *69*(suppl_1), S64–S70. https://doi.org/10.1111/j.1753-4887.2011.00435.x

Black, R. E., Victora, C. G., Walker, S. P., Bhutta, Z. A., Christian, P., De Onis, M., Ezzati, M., Grantham-McGregor, S., Katz, J., Martorell, R., & Uauy, R. (2013). Maternal and child undernutrition and overweight in low-income and middle-income countries. *The Lancet*, *382*(9890), 427–451. https://doi.org/10.1016/S0140-6736(13)60937-X

Blaustein-Rejto, D., Blomqvist, L., McNamara, J., & de Kirby, K. (2019). *Achieving peak Pasture: Shrinking Pasture's footprint by spreading the livestock revolution*. The Breakthrough Institute. https://s3.us-east-2.amazonaws.com/uploads.thebreakthrough.org/articles/achieving-peak-pasture/Pasture_Report_FINAL%20%281%29.pdf

Blesso, C., & Fernandez, M. (2018). Dietary cholesterol, serum lipids, and heart disease: Are eggs working for or against you? *Nutrients*, *10*(4), 426. https://doi.org/10.3390/nu10040426

Bogin, B. (2011).! Kung nutritional status and the original "affluent society"—A new analysis. *Anthropologischer Anzeiger*, *68*(4), 349–366. https://doi.org/10.1127/0003-5548/2011/0148

Bouvard, V., Loomis, D., Guyton, K. Z., Grosse, Y., Ghissassi, F. E., Benbrahim-Tallaa, L., Guha, N., Mattock, H., & Straif, K. (2015). Carcinogenicity of consumption of red and processed meat. *The Lancet Oncology*, *16*(16), 1599–1600. https://doi.org/10.1016/S1470-2045(15)00444-1

Brashares, J. S., Abrahms, B., Fiorella, K. J., Golden, C. D., Hojnowski, C. E., Marsh, R. A., McCauley, D. J., Nuñez, T. A., Seto, K., & Withey, L. (2014). Wildlife decline and social conflict. *Science, 345*(6195), 376–378. https://doi.org/10.1126/science.1256734

Budge, S., Parker, A. H., Hutchings, P. T., & Garbutt, C. (2019). Environmental enteric dysfunction and child stunting. *Nutrition Reviews, 77*(4), 240–253. https://doi.org/10.1093/nutrit/nuy068

Cadinu, L. A., Barra, P., Torre, F., Delogu, F., & Madau, F. A. (2020). Insect rearing: Potential, challenges, and circularity. *Sustainability, 12*(11), 4567. https://doi.org/10.3390/su12114567

Cambra-López, M., Aarnink, A. J. A., Zhao, Y., Calvet, S., & Torres, A. G. (2010). Airborne particulate matter from livestock production systems: A review of an air pollution problem. *Environmental Pollution, 158*(1), 1–17. https://doi.org/10.1016/j.envpol.2009.07.011

Carletto, G., Ruel, M., Winters, P., & Zezza, A. (2015). Farm-level pathways to improved nutritional status: Introduction to the special issue. *The Journal of Development Studies, 51*(8), 945–957. https://doi.org/10.1080/00220388.2015.1018908

Ceballos, G., Ehrlich, P. R., & Raven, P. H. (2020). Vertebrates on the brink as indicators of biological annihilation and the sixth mass extinction. *Proceedings of the National Academy of Sciences, 117*(24), 13596–13602. https://doi.org/10.1073/pnas.1922686117

Chausson, A. M., Rowcliffe, J. M., Escouflaire, L., Wieland, M., & Wright, J. H. (2019). Understanding the sociocultural drivers of urban bushmeat consumption for behavior change interventions in Pointe Noire, Republic of Congo. *Human Ecology, 47*(2), 179–191. https://doi.org/10.1007/s10745-019-0061-z

Chowdhury, R., Warnakula, S., Kunutsor, S., Crowe, F., Ward, H. A., Johnson, L., Franco, O. H., Butterworth, A. S., Forouhi, N. G., Thompson, S. G., Khaw, K.-T., Mozaffarian, D., Danesh, J., & Di Angelantonio, E. (2014). Association of dietary, circulating, and supplement fatty acids with coronary risk: A systematic review and meta-analysis. *Annals of Internal Medicine, 160*(6), 398. https://doi.org/10.7326/M13-1788

Coad L., Fa J.E., Abernethy K., Van Vliet N., Santamaria C., Wilkie D., El Bizri H.R., Ingram D.J., Cawthorn D-M., & Nasi R. (2019). *Toward a sustainable, participatory and inclusive wild meat sector*. Center for International Forestry Research (CIFOR). https://doi.org/10.17528/cifor/007046

Committee on World Food Security. (2012). *Coming to terms with terminology: Food security, nutrition security, food security and nutrition, food and nutrition security* (CFS 2012/39/4). Food and Agriculture Organization. https://www.ipcinfo.org/fileadmin/user_upload/cfs/doclibrary/CFS%20Terminology%2016%20July%202012_rev2.pdf

Cuesta-Triana, F., Verdejo-Bravo, C., Fernández-Pérez, C., & Martín-Sánchez, F. J. (2019). Effect of milk and other dairy products on the risk of frailty, Sarcopenia, and cognitive performance decline in the elderly: A systematic review. *Advances in Nutrition, 10*, S105–S119. https://doi.org/10.1093/advances/nmy105

Cui, K., Liu, Y., Zhu, L., Mei, X., Jin, P., & Luo, Y. (2019). Association between intake of red and processed meat and the risk of heart failure: A meta-analysis. *BMC Public Health, 19*(1), 354. https://doi.org/10.1186/s12889-019-6653-0

Cumming, O., Arnold, B. F., Ban, R., Clasen, T., Esteves Mills, J., Freeman, M. C., Gordon, B., Guiteras, R., Howard, G., Hunter, P. R., Johnston, R. B., Pickering, A. J., Prendergast, A. J., Prüss-Ustün, A., Rosenboom, J. W., Spears, D., Sundberg, S., Wolf, J., Null, C., … Colford, J. M. (2019). The implications of three major new trials for the effect of water, sanitation and hygiene on childhood diarrhea and stunting: A consensus statement. *BMC Medicine*, *17*(1), 173. https://doi.org/10.1186/s12916-019-1410-x

Daszak, P., Zambrana-Torrelio, C., Bogich, T. L., Fernandez, M., Epstein, J. H., Murray, K. A., & Hamilton, H. (2013). Interdisciplinary approaches to understanding disease emergence: The past, present, and future drivers of Nipah virus emergence. *Proceedings of the National Academy of Sciences*, *110*(supplement_1), 3681–3688. https://doi.org/10.1073/pnas.1201243109

De Bruyn, J., Thomson, P. C., Bagnol, B., Maulaga, W., Rukambile, E., & Alders, R. G. (2017). The chicken or the egg? Exploring bi-directional associations between Newcastle disease vaccination and village chicken flock size in rural Tanzania. *PLoS One*, *12*(11), e0188230. https://doi.org/10.1371/journal.pone.0188230

De Oliveira Silva, R., Barioni, L. G., Hall, J. A. J., Folegatti Matsuura, M., Zanett Albertini, T., Fernandes, F. A., & Moran, D. (2016). Increasing beef production could lower greenhouse gas emissions in Brazil if decoupled from deforestation. *Nature Climate Change*, *6*(5), 493–497. https://doi.org/10.1038/nclimate2916

De Rooij, M. M. T., Smit, L. A. M., Erbrink, H. J., Hagenaars, T. J., Hoek, G., Ogink, N. W. M., Winkel, A., Heederik, D. J. J., & Wouters, I. M. (2019). Endotoxin and particulate matter emitted by livestock farms and respiratory health effects in neighboring residents. *Environment International*, *132*, 105009. https://doi.org/10.1016/j.envint.2019.105009

De Roos, B., Sneddon, A. A., Sprague, M., Horgan, G. W., & Brouwer, I. A. (2017). The potential impact of compositional changes in farmed fish on its health-giving properties: Is it time to reconsider current dietary recommendations? *Public Health Nutrition*, *20*(11), 2042–2049. https://doi.org/10.1017/S1368980017000696

De Souza, R. J., Mente, A., Maroleanu, A., Cozma, A. I., Ha, V., Kishibe, T., Uleryk, E., Budylowski, P., Schünemann, H., Beyene, J., & Anand, S. S. (2015). Intake of saturated and trans unsaturated fatty acids and risk of all cause mortality, cardiovascular disease, and type 2 diabetes: Systematic review and meta-analysis of observational studies. *BMJ*, h3978. https://doi.org/10.1136/bmj.h3978

Development Initiatives. (2018). *2018 Global Nutrition Report: Shining a light to spur action on nutrition*. https://globalnutritionreport.org/reports/global-nutrition-report-2018/

Development Initiatives. (2021). *2021 global nutrition report: The state of global nutrition*. https://globalnutritionreport.org/reports/2021-global-nutrition-report/

Dhakal, S., Wang, L., Antony, L., Rank, J., Bernardo, P., Ghimire, S., Bondra, K., Siems, C., Lakshmanappa, Y. S., Renu, S., Hogshead, B., Krakowka, S., Kauffman, M., Scaria, J., LeJeune, J. T., Yu, Z., & Renukaradhya, G. J. (2019). Amish (Rural) vs. Non-Amish (Urban) infant fecal microbiotas are highly diverse and their transplantation lead to differences in Mucosal immune maturation in a humanized Germfree Piglet Model. *Frontiers in Immunology*, *10*, 1509. https://doi.org/10.3389/fimmu.2019.01509

Dickie, F., Miyamoto, M., & Collins, C. M. (2020). The potential of insect farming to increase food security. In H. Mikkola (Ed.), *Edible Insects*. IntechOpen. https://doi.org/10.5772/intechopen.77835

Dolan, L. C., Matulka, R. A., & Burdock, G. A. (2010). Naturally occurring food toxins. *Toxins*, *2*(9), 2289–2332. https://doi.org/10.3390/toxins2092289

Eaton, J. C., & Iannotti, L. L. (2017). Genome–nutrition divergence: Evolving understanding of the malnutrition spectrum. *Nutrition Reviews*, *75*(11), 934–950. https://doi.org/10.1093/nutrit/nux055

Eaton, J. C., Rothpletz-Puglia, P., Dreker, M. R., Iannotti, L., Lutter, C., Kaganda, J., & Rayco-Solon, P. (2019). Effectiveness of provision of animal-source foods for supporting optimal growth and development in children 6 to 59 months of age. *Cochrane Database of Systematic Reviews, 2019*(5). https://doi.org/10.1002/14651858.CD012818.pub2

Ege, M. J., Mayer, M., Normand, A.-C., Genuneit, J., Cookson, W. O. C. M., Braun-Fahrländer, C., Heederik, D., Piarroux, R., & Von Mutius, E. (2011). Exposure to environmental microorganisms and childhood Asthma. *New England Journal of Medicine, 364*(8), 701–709. https://doi.org/10.1056/NEJMoa1007302

Fa, J. E., Olivero, J., Real, R., Farfán, M. A., Márquez, A. L., Vargas, J. M., Ziegler, S., Wegmann, M., Brown, D., Margetts, B., & Nasi, R. (2015). Disentangling the relative effects of bushmeat availability on human nutrition in central Africa. *Scientific Reports, 5*(1), 8168. https://doi.org/10.1038/srep08168

Fa, J. E., van Vliet, N., & Nasi, R. (2016). Bushmeat, food security, and conservation in African rainforests. In A. A. Aguirre & R. Sukumar (Eds.), *Tropical conservation: Perspectives on local and global priorities* (pp. 331–334). Oxford University Press.

Fagre, A. C., Cohen, L. E., Eskew, E. A., Farrell, M., Glennon, E., Joseph, M. B., Frank, H. K., Ryan, S. J., Carlson, C. J., & Albery, G. F. (2022). Assessing the risk of human-to-wildlife pathogen transmission for conservation and public health. *Ecology Letters, 25*(6), 1534–1549. https://doi.org/10.1111/ele.14003

FAO, IFAD, UNICEF, WFP, & WHO. (2022). *The State of Food Security and Nutrition in the World 2022. Repurposing food and agricultural policies to make healthy diets more affordable*. FAO. https://doi.org/10.4060/cc0639en

Food and Agriculture Organization. (2018). *Dietary assessment: A resource guide to method selection and application in low resource settings*. https://www.fao.org/3/i9940en/I9940EN.pdf

Food and Agriculture Organization. (2022a). *Contribution of terrestrial animal source food to healthy diets for improved nutrition and health outcomes—Key messages*. FAO. https://doi.org/10.4060/cc3912en

Food and Agriculture Organization. (2022b). *FishStatJ - Software for fishery and aquaculture statistical time series—Fisheries and aquaculture*. https://www.fao.org/fishery/en/statistics/software/fishstatj

Food and Agriculture Organization. (2022c). *The state of world fisheries and aquaculture 2022. Towards blue transformation*. FAO. https://doi.org/10.4060/cc0461en

Food and Agriculture Organization. (2023, October 27). *FAOSTAT*. https://www.fao.org/faostat/en/#data/FBS

Forgie, S. E., Keenliside, J., Wilkinson, C., Webby, R., Lu, P., Sorensen, O., Fonseca, K., Barman, S., Rubrum, A., Stigger, E., Marrie, T. J., Marshall, F., Spady, D. W., Hu, J., Loeb, M., Russell, M. L., & Babiuk, L. A. (2011). Swine outbreak of pandemic Influenza A virus on a Canadian Research Farm supports human-to-swine transmission. *Clinical Infectious Diseases, 52*(1), 10–18. https://doi.org/10.1093/cid/ciq030

Fournié, G., Guitian, J., Desvaux, S., Cuong, V. C., Dung, D. H., Pfeiffer, D. U., Mangtani, P., & Ghani, A. C. (2013). Interventions for avian influenza A (H5N1) risk management in live bird market networks. *Proceedings of the National Academy of Sciences, 110*(22), 9177–9182. https://doi.org/10.1073/pnas.1220815110

Frankenberger, T., Langworthy, M., Spangler, T., & Nelson, S. (2012). *Enhancing resilience to food security shocks*. TANGO International, Inc. https://www.technicalconsortium.org/wp-content/uploads/2014/05/Enhancing-Resilience-to-Food-Security-Shocks_White-paper1.pdf

Gahukar, R. T. (2016). Edible insects farming: efficiency and impact on family livelihood, food security, and environment compared with livestock and crops. In *Insects as sustainable food ingredients* (pp. 85–111). Elsevier. https://doi.org/10.1016/B978-0-12-802856-8.00004-1

Garlock, T., Asche, F., Anderson, J., Ceballos-Concha, A., Love, D. C., Osmundsen, T. C., & Pincinato, R. B. M. (2022). Aquaculture: The missing contributor in the food security agenda. *Global Food Security*, *32*, 100620. https://doi.org/10.1016/j.gfs.2022.100620

Gerber, P. J., Steinfeld, H., Henderson, B., Mottet, A., Opio, C., Dijkman, J., Falcucci, A., & Tempio, G. (2013). *Tackling climate change through livestock: A global assessment of emissions and mitigation opportunities*. FAO. https://www.fao.org/3/i3437e/i3437e.pdf

Gibson, R. S., Perlas, L., & Hotz, C. (2006). Improving the bioavailability of nutrients in plant foods at the household level. *Proceedings of the Nutrition Society*, *65*(2), 160–168. https://doi.org/10.1079/PNS2006489

Gil, Á., Martinez de Victoria, E., & Olza, J. (2015). Indicadores de evaluación de la calidad de la dieta. *Nutricion Hospitalaria*, *31*(3), 128–144. https://doi.org/10.3305/nh.2015.31.sup3.8761

Godde, C. M., Mason-D'Croz, D., Mayberry, D. E., Thornton, P. K., & Herrero, M. (2021). Impacts of climate change on the livestock food supply chain; a review of the evidence. *Global Food Security*, *28*, 100488. https://doi.org/10.1016/j.gfs.2020.100488

Golden, C. D., Allison, E. H., Cheung, W. W. L., Dey, M. M., Halpern, B. S., McCauley, D. J., Smith, M., Vaitla, B., Zeller, D., & Myers, S. S. (2016). Nutrition: Fall in fish catch threatens human health. *Nature*, *534*(7607), 317–320. https://doi.org/10.1038/534317a

Golden, C. D., & Comaroff, J. (2015). Effects of social change on wildlife consumption taboos in northeastern Madagascar. *Ecology and Society*, *20*(2), art41. https://doi.org/10.5751/ES-07589-200241

Golden, C. D., Fernald, L. C. H., Brashares, J. S., Rasolofoniaina, B. J. R., & Kremen, C. (2011). Benefits of wildlife consumption to child nutrition in a biodiversity hotspot. *Proceedings of the National Academy of Sciences*, *108*(49), 19653–19656. https://doi.org/10.1073/pnas.1112586108

Golden, C. D., Gupta, A. C., Vaitla, B., & Myers, S. S. (2016). Ecosystem services and food security: Assessing inequality at community, household and individual scales. *Environmental Conservation*, *43*(4), 381–388. https://doi.org/10.1017/S0376892916000163

Grace, D. (2015a). Food safety in low and middle income countries. *International Journal of Environmental Research and Public Health*, *12*(9), 10490–10507. https://doi.org/10.3390/ijerph120910490

Grace, D. (2015b). *Review of evidence on antimicrobial resistance and animal agriculture in developing countries*. Evidence on Demand. https://doi.org/10.12774/eod_cr.june2015.graced

Grace, D., Dominguez Salas, P., Alonso, S., Lannerstad, M., Muunda, E. M., Ngwili, N., Omar, A., Khan, M., & Otobo, E. (2018). *The influence of livestock-derived foods on nutrition during the first 1,000 days of life* (ILRI Research Report 44). ILRI.

Grace, D., Gilbert, J., Randolph, T., & Kang'ethe, E. (2012). The multiple burdens of zoonotic disease and an ecohealth approach to their assessment. *Tropical Animal Health and Production*, *44*(S1), 67–73. https://doi.org/10.1007/s11250-012-0209-y

Grünberger, K. (2014). *Estimating food consumption patterns by reconciling food balance sheets and household budget surveys*. Food and Agriculture Organization of the United Nations. https://www.fao.org/3/i4315e/I4315E.pdf

Gupta, S. (2016). Brain food: Clever eating. *Nature, 531*(7592), S12–S13. https://doi.org/10.1038/531S12a

Hamley, S. (2017). The effect of replacing saturated fat with mostly n-6 polyunsaturated fat on coronary heart disease: A meta-analysis of randomised controlled trials. *Nutrition Journal, 16*(1), 30. https://doi.org/10.1186/s12937-017-0254-5

Hartung, J. (2013). A short history of livestock production. In A. Aland & T. Banhazi (Eds.), *Livestock housing* (pp. 21–34). Brill | Wageningen Academic. https://doi.org/10.3920/978-90-8686-771-4_01

Headey, D., Heidkamp, R., Osendarp, S., Ruel, M., Scott, N., Black, R., Shekar, M., Bouis, H., Flory, A., Haddad, L., & Walker, N. (2020). Impacts of COVID-19 on childhood malnutrition and nutrition-related mortality. *The Lancet, 396*(10250), 519–521. https://doi.org/10.1016/S0140-6736(20)31647-0

Henchion, M., Hayes, M., Mullen, A., Fenelon, M., & Tiwari, B. (2017). Future protein supply and demand: Strategies and factors influencing a sustainable equilibrium. *Foods, 6*(7), 53. https://doi.org/10.3390/foods6070053

Herforth, A., & Ballard, T. J. (2016). Nutrition indicators in agriculture projects: Current measurement, priorities, and gaps. *Global Food Security, 10*, 1–10. https://doi.org/10.1016/j.gfs.2016.07.004

Herforth, A., & Harris, J. (2014). Understanding and applying primary pathways and principles. Brief #1. *USAID/Strengthening Partnerships, Results, and Innovations in Nutrition Globally (SPRING) project*. https://www.spring-nutrition.org/sites/default/files/publications/briefs/spring_understandingpathways_brief_1.pdf

Herrero, M., Grace, D., Njuki, J., Johnson, N., Enahoro, D., Silvestri, S., & Rufino, M. C. (2013). The roles of livestock in developing countries. *Animal, 7*, 3–18. https://doi.org/10.1017/S1751731112001954

Herrero, M., Henderson, B., Havlík, P., Thornton, P. K., Conant, R. T., Smith, P., Wirsenius, S., Hristov, A. N., Gerber, P., Gill, M., Butterbach-Bahl, K., Valin, H., Garnett, T., & Stehfest, E. (2016). Greenhouse gas mitigation potentials in the livestock sector. *Nature Climate Change, 6*(5), 452–461. https://doi.org/10.1038/nclimate2925

High Level Panel of Experts on Food Security and Nutrition of the Committee on World Food Security. (2020). *Food security and nutrition: Building a global narrative towards 2030* (HLPE 15). Food and Agriculture Organization. https://openknowledge.fao.org/server/api/core/bitstreams/8357b6eb-8010-4254-814a-1493faaf4a93/content

Hirvonen, K., Bai, Y., Headey, D., & Masters, W. A. (2020). Affordability of the EAT–Lancet reference diet: A global analysis. *The Lancet Global Health, 8*(1), e59–e66. https://doi.org/10.1016/S2214-109X(19)30447-4

Hlaing, L. M., Fahmida, U., Htet, M. K., Utomo, B., Firmansyah, A., & Ferguson, E. L. (2016). Local food-based complementary feeding recommendations developed by the linear programming approach to improve the intake of problem nutrients among 12–23-month-old Myanmar children. *British Journal of Nutrition, 116*(S1), S16–S26. https://doi.org/10.1017/S000711451500481X

Hoffmann, S., Devleesschauwer, B., Aspinall, W., Cooke, R., Corrigan, T., Havelaar, A., Angulo, F., Gibb, H., Kirk, M., Lake, R., Speybroeck, N., Torgerson, P., & Hald, T. (2017). Attribution of global foodborne disease to specific foods: Findings from a World Health Organization structured expert elicitation. *PLoS One, 12*(9), e0183641. https://doi.org/10.1371/journal.pone.0183641

Holmes, E. C., Goldstein, S. A., Rasmussen, A. L., Robertson, D. L., Crits-Christoph, A., Wertheim, J. O., Anthony, S. J., Barclay, W. S., Boni, M. F., Doherty, P. C., Farrar, J., Geoghegan, J. L., Jiang, X., Leibowitz, J. L., Neil, S. J. D., Skern, T., Weiss, S. R., Worobey, M., Andersen, K. G., … Rambaut, A. (2021). The origins of SARS-CoV-2: A critical review. *Cell*, *184*(19), 4848–4856. https://doi.org/10.1016/j.cell.2021.08.017

Iannotti, L. (2021). *Livestock-derived foods and sustainable healthy diets*. UN Nutrition. https://www.unnutrition.org/wp-content/uploads/Presentation-Livestock-derived-foods-and-sustainable-healthy-diets.pdf

Iannotti, L. L., Lutter, C. K., Stewart, C. P., Gallegos Riofrío, C. A., Malo, C., Reinhart, G., Palacios, A., Karp, C., Chapnick, M., Cox, K., & Waters, W. F. (2017). Eggs in early complementary feeding and child growth: A randomized controlled trial. *Pediatrics*, *140*(1), e20163459. https://doi.org/10.1542/peds.2016-3459

Ingram, D. J., Coad, L., Milner-Gulland, E. J., Parry, L., Wilkie, D., Bakarr, M. I., Benítez-López, A., Bennett, E. L., Bodmer, R., Cowlishaw, G., El Bizri, H. R., Eves, H. E., Fa, J. E., Golden, C. D., Iponga, D. M., Minh, N. V., Morcatty, T. Q., Mwinyihali, R., Nasi, R., … Abernethy, K. (2021). Wild meat is still on the menu: Progress in wild meat research, policy, and practice from 2002 to 2020. *Annual Review of Environment and Resources*, *46*(1), 221–254. https://doi.org/10.1146/annurev-environ-041020-063132

IPES- Food. (2017). *Unravelling the Food-Health Nexus: Addressing practices, political economy, and power relations to build healthier food systems*. The Global Alliance for the Future of Food and IPES-Food. https://www.ipes-food.org/_img/upload/files/Health_FullReport(1).pdf

Johnson, C. K., Hitchens, P. L., Pandit, P. S., Rushmore, J., Evans, T. S., Young, C. C. W., & Doyle, M. M. (2020). Global shifts in mammalian population trends reveal key predictors of virus spillover risk. *Proceedings of the Royal Society B: Biological Sciences*, *287*(1924), 20192736. https://doi.org/10.1098/rspb.2019.2736

Johnson, M. C. (2011). Lobbying for trade barriers: A comparison of poultry producers' success in Cameroon, Senegal and Ghana. *The Journal of Modern African Studies*, *49*(4), 575–599. https://doi.org/10.1017/S0022278X11000486

Jones, K. E., Patel, N. G., Levy, M. A., Storeygard, A., Balk, D., Gittleman, J. L., & Daszak, P. (2008). Global trends in emerging infectious diseases. *Nature*, *451*(7181), 990–993. https://doi.org/10.1038/nature06536

Joy, A., Dunshea, F. R., Leury, B. J., Clarke, I. J., DiGiacomo, K., & Chauhan, S. S. (2020). Resilience of small ruminants to climate change and increased environmental temperature: A review. *Animals*, *10*(5), 867. https://doi.org/10.3390/ani10050867

Katabazi, L. V. (2018). *Analysis of the household village chicken feeding practices across the seasons in Central Tanzania* [Master of Science]. University of Sydney.

Katz, D. L., Doughty, K. N., Geagan, K., Jenkins, D. A., & Gardner, C. D. (2019). Perspective: The Public Health Case for modernizing the definition of protein quality. *Advances in Nutrition*, *10*(5), 755–764. https://doi.org/10.1093/advances/nmz023

Kiswaga, G. (2019). *Participatory epidemiology of Newcastle disease across the Rift Valley in Manyoni district, Tanzania* [Master of Science, Sokoine University]. https://www.suaire.sua.ac.tz/server/api/core/bitstreams/57c7b919-0290-44fd-9a3d-c41e666053c9/content

Komarek, A. M., Dunston, S., Enahoro, D., Godfray, H. C. J., Herrero, M., Mason-D'Croz, D., Rich, K. M., Scarborough, P., Springmann, M., Sulser, T. B., Wiebe, K., & Willenbockel, D. (2021). Income, consumer preferences, and the future of livestock-derived food demand. *Global Environmental Change*, *70*, 102343. https://doi.org/10.1016/j.gloenvcha.2021.102343

Kuipers, R. S., Joordens, J. C. A., & Muskiet, F. A. J. (2012). A multidisciplinary reconstruction of Palaeolithic nutrition that holds promise for the prevention and treatment of diseases of civilisation. *Nutrition Research Reviews*, *25*(1), 96–129. https://doi.org/10.1017/S0954422412000017

Lelieveld, J., Evans, J. S., Fnais, M., Giannadaki, D., & Pozzer, A. (2015). The contribution of outdoor air pollution sources to premature mortality on a global scale. *Nature*, *525*(7569), 367–371. https://doi.org/10.1038/nature15371

Leroy, E. M., Rouquet, P., Formenty, P., Souquière, S., Kilbourne, A., Froment, J.-M., Bermejo, M., Smit, S., Karesh, W., Swanepoel, R., Zaki, S. R., & Rollin, P. E. (2004). Multiple Ebola virus transmission events and rapid decline of Central African Wildlife. *Science*, *303*(5656), 387–390. https://doi.org/10.1126/science.1092528

Leroy, F., Abraini, F., Beal, T., Dominguez-Salas, P., Gregorini, P., Manzano, P., Rowntree, J., & Van Vliet, S. (2022). Animal board invited review: Animal source foods in healthy, sustainable, and ethical diets – An argument against drastic limitation of livestock in the food system. *Animal*, *16*(3), 100457. https://doi.org/10.1016/j.animal.2022.100457

Leroy, J. L., Ruel, M. T., & Olney, D. K. (2020). *Measuring the impact of agriculture programs on diets and nutrition*. International Food Policy Research Institute. https://doi.org/10.2499/p15738coll2.133954

Li, M., Jia, N., Lenzen, M., Malik, A., Wei, L., Jin, Y., & Raubenheimer, D. (2022). Global food-miles account for nearly 20% of total food-systems emissions. *Nature Food*, *3*(6), 445–453. https://doi.org/10.1038/s43016-022-00531-w

Lim, X. (2021). Microplastics are everywhere—But are they harmful? *Nature*, *593*(7857), 22–25. https://doi.org/10.1038/d41586-021-01143-3

Ma, W., Kahn, R. E., & Richt, J. A. (2008). The pig as a mixing vessel for influenza viruses: Human and veterinary implications. *Journal of Molecular and Genetic Medicine: An International Journal of Biomedical Research*, *3*(1), 158–166.

Magoke, G. Z. (2021). *Assessing susceptibility of village chickens to aflatoxin exposure and contamination of village grains and village chicken products in Tanzania* [Doctor of Philosophy, University of Sydney]. https://ses.library.usyd.edu.au/handle/2123/27405

Mah, E., Chen, C.-Y. O., & Liska, D. J. (2020). The effect of egg consumption on cardiometabolic health outcomes: An umbrella review. *Public Health Nutrition*, *23*(5), 935–955. https://doi.org/10.1017/S1368980019002441

Maxwell, S. L., Fuller, R. A., Brooks, T. M., & Watson, J. E. M. (2016). Biodiversity: The ravages of guns, nets and bulldozers. *Nature*, *536*(7615), 143–145. https://doi.org/10.1038/536143a

Mejos, K. K., Ignacio, M. S., Jayasuriya, R., & Arcot, J. (2021). Use of linear programming to develop complementary feeding recommendations to improve nutrient adequacy and dietary diversity among breastfed children in the Rural Philippines. *Food and Nutrition Bulletin*, *42*(2), 274–288. https://doi.org/10.1177/0379572121998125

Mekonnen, M. M., & Hoekstra, A. Y. (2012). A global assessment of the water footprint of farm animal products. *Ecosystems*, *15*(3), 401–415. https://doi.org/10.1007/s10021-011-9517-8

Messenger, A. M., Barnes, A. N., & Gray, G. C. (2014). Reverse Zoonotic disease transmission (Zooanthroponosis): A systematic review of seldom-documented human biological threats to animals. *PLoS One*, *9*(2), e89055. https://doi.org/10.1371/journal.pone.0089055

Mottet, A., De Haan, C., Falcucci, A., Tempio, G., Opio, C., & Gerber, P. (2017). Livestock: On our plates or eating at our table? A new analysis of the feed/food debate. *Global Food Security*, *14*, 1–8. https://doi.org/10.1016/j.gfs.2017.01.001

Mottet, A., Henderson, B., Opio, C., Falcucci, A., Tempio, G., Silvestri, S., Chesterman, S., & Gerber, P. J. (2017). Climate change mitigation and productivity gains in livestock supply chains: Insights from regional case studies. *Regional Environmental Change, 17*(1), 129–141. https://doi.org/10.1007/s10113-016-0986-3

Murphy, S. P., & Allen, L. H. (2003). Nutritional importance of animal source foods. *The Journal of Nutrition, 133*(11), 3932S–3935S. https://doi.org/10.1093/jn/133.11.3932S

Muthayya, S., Rah, J. H., Sugimoto, J. D., Roos, F. F., Kraemer, K., & Black, R. E. (2013). The global hidden hunger indices and maps: An advocacy tool for action. *PLoS ONE, 8*(6), e67860. https://doi.org/10.1371/journal.pone.0067860

Nasi, R., Taber, A., & Van Vliet, N. (2011). Empty forests, empty stomachs? Bushmeat and livelihoods in the Congo and Amazon Basins. *International Forestry Review, 13*(3), 355–368. https://doi.org/10.1505/146554811798293872

Neumann, C. G., Murphy, S. P., Gewa, C., Grillenberger, M., & Bwibo, N. O. (2007). Meat supplementation improves growth, cognitive, and behavioral outcomes in Kenyan Children1. *The Journal of Nutrition, 137*(4), 1119–1123. https://doi.org/10.1093/jn/137.4.1119

Newton, P., Civita, N., Frankel-Goldwater, L., Bartel, K., & Johns, C. (2020). What is regenerative agriculture? A review of scholar and practitioner definitions based on processes and outcomes. *Frontiers in Sustainable Food Systems, 4*, 577723. https://doi.org/10.3389/fsufs.2020.577723

Nguyen, T. L. T., Hermansen, J. E., & Mogensen, L. (2010). Environmental consequences of different beef production systems in the EU. *Journal of Cleaner Production, 18*(8), 756–766. https://doi.org/10.1016/j.jclepro.2009.12.023

Nielsen, M. R., Meilby, H., Smith-Hall, C., Pouliot, M., & Treue, T. (2018). The importance of wild meat in the Global South. *Ecological Economics, 146*, 696–705. https://doi.org/10.1016/j.ecolecon.2017.12.018

Ojha, S., Bekhit, A. E.-D., Grune, T., & Schlüter, O. K. (2021). Bioavailability of nutrients from edible insects. *Current Opinion in Food Science, 41*, 240–248. https://doi.org/10.1016/j.cofs.2021.08.003

Onyekuru, N. A., Ume, C. O., Ezea, C. P., & Chukwuma Ume, N. N. (2020). Effects of Ebola Virus disease outbreak on bush meat enterprise and environmental health risk behavior among households in South-East Nigeria. *The Journal of Primary Prevention, 41*(6), 603–618. https://doi.org/10.1007/s10935-020-00619-8

Oonincx, D. G. A. B., & De Boer, I. J. M. (2012). Environmental impact of the production of mealworms as a protein source for humans – A life cycle assessment. *PLoS ONE, 7*(12), e51145. https://doi.org/10.1371/journal.pone.0051145

Paige, S. B., Frost, S. D. W., Gibson, M. A., Jones, J. H., Shankar, A., Switzer, W. M., Ting, N., & Goldberg, T. L. (2014). Beyond Bushmeat: Animal contact, injury, and Zoonotic disease risk in Western Uganda. *EcoHealth, 11*(4), 534–543. https://doi.org/10.1007/s10393-014-0942-y

Parodi, A., Gerrits, W. J. J., Van Loon, J. J. A., De Boer, I. J. M., Aarnink, A. J. A., & Van Zanten, H. H. E. (2021). Black soldier fly reared on pig manure: Bioconversion efficiencies, nutrients in the residual material, greenhouse gas and ammonia emissions. *Waste Management, 126*, 674–683. https://doi.org/10.1016/j.wasman.2021.04.001

Pinstrup-Andersen, P. (2007). Agricultural research and policy for better health and nutrition in developing countries: A food systems approach. *Agricultural Economics, 37*(s1), 187–198. https://doi.org/10.1111/j.1574-0862.2007.00244.x

Pokharel, S., Shrestha, P., & Adhikari, B. (2020). Antimicrobial use in food animals and human health: Time to implement 'One Health' approach. *Antimicrobial Resistance & Infection Control, 9*(1), 181. https://doi.org/10.1186/s13756-020-00847-x

Poore, J., & Nemecek, T. (2018). Reducing food's environmental impacts through producers and consumers. *Science, 360*(6392), 987–992. https://doi.org/10.1126/science.aaq0216

Popkin, B. M., Corvalan, C., & Grummer-Strawn, L. M. (2020). Dynamics of the double burden of malnutrition and the changing nutrition reality. *The Lancet, 395*(10217), 65–74. https://doi.org/10.1016/S0140-6736(19)32497-3

Popkin, B. M., & Ng, S. W. (2022). The nutrition transition to a stage of high obesity and noncommunicable disease prevalence dominated by ultra-processed foods is not inevitable. *Obesity Reviews, 23*(1), e13366. https://doi.org/10.1111/obr.13366

Ramilan, T., Kumar, S., Haileslassie, A., Craufurd, P., Scrimgeour, F., Kattarkandi, B., & Whitbread, A. (2022). Quantifying farm household resilience and the implications of livelihood heterogeneity in the semi-arid tropics of India. *Agriculture, 12*(4), 466. https://doi.org/10.3390/agriculture12040466

Ramsden, C. E., Zamora, D., Majchrzak-Hong, S., Faurot, K. R., Broste, S. K., Frantz, R. P., Davis, J. M., Ringel, A., Suchindran, C. M., & Hibbeln, J. R. (2016). Re-evaluation of the traditional diet-heart hypothesis: Analysis of recovered data from Minnesota Coronary Experiment (1968–73). *BMJ*, i1246. https://doi.org/10.1136/bmj.i1246

Randolph, T. F., Schelling, E., Grace, D., Nicholson, C. F., Leroy, J. L., Cole, D. C., Demment, M. W., Omore, A., Zinsstag, J., & Ruel, M. (2007). Invited review: Role of livestock in human nutrition and health for poverty reduction in developing countries. *Journal of Animal Science, 85*(11), 2788–2800. https://doi.org/10.2527/jas.2007-0467

Raymond, J., Agaba, M., Mollay, C., Rose, J. W., & Kassim, N. (2017). Analysis of nutritional adequacy of local foods for meeting dietary requirements of children aged 6–23 months in rural central Tanzania. *Archives of Public Health, 75*(1), 60. https://doi.org/10.1186/s13690-017-0226-4

Redmond, I., Aldred, T., Jedamzik, K., & Westwood, M. (2006). *Recipes for survival: Controlling the bushmeat trade*. Ape Alliance and World Society for the Protection of Animals. https://www.academia.edu/43704242/Recipes_for_Survival_Controlling_the_Bushmeat_Trade_Ape_Alliance_report_funded_by_WSPA

Ripple, W. J., Abernethy, K., Betts, M. G., Chapron, G., Dirzo, R., Galetti, M., Levi, T., Lindsey, P. A., Macdonald, D. W., Machovina, B., Newsome, T. M., Peres, C. A., Wallach, A. D., Wolf, C., & Young, H. (2016). Bushmeat hunting and extinction risk to the world's mammals. *Royal Society Open Science, 3*(10), 160498. https://doi.org/10.1098/rsos.160498

Ritchie, H., & Roser, M. (2024). Land use. *Our World in Data*. https://ourworldindata.org/land-use

Rojas-Downing, M. M., Nejadhashemi, A. P., Harrigan, T., & Woznicki, S. A. (2017). Climate change and livestock: Impacts, adaptation, and mitigation. *Climate Risk Management, 16*, 145–163. https://doi.org/10.1016/j.crm.2017.02.001

Roman, M., Roman, K., & Roman, M. (2021). Spatial variation in particulate emission resulting from animal farming in Poland. *Agriculture, 11*(2), 168. https://doi.org/10.3390/agriculture11020168

Rong, Y., Chen, L., Zhu, T., Song, Y., Yu, M., Shan, Z., Sands, A., Hu, F. B., & Liu, L. (2013). Egg consumption and risk of coronary heart disease and stroke: Dose-response meta-analysis of prospective cohort studies. *BMJ, 346*(jan07 2), e8539–e8539. https://doi.org/10.1136/bmj.e8539

Salmon, G. R., Marshall, K., Tebug, S. F., Missohou, A., Robinson, T. P., & MacLeod, M. (2018). The greenhouse gas abatement potential of productivity improving measures applied to cattle systems in a developing region. *Animal, 12*(4), 844–852. https://doi.org/10.1017/S1751731117002294

Sattar, A. A., Mahmud, R., Mohsin, Md . A. S., Chisty, N. N., Uddin, Md . H., Irin, N., Barnett, T., Fournie, G., Houghton, E., & Hoque, Md . A. (2021). COVID-19 impact on poultry production and distribution networks in Bangladesh. *Frontiers in Sustainable Food Systems*, *5*, 714649. https://doi.org/10.3389/fsufs.2021.714649

Schenck, M., Nsame Effa, E., Starkey, M., Wilkie, D., Abernethy, K., Telfer, P., Godoy, R., & Treves, A. (2006). Why people eat bushmeat: Results from two-choice, taste tests in Gabon, Central Africa. *Human Ecology*, *34*(3), 433–445. https://doi.org/10.1007/s10745-006-9025-1

Schönfeldt, H., Pretorius, B., & Hall, N. (2014). The impact of animal source food products on human nutrition and health. *South African Journal of Animal Science*, *43*(3), 394. https://doi.org/10.4314/sajas.v43i3.11

Schreefel, L., Schulte, R. P. O., De Boer, I. J. M., Schrijver, A. P., & Van Zanten, H. H. E. (2020). Regenerative agriculture – The soil is the base. *Global Food Security*, *26*, 100404. https://doi.org/10.1016/j.gfs.2020.100404

Smith, J., Sones, K., Grace, D., MacMillan, S., Tarawali, S., & Herrero, M. (2013). Beyond milk, meat, and eggs: Role of livestock in food and nutrition security. *Animal Frontiers*, *3*(1), 6–13. https://doi.org/10.2527/af.2013-0002

Song, M.-S., Lee, J. H., Pascua, P. N. Q., Baek, Y. H., Kwon, H., Park, K. J., Choi, H.-W., Shin, Y.-K., Song, J.-Y., Kim, C.-J., & Choi, Y.-K. (2010). Evidence of human-to-swine transmission of the pandemic (H1N1) 2009 Influenza Virus in South Korea. *Journal of Clinical Microbiology*, *48*(9), 3204–3211. https://doi.org/10.1128/JCM.00053-10

Spira, C., Raveloarison, R., Cournarie, M., Strindberg, S., O'Brien, T., & Wieland, M. (2021). Assessing the prevalence of protected species consumption by rural communities in Makira Natural Park, Madagascar, through the unmatched count technique. *Conservation Science and Practice*, *3*(7), e441. https://doi.org/10.1111/csp2.441

Stanton, A. V., Leroy, F., Elliott, C., Mann, N., Wall, P., & De Smet, S. (2022). 36-fold higher estimate of deaths attributable to red meat intake in GBD 2019: Is this reliable? *The Lancet*, *399*(10332), e23–e26. https://doi.org/10.1016/S0140-6736(22)00311-7

Stein, M. M., Hrusch, C. L., Gozdz, J., Igartua, C., Pivniouk, V., Murray, S. E., Ledford, J. G., Marques Dos Santos, M., Anderson, R. L., Metwali, N., Neilson, J. W., Maier, R. M., Gilbert, J. A., Holbreich, M., Thorne, P. S., Martinez, F. D., Von Mutius, E., Vercelli, D., Ober, C., & Sperling, A. I. (2016). Innate immunity and asthma risk in Amish and Hutterite farm children. *New England Journal of Medicine*, *375*(5), 411–421. https://doi.org/10.1056/NEJMoa1508749

Steinfeld, H. & FAO (Eds.). (2006). *Livestock's long shadow: Environmental issues and options*. Food and Agriculture Organization of the United Nations.

Stewart, C. P., Caswell, B., Iannotti, L., Lutter, C., Arnold, C. D., Chipatala, R., Prado, E. L., & Maleta, K. (2019). The effect of eggs on early child growth in rural Malawi: The Mazira Project randomized controlled trial. *The American Journal of Clinical Nutrition*, *110*(4), 1026–1033. https://doi.org/10.1093/ajcn/nqz163

Tang, G. (2010). Bioconversion of dietary provitamin A carotenoids to vitamin A in humans. *The American Journal of Clinical Nutrition*, *91*(5), 1468S–1473S. https://doi.org/10.3945/ajcn.2010.28674G

Tanga, C. M., Egonyu, J. P., Beesigamukama, D., Niassy, S., Emily, K., Magara, H. J., Omuse, E. R., Subramanian, S., & Ekesi, S. (2021). Edible insect farming as an emerging and profitable enterprise in East Africa. *Current Opinion in Insect Science*, *48*, 64–71. https://doi.org/10.1016/j.cois.2021.09.007

Thornton, P. K. (2010). Livestock production: Recent trends, future prospects. *Philosophical Transactions of the Royal Society B: Biological Sciences*, *365*(1554), 2853–2867. https://doi.org/10.1098/rstb.2010.0134

Thumbi, S. M., Njenga, M. K., Marsh, T. L., Noh, S., Otiang, E., Munyua, P., Ochieng, L., Ogola, E., Yoder, J., Audi, A., Montgomery, J. M., Bigogo, G., Breiman, R. F., Palmer, G. H., & McElwain, T. F. (2015). Linking human health and livestock health: A "One-Health" platform for integrated analysis of human health, livestock health, and economic welfare in livestock dependent communities. *PLoS One*, *10*(3), e0120761. https://doi.org/10.1371/journal.pone.0120761

Tian, H., Zhou, S., Dong, L., Van Boeckel, T. P., Cui, Y., Newman, S. H., Takekawa, J. Y., Prosser, D. J., Xiao, X., Wu, Y., Cazelles, B., Huang, S., Yang, R., Grenfell, B. T., & Xu, B. (2015). Avian influenza H5N1 viral and bird migration networks in Asia. *Proceedings of the National Academy of Sciences*, *112*(1), 172–177. https://doi.org/10.1073/pnas.1405216112

UNICEF. (2021). *UNICEF conceptual framework on maternal and child nutrition*. UNICEF. https://www.unicef.org/media/113291/file/UNICEF Conceptual Framework.pdf

United Nations Department of Economic and Social Affairs. (2023). *The sustainable development goals report 2023: Special edition*. United Nations. https://doi.org/10.18356/9789210024914

Van Heuverswyn, F., & Peeters, M. (2007). The origins of HIV and implications for the global epidemic. *Current Infectious Disease Reports*, *9*(4), 338–346. https://doi.org/10.1007/s11908-007-0052-x

Van Zanten, H. H. E., Van Ittersum, M. K., & De Boer, I. J. M. (2019). The role of farm animals in a circular food system. *Global Food Security*, *21*, 18–22. https://doi.org/10.1016/j.gfs.2019.06.003

Von Mutius, E., & Vercelli, D. (2010). Farm living: Effects on childhood asthma and allergy. *Nature Reviews Immunology*, *10*(12), 861–868. https://doi.org/10.1038/nri2871

Wang, L.-F., & Eaton, B. T. (2007). Bats, civets and the emergence of SARS. In J. E. Childs, J. S. Mackenzie, & J. A. Richt (Eds.), *Wildlife and emerging zoonotic diseases: The biology, circumstances and consequences of cross-species transmission* (Vol. 315, pp. 325–344). Springer Berlin Heidelberg. https://doi.org/10.1007/978-3-540-70962-6_13

Wang, Y., Lehane, C., Ghebremeskel, K., & Crawford, M. A. (2010). Modern organic and broiler chickens sold for human consumption provide more energy from fat than protein. *Public Health Nutrition*, *13*(3), 400–408. https://doi.org/10.1017/S1368980009991157

Webb, P., & Kennedy, E. (2014). Impacts of agriculture on nutrition: Nature of the evidence and research gaps. *Food and Nutrition Bulletin*, *35*(1), 126–132. https://doi.org/10.1177/156482651403500113

Weber, K. T., & Horst, S. (2011). Desertification and livestock grazing: The roles of sedentarization, mobility and rest. *Pastoralism: Research, Policy and Practice*, *1*(1), 19. https://doi.org/10.1186/2041-7136-1-19

Westhoek, H., Rood, T., van den Berg, M., Janse, J., Nijdam, D., Reudink, M., & Stehfest, E. (2011). *The Protein Puzzle: The consumption and production of meat, dairy and fish in the European Union*. PBL Netherlands Environmental Assessment Agency. https://www.pbl.nl/en

Wilkie, D. S., Wieland, M., Boulet, H., Le Bel, S., Van Vliet, N., Cornelis, D., BriacWarnon, V., Nasi, R., & Fa, J. E. (2016). Eating and conserving bushmeat in Africa. *African Journal of Ecology*, *54*(4), 402–414. https://doi.org/10.1111/aje.12392

Wilkinson, J. (2006). Fish: A global value chain driven onto the rocks. *Sociologia Ruralis*, *46*(2), 139–153. https://doi.org/10.1111/j.1467-9523.2006.00408.x

Willett, W., Rockström, J., Loken, B., Springmann, M., Lang, T., Vermeulen, S., Garnett, T., Tilman, D., DeClerck, F., Wood, A., Jonell, M., Clark, M., Gordon, L. J., Fanzo, J., Hawkes, C., Zurayk, R., Rivera, J. A., De Vries, W., Majele Sibanda, L., …

Murray, C. J. L. (2019). Food in the Anthropocene: The EAT–Lancet Commission on healthy diets from sustainable food systems. *The Lancet*, *393*(10170), 447–492. https://doi.org/10.1016/S0140-6736(18)31788-4

Wingett, K., Allman-Farinelli, M., & Alders, R. (2019). Food loss and nutrition security: Reviewing pre-consumer loss in Australian sheep meat value chains using a planetary health framework. *CABI Reviews*, 1–12. https://doi.org/10.1079/PAVSNNR201813033

Wong, J. T., Lane, J. K., Allan, F. K., Vidal, G., Vance, C., Donadeu, M., Jackson, W., Nwankpa, V., Abera, S., Mekonnen, G. A., Kebede, N., Admassu, B., Amssalu, K., Lemma, A., Fentie, T., Smith, W., & Peters, A. R. (2022). Reducing calf mortality in Ethiopia. *Animals*, *12*(16), 2126. https://doi.org/10.3390/ani12162126

World Health Organization. (2011). *Haemoglobin concentrations for the diagnosis of anaemia and assessment of severity*. Vitamin and Mineral Information System, WHO. https://apps.who.int/iris/bitstream/handle/10665/85839/WHO_NMH_NHD_MNM_11.1_eng.pdf

World Health Organization. (2014). *WHO initiative to estimate the global burden of foodborne diseases: Fourth formal meeting of the Foodborne Disease Burden Epidemiology Reference Group (FERG): Sharing New Results, Making Future Plans, and Preparing Ground for the Countries*. World Health Organization; WHO IRIS. https://iris.who.int/handle/10665/159844

World Health Organization. (2023, October 12). *Climate change*. https://www.who.int/news-room/fact-sheets/detail/climate-change-and-health

World Health Organization. (2024a). *The WHO child growth standards*. https://www.who.int/tools/child-growth-standards

World Health Organization. (2024b). *WHO EMRO | Double burden of nutrition | Nutrition site*. World Health Organization - Regional Office for the Eastern Mediterranean. https://www.emro.who.int/nutrition/double-burden-of-nutrition/index.html

World Health Organization. (2024c, March 1). *Fact sheets—Malnutrition*. https://www.who.int/news-room/fact-sheets/detail/malnutrition

World Health Organization. (2024d, March 1). *Obesity and overweight*. https://www.who.int/news-room/fact-sheets/detail/obesity-and-overweight

World Organisation for Animal Health. (2020). *Guidance on working with farmed animals of species susceptible to infection with SARS-CoV-2, Version 1.2*. WOAH. https://www.woah.org/app/uploads/2021/03/draft-oie-guidance-farmed-animals-cleanms05-11.pdf

World Organisation for Animal Health. (2022). *SARS-CoV-2 in animals—Situation report 13*. WOAH. https://www.woah.org/app/uploads/2022/06/sars-cov-2-situation-report-13.pdf

Xue, L., Prass, N., Gollnow, S., Davis, J., Scherhaufer, S., Östergren, K., Cheng, S., & Liu, G. (2019). Efficiency and carbon footprint of the German Meat Supply Chain. *Environmental Science & Technology*, *53*(9), 5133–5142. https://doi.org/10.1021/acs.est.8b06079

Young, H. S., McCauley, D. J., Galetti, M., & Dirzo, R. (2016). Patterns, causes, and consequences of anthropocene defaunation. *Annual Review of Ecology, Evolution, and Systematics*, *47*(1), 333–358. https://doi.org/10.1146/annurev-ecolsys-112414-054142

Zhang, X., Chen, X., Xu, Y., Yang, J., Du, L., Li, K., & Zhou, Y. (2021). Milk consumption and multiple health outcomes: Umbrella review of systematic reviews and meta-analyses in humans. *Nutrition & Metabolism*, *18*(1), 7. https://doi.org/10.1186/s12986-020-00527-y

8 Addressing Disaster Management for the Global Health Practitioner

Joann M. Lindenmayer, John P. Bourgeois, and Jackson Tzu-ie Zee

INTRODUCTION

Many "wicked" problems face all living flora and fauna on planet Earth. Of the estimated 8.7 million (±1.3 million) (Zimmer, 2011) species living here, only one is responsible for these crises, and that is ours. Of the health challenges facing us, those caused by climate change dwarf the rest, even the global severe acute respiratory syndrome coronavirus 2 (SARS-CoV-2) pandemic. But climate change itself is a symptom of fossil fuel use, population growth, unchecked land and natural resource consumption, and ecological poisoning. Climate change is also directly responsible for an increase in the incidence and severity of disasters, including those caused by infectious diseases.

This chapter is addressed to the Global Health Practitioner who works, or desires to work, in disaster environments and addresses what the practitioner needs to know about disasters to respond to them effectively. Because the Global Health Practitioner may be an expert in one of several different disaster-related disciplinary fields or a member of a community that is preparing for future disasters, we first define what we mean by the term "Global Health Practitioner" and, second, review disaster management terminology to ensure that all practitioners – regardless of disciplinary expertise or experience – share a common language. The third section reviews the epidemiology of disasters with a focus on the risk factors that predispose populations to harm from disasters, highlighting infectious disease as the cause of a disaster or where it occurs secondary to a primary disaster. The fourth section presents a general outline of the disaster management cycle. The fifth section highlights several infectious disease disasters and lessons learned from them. The final section expands disasters beyond their impact on people, turning to the promise of One Health for extending disaster management beyond human populations.

DOI: 10.1201/9781003232223-8

WHO IS THE GLOBAL HEALTH PRACTITIONER?

Today many people in the world are doing their parts to mitigate the impact of disasters. These individuals include political leaders, members of international and national disaster relief organizations, non-governmental organizations (NGOs) and advocacy groups, professional organizations of all kinds, the private sector, and volunteer members of communities. All disasters involve a range of health professionals who strive collectively to protect and improve the health and survival of people, other animals, and nature before, during, and after disasters. These individuals are located throughout the world and their expertise is rooted in diverse cultures and societies. Collectively, they constitute an informal network of organizations and individuals that addresses specific disasters. Although we turn our attention to disasters in general, we are mindful that the greater challenge before us is to marshal those joint efforts to confront climate change, a disaster of global proportions. These are "Global Health Practitioners."

We define a Global Health Practitioner as an individual who applies technical, communication, organizational, and/or management skills in collaboration with other individuals and groups to improve the health and resilience of people, other animals, and nature, while working to address and mitigate the impact of disasters.

DEVELOPING A COLLECTIVE UNDERSTANDING OF DISASTER TERMINOLOGY

The terminology of disaster management can be confusing. For this reason, we clarify our use of common terms. A hazard is a process, phenomenon, or human activity that may cause loss of life, injury, or other health impacts, property damage, social and economic disruption, or environmental degradation (United Nations Office for Disaster Risk Reduction, 2017). Hazards include not only acute weather events such as hurricanes, tornadoes, heat waves, floods, and cold weather, but also slow-onset conditions such as droughts. Risk is the degree to which a hazard may be expected to cause harm. Global warming is the long-term heating of Earth's surface observed since the pre-industrial period (between 1850 and 1900) that is due to human activities, primarily the burning of fossil fuels, which increases heat-trapping greenhouse gas levels in Earth's atmosphere (NASA, n.d.). "Climate" refers to the long-term (at least 30 years) regional or global average of temperature, humidity, and rainfall patterns over seasons, years, or decades, and "climate change" refers to long-term shifts in temperatures and weather patterns (United Nations, n.d.). "Weather" refers to atmospheric conditions that occur locally over short periods of time – from minutes to hours or days.

As an example, Super-typhoon Haiyan, also known as Super-typhoon Yolanda, was one of the most powerful tropical cyclones ever recorded. This 2013 weather event created a hazard, posing enormous risk to occupants of the Philippines and Leyte Island in particular, and caused a disaster that led to widespread death and destruction. While no one such hazard or event can be attributed to climate change, Haiyan occurred in the context of increasing frequency and severity of other such events, and, for that reason, its occurrence is linked to climate change caused by global warming. We use the word "disaster" in this chapter to refer to a hazard that has caused numerous deaths, injuries, a scarcity of basic survival resources, such as food, water, and shelter, and loss of property.

EPIDEMIOLOGY OF DISASTERS

The classification of a disaster as "natural" should be applied cautiously; calling events natural disasters does not absolve our species of responsibility for their prevention or control. Furthermore, disasters worldwide are becoming more complex due to the increasing number of social, economic, political, and cultural factors that must be considered in managing them (The World Bank, 2015).

The Centre for Research in the Epidemiology of Disasters (CRED; UCLouvain, 2024) at the Université Catholique de Louvain in Brussels, Belgium, hosts the EM-DAT database (CRED, n.d.), the most widely recognized global database of disasters worldwide. CRED has been monitoring disaster occurrences and impacts on human populations and livelihoods since 1900. The initiative counts among its collaborators the United Nations Department of Humanitarian Affairs (UN-DHA), the European Union Humanitarian Office (ECHO), the International Federation of Red Cross and Red Crescent Societies (IFRC), and the United States Agency for International Development's Office of Foreign Disaster Assistance (USAID-OFDA). EM-DAT includes all disasters that meet the following criteria: 10 or more people dead; 100 or more people requiring immediate assistance for basic survival needs such as food, water, shelter, sanitation, and medical assistance; declaration of a state of emergency by a leader of the government in an affected country (or, in some cases, a state); or a call for international assistance by the national government in an affected country (CRED, n.d.).

CRED categorizes disasters into those that are natural and those that are technological, but this chapter considers only those that are natural. CRED defines the following subcategories of natural disasters: geophysical, meteorological, hydrological, climatological, biological, and extraterrestrial. We will focus our discussion primarily, but not solely, on natural disasters of biological origin, which CRED defines as a hazard caused by exposure to live organisms and their toxic substances (e.g., venom, mold) or vector-borne diseases. Examples are venomous wildlife and insects, poisonous plants, and mosquitoes carrying disease-causing agents such as parasites, bacteria, or viruses (e.g., malaria). This category includes epidemics, insect infestations, and animal accidents. As a natural disaster of biological origin, the coronavirus disease (COVID)-19 pandemic is considered a "profound tragedy and a massive global failure at multiple levels" despite years of preparedness planning by nearly all countries and an untold amount of funding dedicated to its prevention (Sachs et al., 2022).

The incidence of natural disasters has been examined at both temporal and spatial scales.

With respect to temporal incidence, there is ample evidence that the frequency of natural disasters has increased over time since monitoring efforts began. In one study that examined reported disaster occurrence in the period from 1950 to 1979, 1,779 natural disasters occurred, an average of 59.3/year, followed in the period 1980 to 2015 by 11,494 disasters, with an annual average of 319.3/year, representing an increase of 438.4% (Shen & Hwang, 2019) in the average annual incidence of all types of disasters. Incidence of disasters in all six natural disaster subgroups increased gradually for that period, but the reported number of disasters was generally higher for both hydrological and meteorological hazards than for biological, climatological, geophysical, and extraterrestrial ones. Another study (Keim, 2020) examined a subset of natural disasters that focused on the incidence of extreme weather events (EWEs), defined as events caused by climatological, hydrological, and meteorological hazards that include cyclones, droughts, floods, heat waves, landslides, cold weather, and storms, all of which are influenced by the global climate. In that study, which examined EWEs during the period 1969–2018, 10,009 EWE disasters occurred, among which floods and storms comprised 47% and 30% of all EWE disasters, respectively, followed by landslides (7%), drought (6%), cold (4%), wildlife (4%), and heat wave (2%). The study suggests with a high degree of confidence that the annual incidence of EWE disasters is increasing even as the study reported no significant changes in numbers of people killed, injured, or affected over the same period. The most recent data on disaster incidence in the first half of 2022 provisional data recorded 187 disasters in 79 countries, close to the median calculated for the same period (UCLouvain et al., 2022).

With respect to spatial patterns of disasters during the period 1900–2015 (Shen & Hwang, 2019), natural disaster occurrence was highest in the world's larger or most populous countries such as the United States, China, India, the Philippines, Indonesia, Bangladesh, Japan, Mexico, Australia, and Brazil, which together captured 35.1% of all disasters, whereas countries where incidence was lowest included small or less-populated territories such as French Guiana, the Virgin Islands, Finland, and Macau, among others (Shen & Hwang, 2019). This study was published prior to the COVID-19 pandemic, which, as a disaster of global proportions, challenged the entire structure of the traditional disaster management cycle and has no equal in terms of spatial extent and impact on human populations. Measures adopted worldwide to respond to the pandemic prompted developed and developing countries alike to join efforts to boost international resilience (Shen & Hwang, 2019).

Lower- and middle-income countries (LMICs) (The World Bank, n.d.-b) are disproportionately affected by disasters and least able to recover from them. The World Bank list of 54 LMICs includes 18 in Africa, 17 in Asia, seven in the Middle East, six in Oceania, four in South America, and one each in Europe and North America. For this reason, we emphasize the epidemiological challenges disasters pose in countries on the continents of Asia and Africa.

Between 1900 and 2022, 6,619 disasters occurred in Asia and 3,041 in Africa. Comparing disaster subtypes between these two continents, the largest discrepancies existed for storms and earthquakes, which were considerably greater in Asia, and for epidemics and droughts, which were considerably greater in Africa (Figure 8.1) – note that extreme T represents extreme temperature. Comparing the source of epidemics, bacterial origins predominate in Africa and viral origins in Asia (Figure 8.2).

Disaster severity has historically been based solely on the human impact of disasters and includes mortality; morbidity; total number of people affected who require basic survival needs such as food, water, shelter, sanitation, and immediate medical assistance, and a measure of total estimated damages and economic losses directly or indirectly related to the disaster in thousands of U.S. dollars valued in the year of occurrence, unadjusted for inflation. Using these figures as proxies for severity of disasters, then, from 1950 to 2015, although the average annual number of disaster-related human deaths decreased by 55.4%, presumably due to the availability of improved disaster prediction capabilities and country capacities to respond, the average number of injuries to people increased by 285.8%, and the average annual value of estimated damages increased by 2,795.3% (CRED, n.d.). A new universal classification system for disaster severity has been developed that considers these quantitative measures as well as qualitative measures of severity based on terminology applied at the time of the event (Caldera & Wirasinghe, 2022). No measures of impact take other animals into account except as people may risk injury and death by refusing to evacuate without a family pet or as the value of livestock is included in economic loss estimates, and none consider the impact on habitat other than the built environment.

Since 2005, when Hurricane Katrina devastated the U.S. Gulf coast, and as recognition has grown that climate change has exerted profound impacts on communities that rely heavily on livestock, certain countries have tried to address the

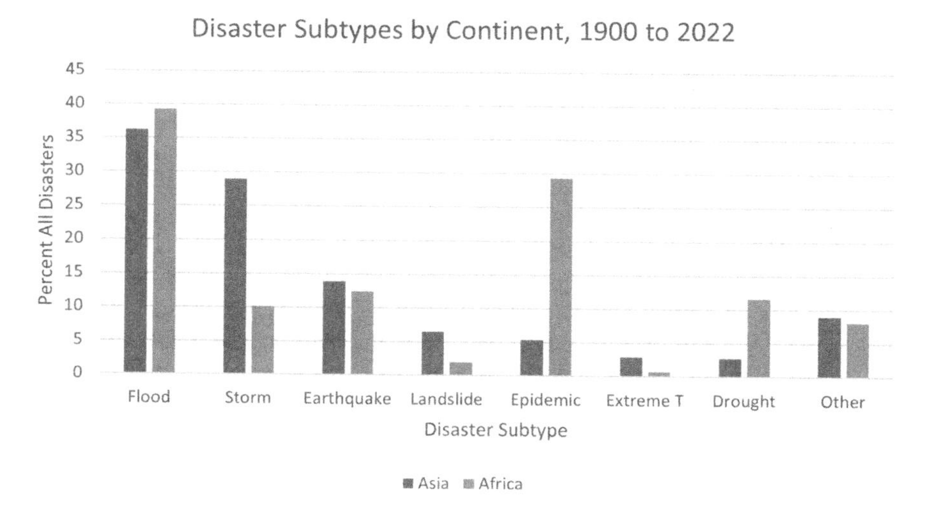

FIGURE 8.1 Comparison of disaster subtypes between Asia and Africa from 1900 to 2022.

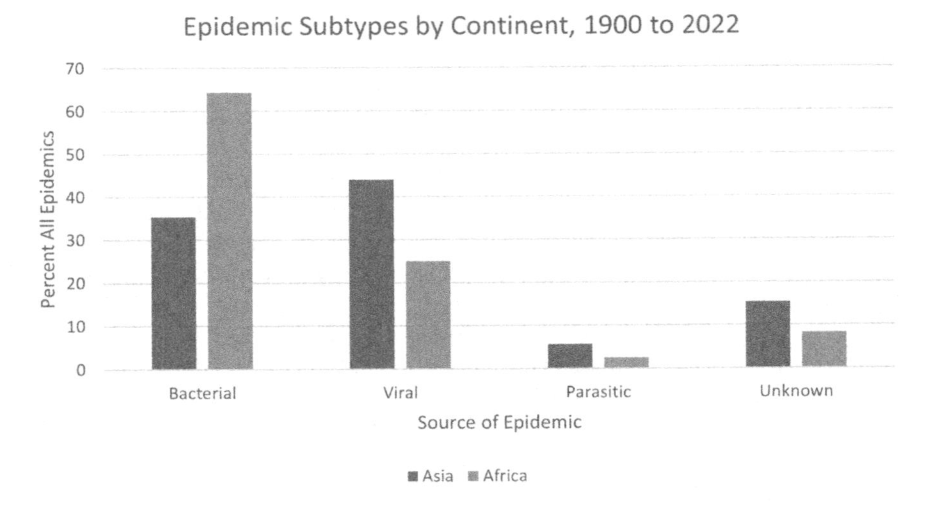

FIGURE 8.2 Comparison of the source of epidemics from disasters in Asia and Africa from 1900 to 2022.

needs of companion animals and livestock in disasters. However, these animals are included in disaster management efforts only because they are considered property.

Features of Disasters That Predispose Human Populations to Harms

The UN Office for Disaster Risk Reduction (UNDRR) defines a disaster as "a serious disruption of the functioning of a community or a society at any scale due to hazardous events interaction with conditions of exposures, vulnerability, and capacity, leading to one or more of the following: human, material, economic and environmental losses, and impact that may test or exceed the capacity of a community or society to cope using its own resources" (United Nations Office for Disaster Risk Reduction, 2023). This definition identifies several characteristics of an event that can be used to anticipate whether it will precipitate a disaster for human populations. However, it must be noted that the UNDRR's definition of a disaster is distinctly anthropocentric because reference to "community" or "society" is understood to refer solely to human communities and societies and not to societies of other animal populations or to communities of species that comprise ecosystems. While disasters always affect companion animals and livestock that are "owned," as well as free-roaming wildlife where people depend on them for food, income, or ecotourism, animal species and ecosystems that do not obviously benefit people are rarely considered when disaster impact is measured or when disaster management efforts are undertaken.

By the UNDRR definition, whether a specific event leads to a disaster is conditional upon the nature of the event as well as the extent to which human populations are exposed to it, vulnerable to it, and can cope with the impact of the event.

What constitutes a disaster is not as straightforward as most people think. Based on the CRED disaster criteria, if there is no obvious exposure of people to a potential hazard, then we do not consider the event to be a disaster. The impact of these types of events can range from no obvious impact at all, such as storms confined to the vast expanse of an ocean, to a bushfire that wiped out between 60 and 84 million acres of land and killed an estimated three billion wild animals (Readfearn & Morton, 2020). For that reason, other factors must be considered when defining what constitutes a disaster.

Although a disaster is defined as a situation or event which overwhelms a community or society's capacity to cope, it may be understood as a vulnerability, that is, where the ability to cope with an event is surpassed regardless of the disaster's nature or its severity (Kelman, 2011). This is an important point because it implies that in these situations, disaster preparedness efforts were either absent or insufficient to address the scope of the disaster, its severity, or both. We must ask why that is the case.

Lastly, the capacity of a community or society to cope with the impact of a hazard is highly dependent on its internal resources and on the willingness of external entities to deliver aid and assistance. The World Bank categorizes low- and middle-income economy levels using the World Bank Atlas method (The World Bank, 2024a) for the 2023 fiscal year as follows: low-income economies are those with a Gross National Income (GNI) per capita of $1,085 or less in 2021; lower-middle-income economies are those with a GNI per capita between $1,086 and $4,255; and upper-middle-income economies are those with a GNI per capita between $4,256 and $13,205. High-income economies are those with a GNI per capita of $13,206 or more (The World Bank, 2024b). Countries with lower income, less educational attainment, less robust financial systems, less transparency, and smaller governments have been shown to experience greater impacts from natural disasters than have those with higher incomes (Toya & Skidmore, 2007). In many of these countries, disaster management is often limited to reactive response rather than proactive measures that might reduce vulnerability to natural disasters, leading to separation of disaster management from the country's development goals (Brower et al., 2014).

Several other features determine how severely natural disasters impact populations. These include the following:

- Actual location and scale of the disaster (local, regional, national, international). The impact of an event on people and their environments can be expected to be greater in urban or peri-urban areas that are densely populated and where weather event control measures such as green infrastructure, porous surfaces and other nature-based solutions are scarce, or where people in LMICs have no choice but to live in coastal areas or other areas prone to disruption (Dewan, 2015).
- The time frame over which a disaster begins and evolves (acute, intermediate, long term). Droughts are slow-onset events that play out over

long time periods and may give people, their companion animals, and livestock sufficient time to adapt to drought conditions or to relocate, provided they have no mobility issues and suitable quarters, or habitats are located nearby, but they constitute disasters for nomadic and semi-nomadic peoples and their animals in sub-Saharan Africa and elsewhere (Giannini, 2015).

- Whether an area could, based on historical record, regularity of previous disasters, or modeling projections, anticipate the timing of future disasters or whether disasters are unusual and unexpected events that occur irregularly. In many countries, hurricane or typhoon seasons are routine and expected, although the severity of these events has become greater than could be anticipated from the historical record alone. An example of this is Super-typhoon Haiyan, which caused great loss of life and property damage in the Philippines in 2013 (Singer, 2014).
- Finally, disasters are not always one-off events and interactions may occur between them as, for example, when floods are followed closely by outbreaks of communicable diseases among people or livestock (Walton & Ivers, 2011).

Features of Disasters That Predispose Human Populations to Infectious Disease Epidemics and Outbreaks

Infectious diseases, whether bacterial, viral, or parasitic, occur as standalone disasters or as consequences of a preceding disaster. They can be emerging diseases, re-emerging diseases in locations where it was previously controlled, or increases in endemic disease over expected levels.

Developing countries are disproportionately affected by epidemics and outbreaks because they lack the resources, infrastructure, and preparedness plans to respond effectively to outbreaks of infectious diseases even when not challenged by a disaster (J. Watson et al., 2006; Kouadio et al., 2012; J. T. Watson et al., 2007; Talisuna et al., 2020). Furthermore, the occurrence of disasters of any kind can wipe out development efforts, leaving LMICs in worse condition than existed prior to the disaster event. We turn now to examining examples of standalone epidemics and others that occurred following a prior natural disaster.

Standalone Epidemics/Outbreaks

Ebola Virus Disease (EVD)

The 2014 Ebola outbreak in West Africa, which reported 11,310 deaths and 28,616 confirmed, probable, or suspected cases as of the end of the outbreak in April 2016 (Dahl et al., 2016), originated with the index case in Guinea in December 2013. The outbreak spread to eight other nations, including five in Africa (Liberia, Sierra Leone, Nigeria, Mali, and Senegal). Guinea, Liberia, and Sierra Leone were the most severely affected nations where years of developmental gains were reversed, whereas Nigeria, Senegal, and Ghana were able to successfully control the virus and prevent transmission of EVD. Delayed identification of cases and lack of disaster preparedness in countries with weak public health and healthcare systems following years of civil war, coupled with inadequate healthcare personnel and ineffective healthcare funding systems, fueled the epidemic (O. O. Oleribe et al., 2015) in the most affected countries.

EVD prevention and control efforts failed utterly in Liberia. Contributing factors included limited government and healthcare infrastructure and lack of preparedness planning after two civil wars and political instability, after which only 51 of 293 public health facilities (17%) survived and only 30 physicians were left to serve more than three million people (1 physician/100,000 people). Of the remaining healthcare facilities, most had no cell phone or radio capability. Distrust of the Sirleaf administration by the public because of nepotism, a stagnant education system, and high unemployment rates led to neglect of local knowledge that could have minimized the impact had these local solutions been implemented. However, the lack of public support for proposed national solutions exacerbated the outbreak (Southall et al., 2017). Similar factors – inadequate human resources, inadequate budgetary allocation to health, and poor leadership and management – have been cited as reasons for failure to address public health challenges more generally in Africa (O. E. Oleribe et al., 2019).

It is instructive, then, to consider not only why some countries failed to prevent and control the epidemic, but why others were able to do so despite having economic and social relationships with affected nations, an approach known as Positive Deviance (Positive Deviance Collaborative, 2020). In Nigeria, following years of public health training support by the Centers for Disease Control and Prevention, the Nigeria Field Epidemiology and Laboratory Training Program (NFELTP) was mobilized to conduct surveillance, investigate, manage, and control the epidemic, assuring a reactive and prompt response. In 1998, Nigeria established an Integrated Disease Surveillance and Response Program that, along with the Nigerian Center for Disease Control Prevention, the responsiveness of the Lagos State Ministry of Health, and previous experience with Lassa Fever, was key to preventing and controlling EVD transmission following the first case. The Nigerian health system was not prepared for the epidemic, but volunteer HealthCare Workers (HCWs) were deployed to the field for contact tracing, monitoring, and isolation of cases; funds for other healthcare activities were redirected to EVD prevention; and rapid and timely communication with the public alerting it to signs, symptoms, and acceptable practices for EVD all contributed to the success of prevention and control efforts. In Senegal, extensive educational campaigns in the months prior to the identification of the first case, the presence of an Institut Pasteur Laboratory in Dakar, and distribution of mobile phones for reporting suspected cases were instrumental to keeping the virus at bay. Ghana's interministerial supervisory effort of Port Health, immigration and security services, and the Ghana Health Service, along with national and regional technical coordinating committees and public sensitization efforts, were likewise key to the country's success.

COVID-19

A disaster on a global scale, as of April 1, 2023, there were 7,727,905 reported COVID-19 deaths, and 18,572,492 total deaths attributable to COVID-19 were reported by the Institute for Health Metrics and Evaluation (n.d.). The epidemic exposed not only the inadequacies of individual countries and governments to prevent and control transmission of the virus, but also highlighted the failure of the international community to cooperate on many levels. In July 2020, the Lancet COVID-19 Commission (2022) was established to develop recommendations on how to best suppress the epidemic, address the related humanitarian crises, and rebuild an inclusive, fair, and sustainable world. The final report of The Commission attributes our collective failure to control the epidemic to the following factors:

- Lack of timely notification of the initial outbreak;
- Delays in acknowledging airborne exposure pathway;
- Failure to coordinate suppression strategies among countries;
- Failure of governments to examine evidence and adopt best prevention and control practices;
- Shortfall of global funding for LMICs;
- Failure to ensure adequate global supplies and equitable distribution of key commodities;
- Lack of timely, accurate, and systematic data;
- Poor enforcement of biosafety regulations prior to the epidemic;
- Failure to combat disinformation; and
- Lack of global and national safety nets for vulnerable populations.

To these, we add the failure to prevent the transmission of the virus from animal to human populations at the outset, a consideration that will require systems-based approaches and a long-term political horizon to address.

Epidemics and Outbreaks Following Natural Disasters

The risk for epidemics and outbreaks following a disaster is presumed to be high, but the epidemiology of disasters does not bear this out. Flooding is the disaster subtype most frequently cited as linked to a subsequent outbreak of infectious disease, but the association is attributable rather to population displacement and not to the flood or other disaster itself (Ahern et al., 2005), and it is higher in LMICs than in other countries (O. E. Oleribe et al., 2019). Diarrheal disease outbreaks following heavy rains or floods are caused by a variety of pathogens including *Vibrio cholerae*, *Escherichia coli*, *Salmonella enterica*, and *Cryptosporidium parvum*, among others, attributed to people drinking contaminated water where clean, potable water has not been provided either on site or in displaced persons camps, or to lack of sanitation. In both Asia and Africa from 1900 to 2022, most bacterial disease outbreaks were attributed specifically to *V. cholerae*, 80 of 127 (70.9%) and 375 of 573 (65.4%) outbreaks on these continents, respectively.

Crowding among displaced persons in temporary settlements facilitates transmission of vaccine-preventable diseases such as measles, where immunization coverage is low, particularly among children, and from 1900 to 2022, measles was attributed to 4/360 (1.1%) outbreaks in Asia and 27/891 (3.0%) outbreaks in Africa. *Neisseria meningitidis* outbreaks have occurred where adequate prophylaxis has not been provided, and acute respiratory infections attributable to several pathogens have occurred where people must live and cook in small quarters and where access to antimicrobial agents is limited or unavailable (O. E. Oleribe et al., 2019).

THE DISASTER MANAGEMENT CYCLE – AN OVERVIEW

Disasters are not new. Although they are most often viewed as occurrences that are not "normal," in fact, some disasters are integral processes of sustainable ecosystems because they serve important ecological functions (Tilman, 1996).

The most obvious manifestation of a disaster is in the aftermath when efforts are mounted to respond to its human victims and their needs. Because this is the most visible of the disaster management phases, it limits public awareness of the other phases of disaster management, which happen in more controlled circumstances and often out of sight of the public. However, the other phases of the disaster management cycle are equally, if not more, important than the response (Figure 8.3).

FIGURE 8.3 Disaster management cycle.

Disaster management illustrates the ongoing process by which governments, businesses, and civil society plan to reduce or avoid potential losses posed by hazards, ensure prompt and appropriate assistance to victims of disasters, and achieve rapid and effective recovery. Appropriate actions at all points in the cycle lead to reduced vulnerability to the next disaster. The complete disaster management cycle includes the shaping of public policies and plans that either modify the causes of disasters or mitigate their effects on people, property, and infrastructure (Warfield, 2022).

Worldwide, but particularly in LMICs, people involved in disaster management often rely on the traditional disaster management cycle without recognizing that each disaster is unique and requires flexibility on the part of leadership as well as the application of tailored management strategies to be successful (Sawalha, 2020). The cycle comprises four phases, which are not temporally distinct and frequently overlap. Each phase of the cycle requires the application of different but related knowledge, skills, abilities, and behaviors to be effective. The following are descriptions of the phases along with some unique features of each, and key considerations for effective management.

Disaster Preparedness

Disaster preparedness is the cornerstone of effective disaster response, recovery, and mitigation. It includes plans or preparations made in advance of an emergency that help governments, communities, and individuals to be prepared for a disaster. Such preparations might include developing evacuation plans in communities susceptible to flooding; stocking of food, water, vaccines, and antimicrobial drugs; or preparation by healthcare services to address physical and mental health needs of affected persons in the intermediate to long-term period following a disaster, particularly among women, children, the elderly, the disabled and in rural communities (Cartwright et al., 2017).

Government entities at all levels, particularly Ministries and offices of Health and Civil Defense, are responsible for disaster preparedness planning. Disasters have caught many countries, particularly LMICs, unprepared or unable to handle the costs of a disaster, and for that reason the most important preparedness step for governments to take is to pre-commit to financing disasters before they happen so that they need not default to "begging bowl" appeals to humanitarian agencies in the aftermath (Clarke & Dercon, 2016). Memoranda of Understanding with other governments, international development organizations, such as the World Bank, and NGOs can help to establish mechanisms of financial and technical assistance before disasters occur, and they can also help to eliminate the delay these partners experience in requesting and obtaining permission to enter an affected country (Dixit, 2004).

Another important consideration is described in both the Hyogo Framework for Action (2005–2015) (International Strategy for Disaster Reduction, 2005) and the Sendai Framework for Disaster Risk Reduction (2015–2030) (United Nations,

2015), which call for more community engagement, participation, and collaboration at multiple levels. Community members can help to "ground truth" information provided by authorities with observations and measurements from the field, and they can also share culturally and socially appropriate suggestions for response. Thus, the preparedness phase of disaster management should include voices from the community in the planning stages and in each subsequent phase of the disaster management cycle.

Response

Disaster response includes any actions taken during or immediately following an emergency, including efforts to save lives and prevent further property damage. Ideally, disaster response involves putting already established disaster preparedness plans into motion. This is the phase of disaster management best known and understood by the public, and it's often all that the public sees. Examples include providing medical treatment, shelter, food, and safe water to affected individuals and communities (Taft-Morales, 2011).

People worldwide still rely primarily on the military, emergency personnel, local authorities, and disaster management agencies during a disaster, and authorities rarely consider voluntary responders, otherwise known as "zero-responders," as key players and first responders before, during, or after major incidents (Cocking, 2013). In the Philippines and many other developing countries, disaster management focused historically on the military and national police as central actors (Brower et al., 2014), although the country's 2010 Disaster Risk Reduction and Management Act expanded the National Disaster Risk Reduction and Management Council to include civil society organizations. However, particularly in the response to a disaster, it's critical to include voices from affected communities, especially those who are vulnerable (Donner & Rodriguez, n.d.), as they often have valuable information to share that could facilitate more effective responses.

Current approaches to disaster management in many LMICs have been criticized as overly technical and decision-centralized (Petterson & Ray-Bennett, 2018), resulting in documentation replete with technical details that no one reads in the immediate aftermath of a disaster when time is of the essence and flexibility and innovation are required of leadership (Strelec, 2010). This may prove challenging in countries with top-down decision-making and an over-reliance on procedures and protocols that can slow down a response or make it less than effective (Sawalha, 2020). Furthermore, in many countries, as was seen in the COVID-19 pandemic, other outbreaks, and epidemics, the rapid proliferation and distribution of misinformation can lead to distrust of knowledgeable sources, leading to delays or reduced effectiveness to humanitarian needs during the response phase (Sachs et al., 2022).

Recovery

Recovery from a disaster occurs only after damages have been assessed and involves taking actions to return the affected community to its pre-disaster state or better, making it less vulnerable to future hazards. Examples include reconstruction of houses and rehabilitation of community infrastructure, including ensuring employment for residents in road construction and solid waste management through cash for work programs (Vidyattama et al., 2021).

While NGO support is critical to effective disaster response, the organizational goals of many such organizations include raising funds for future disaster responses, and they do not typically include funding for longer term activities in affected countries, nor do they often have the staff equipped to do so. NGOs often leave a country during the late response–early recovery phases and they cannot necessarily be counted upon to participate in or support long-term recovery and mitigation efforts.

Values and ethics, including respect for people, their beliefs, and values, come into play throughout the disaster management cycle but particularly during immediate response and recovery efforts. Tensions may flare between responders and affected human populations where cultural values differ or where political tensions between responding and affected countries predate the occurrence of a disaster (Subedi & Chhetri, 2019).

Mitigation

This phase aims to prevent future emergencies and/or minimize their negative effects. It requires hazard risk analysis skills and the application of strategies to reduce the likelihood that hazards will become disasters. Examples of this include national measures to strengthen building construction in earthquake-prone areas (Dixit, 2004) and moving entire communities where rising sea levels threaten their existence (Flavelle & Irvine, 2022).

Because NGO goals rarely extend beyond response and recovery phases, their missions do not often encompass longer term participation in mitigation efforts for hazard reduction. Furthermore, even if timely after-action reports are written that include lessons learned, these reports are often limited to internal consumption out of concern for their impact on leaders' reputations or on donor funding and are published as gray literature, making them a challenge to find. Yet the survival of people, including small livestock owners, can teeter in the balance if their sources of income and food are not restored in a reasonable amount of time. Calls to compensate livestock holders for the value of their animals often go unheeded or these values are vastly underestimated, as was the case in Indonesia following the Highly Pathogenic Avian Influenza outbreak in 2008 (Scoones & Forster, 2008), but many LMICs are challenged to provide for even basic humanitarian needs and adequate compensation for livestock is simply beyond reach.

Ideally, mitigation measures recognize the need to address distal social-economic factors that contribute to low community resilience in the face of what are likely to be more frequent and severe disasters. An example of the need for systems thinking related to disaster mitigation is that used in Indonesia, which suffered massive damage in the aftermath of an Indian Ocean earthquake and tsunami in 2004. Beyond the heavy toll on human lives, the Indian Ocean earthquake caused an enormous impact on ecosystems including mangroves, coral reefs, forests, coastal wetlands, vegetation, sand dunes and rock formations, animal and plant biodiversity, and groundwater that will affect the region for many years. The spread of solid and liquid waste and industrial chemicals, water pollution, and the destruction of sewage collectors and treatment plants threatened the environment even further (Hannah, 2021). One systems-based disaster risk reduction approach adopted by the Government of Indonesia emphasizes the use of existing vegetation in the area, which has always played a role in the local culture, calls for the active participation of the community, and is relatively less costly compared to other structural disaster mitigation approaches. Furthermore, it has the additional benefit of reinforcing environmental preservation and biodiversity. Mangrove forests became the fortress that protected Kabonga Besar Village in Donggala, Central Sulawesi, from the tsunami caused by the 7.4 magnitude earthquake in 2018 (Yanuarto, 2022).

CASE STUDY: HAITI

On January 12, 2010, a 7.0 magnitude earthquake ravaged Haiti, one of the poorest nations in the world where 88% of people live below the poverty line (The World Bank, n.d.-a), leaving an estimated 250,000 people dead, 300,000 injured, and more than 1.3 million homeless. Camps for internally displaced people were quickly established in the capital of Port-au-Prince. However, ten months later, 60 cases of acute, watery diarrhea were recorded at L' Hôpital de Saint Nicolas, more than 55 miles from the nearest camp but in a location where thousands of displaced people had sought refuge with friends or family. Within 48 hours, the hospital received more than 1,500 additional patients with the same symptoms and the national laboratory confirmed the cause as cholera, which had not previously been reported in Haiti in more than a century. Between October 20 and November 9, Partners in Health recorded 7,159 cases of severe cholera and 161 deaths. Sporadic cases had been reported in the capital, but following a hurricane on November 6, by November 9, the Ministry of Health reported 11,125 hospitalized patients and 724 confirmed deaths (Walton & Ivers, 2011). By the end of the epidemic, more than 600,000 cases of cholera had been reported.

The internal social, economic, and political factors that underlay Haiti's fragile health status at the time of the earthquake and outbreak have been well described but bear repeating here. They include a barely functioning government that was rated as one of the most corrupt in the world; public services (education, sanitation, and healthcare) often provided by private institutions or NGOs but not the government; the government's lack of financial resources, management, and leadership infrastructure to respond effectively; the absence of a standing army, fire, or prehospital services; a small, unprofessional police force; and a long history of political and civil violence (Walton & Ivers, 2011; Kirsch et al., 2012).

External risk factors also contributed to the chaos following the earthquake and failure to control the cholera epidemic. These included the presence of an estimated 2,000 NGOs of which only 400 provided healthcare but many had little previous experience, skills, and tools required to operate in such a complex and chaotic environment, a UN cluster system that was overwhelmed by the thousands of NGOs, and a lack of accountability of international NGOs (INGOs) that, while they had the best interest of the affected population in mind, largely failed to respond to requests from the Haitian population, leading to a disconnect between INGO priorities and the population's needs and concerns (Manilla Arroyo, 2014). Adding to the population's distrust of INGOs, many UN cluster meetings were held in English, limiting the ability of Haitian officials and NGOs to participate, and rumors circulated for months that Nepalese soldiers serving as UN peacekeepers were the source of cholera (Frerichs, 2017); ten months later this was confirmed by genomic analysis, highlighting how rapidly infectious diseases might be brought to disaster sites by international responders (Hendriksen et al., 2011; Orata et al., 2014). Most important to longer term recovery, most NGOs and INGOs left after the initial response and many promises of foreign aid failed to materialize.

While Haiti is among the poorest and most fragile countries in the world, it is not alone, and many other countries are similarly vulnerable to disasters and susceptible to epidemics and outbreaks such as these for similar reasons (World Population Review, 2024).

DISASTERS – NOT JUST A HUMAN HEALTH PROBLEM

The term "One Health" originated from the term "One Medicine" (Schwabe, 1984) and extended the concept of the interdependence between human and other animal health beyond treatment to include prevention (Zinsstag et al., 2011). It has undergone several revisions since then. The most recent definition of One Health was drafted by the One Health High-Level Expert Panel (OHHLEP), comprised of representatives of the Food and Agriculture Organization of the United Nations (FAO), the World Organisation for Animal Health (OIE), the United Nations Environment Programme (UNEP), and the World Health Organization (WHO) and others. The members of these organizations, known collectively as the Quadripartite, represent a broad range of disciplines in science and policy-related sectors relevant to One Health from around the world (World Health Organization, 2021).

As defined elsewhere in the book, the OHHLEP states One Health is "an integrated, unifying approach that aims to sustainably balance and optimize the health of people, animals, and ecosystems. It recognizes the health of humans, domestic and wild animals, plants, and the wider environment (including ecosystems) are closely linked and interdependent. The approach mobilizes multiple sectors, disciplines, and communities at varying levels of society to work together to foster well-being and tackle threats to health and ecosystems, while addressing the collective need for clean water, energy and air, safe and nutritious food, acting on climate change and contributing to sustainable development" (One Health High-Level Expert Panel (OHHLEP) et al., 2022).

By this definition, disasters satisfy the criteria for One Health challenges because they impact people, companion animals, livestock, wildlife, and their habitats, whether urban, semi-urban, or rural. People who own companion animals may put their own lives at risk to save their pets (Mike et al., 2011), and the loss of livestock has long-lasting effects on the livelihoods and economies of individuals, families, and communities (Livestock Emergency Guidelines and Standards, 2023), especially for subsistence-level farmers, people in rural communities, and even for a country's GDP. The loss of wildlife threatens to further erode biodiversity and the ecosystem services that sustain us all (Rondeau et al., 2020), and the loss of habitats, both built and natural, forces people to migrate and other animals to translocate (Chua et al., 2002) (Maldonado et al., 2013), assuming that suitable quarters are located within their mobility range. The loss of natural habitats especially adds further insult to the environment which is already teetering on the brink of destruction on a global scale (United Nations Environment Programme, 2022).

A One Health Approach to Disaster Management

The typical approach to disaster management focuses on the health and well-being of people first, companion animals and livestock second, and wildlife last, if at all. But people in some countries put their own lives at risk to save a beloved pet; small livestock owners, families, and communities cannot survive in the absence of restocking or full compensation for dead animals; and the destruction of wildlife and their habitats adds further insult to the environment which is already teetering on the brink of destruction on a global scale, with implications for human health and well-being as a result of biodiversity loss. While we do not argue that animals should receive priority in disasters, taking steps to include them in the disaster management cycle could contribute to strengthening the resilience of individuals, families, and communities. Because the health and well-being of people, other animals, and nature are interdependent, disasters satisfy all criteria to be considered as One Health challenges. For that reason, several authors have advocated for integrating One Health strategies and principles into emergency and disaster management (Squance et al., 2021; Dalla Villa et al., 2020; Gallagher et al., 2021; Asokan & Vanitha, 2016).

The approach to disaster management therefore should involve public and private entities and members of indigenous and local communities (Hadlos et al., 2022) in all phases of disaster management out of concern for the health, well-being, and survival of people, companion animals, livestock, wildlife, and the integrity of natural habitats and the sustainability of ecosystem services.

In 2009, an outbreak of Highly Pathogenic Avian Influenza occurred in Indonesia and spread to much of Southeast Asia, Africa, and other countries in the world. In response to this and other emerging infectious disease (EID) threats, USAID mounted a 5-year Emerging Pandemic Threats (EPT) Program designed to improve effectiveness and timely responses to outbreaks of EIDs. The RESPOND component of the Program (U.S. Agency for International Development, 2014) improved training capacity for skills necessary to respond to suspected outbreaks, improved cross-sectoral linkages to support coordinated outbreak response, and improved capacity to conduct investigations of suspected outbreaks in countries and areas considered to be at high risk for emergence of viral diseases from wildlife that had the potential to become pandemic threats to human populations.

The One Health Workforce, EPT's successor program, produced the "One Health Core Competency Domains, Subdomains, and Competency Examples" (Frankson et al., 2016). This document describes the competencies needed for the One Health Workforce to successfully address EPTs, but it can easily be repurposed for disasters of all types. Examples of applying the domains to disaster management may be seen in Table 8.1. Although core competencies for One Health were reinterpreted and updated in 2018 (Togami et al., 2018) and 2023 (Laing et al., 2023), they remain essentially the same.

The Global Health Practitioner builds and maintains relationships based on One Health competencies. Because no single individual possesses the expertise,

TABLE 8.1
Applying the Domains of Disaster Management to One Health Examples

Domain	Description	One Health Example
Management	Competencies that enable partners to plan, design, implement, and evaluate programs across disciplines and sector to maximize effectiveness of action and One Health outcomes.	Veterinary Services and Environmental Health entities are included in all phases of the Disaster Management cycle.
Culture and beliefs	Competencies focusing on effective communication and interactions through the understanding of diverse social norms, roles, and practices of individuals, communities, and organizations that impact an intended One Health outcome.	Pets are permitted to shelter alongside their owners, and euthanasia of animals is not performed where religious values oppose it.

(Continued)

TABLE 8.1 (*Continued*)
Applying the Domains of Disaster Management to One Health Examples

Domain	Description	One Health Example
Values and ethics	Competencies that enable partners to identify and respond with respect and fairness across all disciplines and sectors to One Health issues in diverse human, animal, and ecosystem contexts, and promote accountability for the full impact of decisions on the integrated system at local, national, and international levels.	Animal welfare and environmental NGOs and members of affected communities are invited to participate in decision-making in all phases of disaster management, and their perspectives are valued.
Collaboration and partnerships	Competencies that identify, recruit, work with, and sustain the willingness and ability of a diverse range of stakeholders to work effectively to advance One Health.	Memoranda of Understanding are established with humanitarian, animal welfare, and environmental NGOs and civil service organizations during the preparedness phase.
Leadership	Competencies that focus on creating shared visions, championing collaborative solutions through critical and strategic decision-making, and energizing commitment to transdisciplinary approaches for One Health challenges.	Animal health and welfare representatives are included in Incident Command Structure in response to a disaster.
Communication	Competencies that foster effective communication and information sharing across disciplines and sectors.	Phone-based applications permit citizen scientists and others in affected populations to "groundtruth" data from satellites about the extent of a disaster and the impact on human and other animal populations and the environment.
Systems thinking	Competencies that recognize how elements influence and interact with one another within a whole that results from the dynamic interdependencies among human, animal, environmental, and ecological systems, and how these interdependencies affect the relationships among individuals, groups, organizations, and communities.	Mitigation measures include policy changes that preserve natural habitat for threatened and endangered species and limit appropriation of forests and open spaces for private gain.

Adapted from Frankson et al. (2016).

skills, and competencies required to manage all phases of a disaster successfully, One Health competencies apply in all disaster management phases to teams of individuals with different disciplinary, cultural, and social experience and beliefs, regardless of whether disaster management is being undertaken for a locality, region, country, or multiple countries at once.

One last point deserves to be made. There is evidence that as countries' economies develop, they suffer fewer disaster-related deaths and damages than do countries with weaker economies, explained in part by increases in income which generate greater individual demand for safety, including measures to reduce the impact of natural disasters; this demand extends countrywide (Toya & Skidmore, 2007). Furthermore, the impact of disasters can have lasting effects on economic growth among countries with low levels of financial sector development that face borrowing constraints (McDermott et al., 2014). The implications of these findings for lower -and middle-income countries are that resources that might otherwise be dedicated to development are directed instead to disaster management and response, thereby keeping these countries in a perpetual cycle of low economic development and high vulnerability to natural disasters. This also raises the question of whether a One Health approach to disaster management – especially in lower- to middle-income countries – might lessen the impact of natural disasters and permit greater allocation of funds to measures designed to mitigate the impact of future disasters.

REFERENCES

Ahern, M., Kovats, R. S., Wilkinson, P., Few, R., & Matthies, F. (2005). Global health impacts of floods: Epidemiologic evidence. *Epidemiologic Reviews*, *27*(1), 36–46. https://doi.org/10.1093/epirev/mxi004

Asokan, G. V., & Vanitha, A. (2016). Disaster response under One Health in the aftermath of Nepal earthquake, 2015. *Journal of Epidemiology and Global Health*, *7*(1), 91. https://doi.org/10.1016/j.jegh.2016.03.001

Brower, R. S., Magno, F. A., & Dilling, J. (2014). Evolving and implementing a new disaster management paradigm: The case of the Philippines. In N. Kapucu & K. T. Liou (Eds.), *Disaster and development* (pp. 289–313). Springer International Publishing. https://doi.org/10.1007/978-3-319-04468-2_17

Caldera, H. J., & Wirasinghe, S. C. (2022). A universal severity classification for natural disasters. *Natural Hazards*, *111*(2), 1533–1573. https://doi.org/10.1007/s11069-021-05106-9

Cartwright, C., Hall, M., & Lee, A. C. K. (2017). The changing health priorities of earthquake response and implications for preparedness: A scoping review. *Public Health*, *150*, 60–70. https://doi.org/10.1016/j.puhe.2017.04.024

Chua, K. B., Chua, B. H., & Wang, C. W. (2002). Anthropogenic deforestation, El Nino and the emergence of Nipah virus in Malaysia. *Malaysian Journal of Pathology*, *24*(1), 15–21.

Clarke, D. J., & Dercon, S. (2016). *Dull disasters?: How planning ahead will make a difference* (1st ed.). Oxford University Press, Oxford. https://doi.org/10.1093/acprof:oso/9780198785576.001.0001

Cocking, C. (2013). The role of "zero-responders" during 7/7: Implications for the emergency services. *International Journal of Emergency Services*, *2*(2), 79–93. https://doi.org/10.1108/IJES-08-2012-0035

CRED. (n.d.). *EM-DAT - The international disaster database*. Retrieved June 4, 2024, from https://www.emdat.be/

Dahl, B. A., Kinzer, M. H., Raghunathan, P. L., Christie, A., De Cock, K. M., Mahoney, F., Bennett, S. D., Hersey, S., & Morgan, O. W. (2016). CDC's Response to the 2014–2016 Ebola Epidemic—Guinea, Liberia, and Sierra Leone. *MMWR Supplements*, *65*(3), 12–20. https://doi.org/10.15585/mmwr.su6503a3

Dalla Villa, P., Watson, C., Prasarnphanich, O., Huertas, G., & Dacre, I. (2020). Integrating animal welfare into disaster management using an 'all-hazards' approach. *Revue Scientifique et Technique de l'OIE*, *39*(2), 599–613. https://doi.org/10.20506/rst.39.2.3110

Dewan, T. H. (2015). Societal impacts and vulnerability to floods in Bangladesh and Nepal. *Weather and Climate Extremes*, *7*, 36–42. https://doi.org/10.1016/j.wace.2014.11.001

Dixit, A. M. (2004). Promoting safer building construction in Nepal. *13th World Conference on Earthquake Engineering*, Vancouver, British Columbia, Canada.

Donner, W., & Rodriguez, H. (n.d.). *Disaster risk and vulnerability: The role and impact of population and society*. PRB. Retrieved June 4, 2024, from https://www.prb.org/resources/disaster-risk/

Flavelle, C., & Irvine, T. (2022, November 2). Here's where the U.S. is testing a new response to rising seas. *The New York Times*. https://www.nytimes.com/2022/11/02/climate/native-tribes-relocation-climate.html

Frankson, R., Hueston, W., Christian, K., Olson, D., Lee, M., Valeri, L., Hyatt, R., Annelli, J., & Rubin, C. (2016). One Health core competency domains. *Frontiers in Public Health*, *4*. https://doi.org/10.3389/fpubh.2016.00192

Frerichs, R. R. (2017). *Deadly river: Cholera and cover-up in post-earthquake Haiti*. Cornell University Press. https://doi.org/10.7591/9781501703638

Gallagher, C. A., Jones, B., & Tickel, J. (2021). Towards resilience: The one health approach in disasters. In *One Health: The theory and practice of integrated health approaches* (pp. 310–326). CABI Wallingford UK.

Giannini, A. (2015). Climate change comes to the Sahel. *Nature Climate Change*, *5*(8), 720–721. https://doi.org/10.1038/nclimate2739

Hadlos, A., Opdyke, A., & Hadigheh, S. A. (2022). Where does local and indigenous knowledge in disaster risk reduction go from here? A systematic literature review. *International Journal of Disaster Risk Reduction*, *79*, 103160. https://doi.org/10.1016/j.ijdrr.2022.103160

Hannah, C. W. (2021, October 20). *In the aftermath of a Tsunami, mangrove forests in Indonesia sustain lives and livelihoods*. UNDRR. https://www.undrr.org/news/aftermath-tsunami-mangrove-forests-indonesia-protect-lives-and-livelihoods

Hendriksen, R. S., Price, L. B., Schupp, J. M., Gillece, J. D., Kaas, R. S., Engelthaler, D. M., Bortolaia, V., Pearson, T., Waters, A. E., Prasad Upadhyay, B., Devi Shrestha, S., Adhikari, S., Shakya, G., Keim, P. S., & Aarestrup, F. M. (2011). Population genetics of Vibrio cholerae from Nepal in 2010: Evidence on the origin of the Haitian Outbreak. *mBio*, *2*(4), e00157–11. https://doi.org/10.1128/mBio.00157-11

Institute for Health Metrics and Evaluation. (n.d.). *COVID-19 projections*. Institute for Health Metrics and Evaluation. Retrieved June 4, 2024, from https://covid19.healthdata.org/

International Strategy for Disaster Reduction. (2005). *Hyogo framework for action 2005–2015: Building the resilience of nations and communities to disasters*. World Conference on Disaster Reduction. https://www.unisdr.org/2005/wcdr/intergover/official-doc/L-docs/Hyogo-framework-for-action-english.pdf

Keim, M. E. (2020). The epidemiology of extreme weather event disasters (1969–2018). *Prehospital and Disaster Medicine*, *35*(3), 267–271. https://doi.org/10.1017/S1049023X20000461

Kelman, I. (2011). Understanding vulnerability to understand disasters. In *Canadian disaster management textbook, Canada: Canadian risk and hazards network (cap 7)*.

Kirsch, T., Sauer, L., & Guha Sapir, D. (2012). Analysis of the international and US response to the Haiti Earthquake: Recommendations for change. *Disaster Medicine and Public Health Preparedness*, *6*(3), 200–208. https://doi.org/10.1001/dmp.2012.48

Kouadio, I. K., Aljunid, S., Kamigaki, T., Hammad, K., & Oshitani, H. (2012). Infectious diseases following natural disasters: Prevention and control measures. *Expert Review of Anti-Infective Therapy*, *10*(1), 95–104. https://doi.org/10.1586/eri.11.155

Laing, G., Duffy, E., Anderson, N., Antoine-Moussiaux, N., Aragrande, M., Luiz Beber, C., Berezowski, J., Boriani, E., Canali, M., Pedro Carmo, L., Chantziaras, I., Cousquer, G., Meneghi, D., Gloria Rodrigues Sanches Da Fonseca, A., Garnier, J., Hitziger, M., Jaenisch, T., Keune, H., Lajaunie, C., … Häsler, B. (2023). Advancing One Health: Updated core competencies. *CABI One Health*, ohcs20230002. https://doi.org/10.1079/cabionehealth.2023.0002

Livestock Emergency Guidelines and Standards. (2023). *Livestock emergency guidelines and standards*. UK: Practical Action Publishing. http://doi.org/10.3362/9781788532488LEGS

Maldonado, J. K., Shearer, C., Bronen, R., Peterson, K., & Lazrus, H. (2013). The impact of climate change on tribal LEGS communities in the US: Displacement, relocation, and human rights. In J. K. Maldonado, B. Colombi, & R. Pandya (Eds.), *Climate change and indigenous peoples in the United States* (pp. 93–106). Springer International Publishing. https://doi.org/10.1007/978-3-319-05266-3_8

Manilla Arroyo, D. (2014). Blurred lines: Accountability and responsibility in post-earthquake Haiti. *Medicine, Conflict and Survival*, *30*(2), 110–132. https://doi.org/10.1080/13623699.2014.904642

McDermott, T. K. J., Barry, F., & Tol, R. S. J. (2014). Disasters and development: Natural disasters, credit constraints, and economic growth. *Oxford Economic Papers*, *66*(3), 750–773. https://doi.org/10.1093/oep/gpt034

Mike, M., Mike, R., & Lee, C. J. (2011). Katrina's animal legacy: The PETS act. *Journal of Animal Law and Ethics*, *4*(133), 133–160.

NASA. (n.d.). *What is climate change? - NASA Science*. Retrieved June 4, 2024, from https://science.nasa.gov/climate-change/what-is-climate-change/

Oleribe, O. E., Momoh, J., Uzochukwu, B. S., Mbofana, F., Adebiyi, A., Barbera, T., Williams, R., & Taylor Robinson, S. D. (2019). Identifying key challenges facing healthcare systems in Africa and potential solutions. *International Journal of General Medicine*, *12*, 395–403. https://doi.org/10.2147/IJGM.S223882

Oleribe, O. O., Salako, B. L., Ka, M. M., Akpalu, A., McConnochie, M., Foster, M., & Taylor-Robinson, S. D. (2015). Ebola virus disease epidemic in West Africa: Lessons learned and issues arising from West African countries. *Clinical Medicine*, *15*(1), 54–57. https://doi.org/10.7861/clinmedicine.15-1-54

One Health High-Level Expert Panel (OHHLEP), Adisasmito, W. B., Almuhairi, S., Behravesh, C. B., Bilivogui, P., Bukachi, S. A., Casas, N., Cediel Becerra, N., Charron, D. F., Chaudhary, A., Ciacci Zanella, J. R., Cunningham, A. A., Dar, O.,

Debnath, N., Dungu, B., Farag, E., Gao, G. F., Hayman, D. T. S., Khaitsa, M., … Zhou, L. (2022). One Health: A new definition for a sustainable and healthy future. *PLoS Pathogens*, *18*(6), e1010537. https://doi.org/10.1371/journal.ppat.1010537

Orata, F. D., Keim, P. S., & Boucher, Y. (2014). The 2010 Cholera Outbreak in Haiti: How science solved a controversy. *PLoS Pathogens*, *10*(4), e1003967. https://doi.org/10.1371/journal.ppat.1003967

Petterson, M., & Ray-Bennett, N. (2018). Avoidable deaths: A systems failure approach to disaster risk management. *Disaster Prevention and Management: An International Journal*, *27*(2), 271–274. https://doi.org/10.1108/DPM-04-2018-301

Positive Deviance Collaborative. (2020). *Positive deviance collaborative*. Positive Deviance Collaborative. https://positivedeviance.org

Readfearn, G., & Morton, A. (2020, July 28). Almost 3 billion animals affected by Australian bushfires, report shows. *The Guardian*. https://www.theguardian.com/environment/2020/jul/28/almost-3-billion-animals-affected-by-australian-megafires-report-shows-aoe

Rondeau, D., Perry, B., & Grimard, F. (2020). The Consequences of COVID-19 and Other Disasters for Wildlife and Biodiversity. *Environmental and Resource Economics*, *76*(4), 945–961. https://doi.org/10.1007/s10640-020-00480-7

Sachs, J. D., Karim, S. S. A., Aknin, L., Allen, J., Brosbøl, K., Colombo, F., Barron, G. C., Espinosa, M. F., Gaspar, V., Gaviria, A., Haines, A., Hotez, P. J., Koundouri, P., Bascuñán, F. L., Lee, J.-K., Pate, M. A., Ramos, G., Reddy, K. S., Serageldin, I., … Michie, S. (2022). The Lancet Commission on lessons for the future from the COVID-19 pandemic. *The Lancet*, *400*(10359), 1224–1280. https://doi.org/10.1016/S0140-6736(22)01585-9

Sawalha, I. H. (2020). A contemporary perspective on the disaster management cycle. *Foresight*, *22*(4), 469–482. https://doi.org/10.1108/FS-11-2019-0097

Schwabe, C. W. (1984). *Veterinary medicine and human health* (3rd ed). Williams & Wilkins.

Scoones, I., & Forster, P. (2008). *The international response to highly pathogenic Avian Influenza: Science, policy and politics* (STEPS Working Paper 10). STEPS Centre. https://citeseerx.ist.psu.edu/document?repid=rep1&type=pdf&doi=9057e8f06991eaf42792bd2dd50d0540e644fa4f

Shen, G., & Hwang, S. N. (2019). Spatial–Temporal snapshots of global natural disaster impacts Revealed from EM-DAT for 1900–2015. *Geomatics, Natural Hazards and Risk*, *10*(1), 912–934. https://doi.org/10.1080/19475705.2018.1552630

Singer, M. (2014, July 13). *2013 State of the climate: Record-breaking Super Typhoon Haiyan*. Climate.Gov. https://www.climate.gov/news-features/understanding-climate/2013-state-climate-record-breaking-super-typhoon-haiyan

Southall, H. G., DeYoung, S. E., & Harris, C. A. (2017). Lack of cultural competency in international aid responses: The Ebola Outbreak in Liberia. *Frontiers in Public Health*, *5*. https://doi.org/10.3389/fpubh.2017.00005

Squance, H., MacDonald, C., Stewart, C., Prasanna, R., & Johnston, D. M. (2021). Strategies for implementing a one welfare framework into emergency management. *Animals*, *11*(11), 3141. https://doi.org/10.3390/ani11113141

Strelec, J. (2010). *The pros and cons of the ISO certification | Own way*. OwnWayEU. https://www.ownway.eu/articles/the-pros-and-cons-of-the-iso-certification-1/

Subedi, S., & Chhetri, M. B. P. (2019). Impacts of the 2015 Gorkha earthquake: Lessons learnt from Nepal. In Jaime Santos-Reyes (Ed.) *Earthquakes-impact, community vulnerability and resilience*. IntechOpen.

Taft-Morales, M. (2011). *Haiti earthquake: Crisis and response*. DIANE Publishing.

Talisuna, A. O., Okiro, E. A., Yahaya, A. A., Stephen, M., Bonkoungou, B., Musa, E. O., Minkoulou, E. M., Okeibunor, J., Impouma, B., Djingarey, H. M., Yao, N. K. M., Oka, S., Yoti, Z., & Fall, I. S. (2020). Spatial and temporal distribution of infectious disease epidemics, disasters and other potential public health emergencies in the World Health Organisation Africa region, 2016–2018. *Globalization and Health, 16*(1), 9. https://doi.org/10.1186/s12992-019-0540-4

The Lancet COVID-19 Commission. (2022). *The Lancet COVID-19 commission*. Lancet Commission on COVID-19. https://www.thelancet.com/commissions/covid19

The Visual Journalism Team. (2022, August 31). Pakistan floods: Map and satellite photos show extent of devastation. *BBC News*. https://www.bbc.com/news/world-asia-62728678

The World Bank. (n.d.-a). *Overview—Haiti* [Text/HTML]. World Bank. Retrieved June 4, 2024, from https://www.worldbank.org/en/country/haiti/overview

The World Bank. (n.d.-b). *World Bank open data*. Retrieved June 4, 2024, from https://data.worldbank.org

The World Bank. (2015). *The making of a riskier future*. Global Facility for Disaster Reduction and Recovery. https://doi.org/10.1596/978-1-4648-0484-7

The World Bank. (2024a). *The World Bank Atlas method—Detailed methodology – World Bank Data Help Desk*. https://datahelpdesk.worldbank.org/knowledgebase/articles/378832-what-is-the-world-bank-atlas-method

The World Bank. (2024b). *World Bank country and lending groups – World Bank Data Help Desk*. https://datahelpdesk.worldbank.org/knowledgebase/articles/906519-world-bank-country-and-lending-

Tilman, D. (1996). The benefits of natural disasters. *Science, 273*(5281), 1518–1518. https://doi.org/10.1126/science.273.5281.1518

Togami, E., Gardy, J. L., Hansen, G. R., Poste, G. H., Rizzo, D. M., Wilson, M. E., & Mazet, J. A. (2018). Core competencies in one health education: What are we missing? *NAM Perspectives*.

Toya, H., & Skidmore, M. (2007). Economic development and the impacts of natural disasters. *Economics Letters, 94*(1), 20–25. https://doi.org/10.1016/j.econlet.2006.06.020

UCLouvain. (2024). *CRED: Epidemiology of disasters*. UCLouvain. https://uclouvain.be/en/research-institutes/irss/cred-center-of-research-on-the-epidemiology-of-disasters-0.html

UCLouvain, CRED, & USAID. (2022). *Natural hazards and disasters—An overview of the first half of 2022. 68*. https://reliefweb.int/report/world/cred-crunch-newsletter-issue-no-68-september-2022-natural-hazards-disasters-overview-first-half-2022

United Nations. (n.d.). *Press materials—Climate action*. United Nations; United Nations. Retrieved June 4, 2024, from https://www.un.org/en/climatechange/press-materials

United Nations. (2015). *Sendai framework for disaster risk reduction 2015—2030*. https://www.undrr.org/publication/sendai-framework-disaster-risk-reduction-2015-2030

United Nations Environment Programme. (2022, December 20). *COP15 ends with landmark biodiversity agreement*. UNEP. https://www.unep.org/news-and-stories/story/cop15-ends-landmark-biodiversity-agreement

United Nations Office for Disaster Risk Reduction. (2017, February 2). *Sendai framework terminology on disaster risk reduction*. https://www.undrr.org/drr-glossary/terminology

United Nations Office for Disaster Risk Reduction. (2023). *Disaster | UNDRR*. https://www.undrr.org/terminology/disaster

US Agency for International Development. (2014). *Welcome to the RESPOND Project Digital Archive, v2.0*. https://www.respond-project.org/archive/Digital-Archive.html

Vidyattama, Y., Merdikawati, N., & Tadjoeddin, M. Z. (2021). Aceh tsunami: Long-term economic recovery after the disaster. *International Journal of Disaster Risk Reduction*, *66*, 102606. https://doi.org/10.1016/j.ijdrr.2021.102606

Walton, D. A., & Ivers, L. C. (2011). Responding to Cholera in post-Earthquake Haiti. *New England Journal of Medicine*, *364*(1), 3–5. https://doi.org/10.1056/NEJMp1012997

Warfield, C. (2022). *The disaster management cycle*. GDRC. https://www.gdrc.org/uem/disasters/1-dm_cycle.html

Watson, J., Connolly, M., & Gayer, M. (2006). *Communicable diseases following natural disasters: Risk assessment and priority interventions* (WHO/CDS/NTD/DCE/2006.4) [Technical Document]. Programme on Disease Control in Humanitarian Emergencies Communicable Diseases Cluster, WHO. https://cdn.who.int/media/docs/default-source/documents/emergencies/communicable-diseases-following-natural-disasters.pdf?sfvrsn=4a185b2c_2&download=true

Watson, J. T., Gayer, M., & Connolly, M. A. (2007). Epidemics after natural disasters. *Emerging Infectious Diseases*, *13*(1), 1–5. https://doi.org/10.3201/eid1301.060779

World Health Organization. (2021, December 1). *Tripartite and UNEP support OHHLEP's definition of "One Health."* https://www.who.int/news/item/01-12-2021-tripartite-and-unep-support-ohhlep-s-definition-of-one-health

World Population Review. (2024). *Poorest Countries in the World 2024*. https://worldpopulationreview.com/country-rankings/poorest-countries-in-the-world

Yanuarto, T. (2022, April 25). *Indonesia: Disaster prevention and mitigation through vegetation | PreventionWeb*. Prevention Web - UNDRR. https://www.preventionweb.net/news/disaster-prevention-and-mitigation-through-vegetation-indonesias-experience

Zimmer, C. (2011, August 23). How many species? A study says 8.7 million, but it's tricky. *The New York Times*. https://www.nytimes.com/2011/08/30/science/30species.html

Zinsstag, J., Schelling, E., Waltner-Toews, D., & Tanner, M. (2011). From "one medicine" to "one health" and systemic approaches to health and well-being. *Preventive Veterinary Medicine*, *101*(3–4), 148–156. https://doi.org/10.1016/j.prevetmed.2010.07.003

9 Ecosystems and Infectious Disease

Tara E. Stewart Merrill, A. Alonso Aguirre, Brian F. Allan, and Christopher A. Whittier

INTRODUCTION

One Health is united by an emphasis on healthy people, healthy organisms, and healthy ecosystems. Most health practitioners recognize the intrinsic importance of human and animal health, but we are still learning about the value of healthy ecosystems. Ecosystems are complex networks of interacting organisms and habitats and provide numerous benefits to the species that live within them. For example, oyster reefs are a critical ecosystem found in shallow, nearshore marine environments. When oyster reefs are "healthy" – supporting high numbers of oysters – oysters filter large amounts of water, which improves water clarity (Vaughn & Hoellein, 2018). This enhanced water clarity increases the amount of sunlight that can penetrate the water and reach aquatic plants like seagrasses (Vaughn & Hoellein, 2018), which then support large herbivorous organisms like manatees and sea turtles (Jackson, 2001). While the positive connections from bivalve to macrophyte to herbivore are easy to trace within a system, such beneficial links are not confined to just the ecosystem. That is, the benefits of ecosystems can extend even further to reach and benefit people.

Ecosystems provide humans with myriad services. They form the foundation of our food chains, supply us with clean air and water, and yield natural resources like timber (Costanza et al., 1997). In addition to supporting our fundamental needs, ecosystems also enrich our lives. Human recreational activities – whether from hunting, fishing, swimming, or hiking – often occur in a natural setting and data are accumulating that exposure to nature maintains mental health and well-being (Bratman et al., 2012). In the One Health context, ecosystems are also thought to provide an additional essential service: ecosystems may protect health in the face of infectious disease.

Infectious disease is a critical stressor with broad consequences. Human infectious disease can result in morbidity and mortality and reduced lifespan and disability, and these outcomes often have downstream negative effects on societies and economies (Bonds et al., 2010; Murray et al., 2022). Beyond people, infectious disease in farmed and harvested animals leads to substantial economic losses (The World Bank, 2010; Lafferty et al., 2015), and companion animals that live with us can also be adversely affected by infection

DOI: 10.1201/9781003232223-9

(Messenger et al., 2014). Finally, disease represents an ongoing concern for wild organisms, including those deemed vulnerable and threatened by extinction. For instance, *Batrachochytrium dendrobatidis* (a fungal pathogen that causes the disease chytridiomycosis) is a significant and growing threat to amphibians and has been implicated in their global declines (Scheele et al., 2019). Disease can move freely among these contexts, spreading from ecosystems and animals to people and back again, which suggests that maintaining health in one context can result in health in another. Understanding how healthy ecosystems may mitigate disease is therefore a paramount issue for global health practitioners, from epidemiologists and veterinarians to conservation biologists.

We have an idea of what health looks like in a person, plant, or animal, but what is a healthy ecosystem? Ecosystem health is the least understood slice of the One Health pie. Rapport et al. (1985) defined the "ecological distress syndrome" as a series of unprecedented environmental changes that lead to health concerns for humans, species, and ecosystems. Ecological health is defined as a continuous state of flux and a complex interrelationship between health and those ecological concerns. It includes linkages among changes in habitat structure and land change; emergence and resurgence of infectious and non-infectious diseases including environmental contaminants; maintenance of biodiversity; and sustaining the health of plants and animal communities including humans (Aguirre et al., 2002). The complexity of ecosystems, as well as how much they can vary, can make ecosystem health especially challenging to define or measure. Ecosystems contain huge and varied numbers of species (both eukaryotic and prokaryotic) engaged in a complex web of feeding interactions, and these species vary in body size, structure, and interaction type, with numbers frequently fluctuating over time and space. Given this complexity, it is unlikely that we can derive one comprehensive measure of ecosystem health, or a set of measures that could apply generally across diverse ecosystems. When we think about ecosystem health perhaps the closest we might get is by considering how unperturbed an ecosystem is by anthropogenic influences. In this case, a healthy ecosystem is likely one that contains intact biodiversity, where species have not been extirpated, as well as intact habitat that has not been destroyed, fragmented, or degraded. In the Sections "Biodiversity and Infectious Disease" and "Healthy Environments" we consider how these two core facets – biodiversity and healthy environments – may reduce infectious disease across forms of life.

Importantly, the relationship between an ecosystem and infectious disease is not unidirectional; that is, disease can also impact how an ecosystem operates. For example, amphibians are important players in natural systems because they are predators of diverse invertebrates, provide resources for birds, mammals, and fish, and can move energy across the aquatic-terrestrial boundary (Hocking & Babbitt, 2014). When amphibians decline or are extirpated due to infections like amphibian chytridiomycosis, we also may lose their critical ecological roles. Reducing disease in amphibians may therefore restore the roles they play in ecosystem processes, improving ecosystem health and function. In the Section "Effects of Health on Ecosystems" we explore this feedback and consider how disease reduction in human, agricultural, and wildlife contexts can benefit ecosystems. Finally, in the Section "Nature-Based Solutions" we embrace the connections between ecosystems, environments, biodiversity, and disease and discuss nature-based solutions (NbS) where a single intervention can preserve health across multiple sectors. Throughout this chapter, we refer to parasites and pathogens collectively as organisms that live on or within their hosts, while deriving a net benefit from the host (typically through feeding) and exerting a cost on the host (harm that often manifests as infection symptoms (Stewart & Schnitzer, 2017)). For further information on the diversity of parasitic interactions and strategies, see Lafferty and Kuris (2002).

Biodiversity and Infectious Disease

Biodiversity may limit the spread of parasites and pathogens, reducing infection and disease (Dobson et al., 2006; Keesing et al., 2010; Ostfeld & Keesing, 2012). This potential benefit arises from the intimate connections between parasites and their natural ecosystems. Like all other organisms, parasites have lived in ecosystems throughout their evolutionary history. Because of this, ecosystem properties – including abiotic (non-living) factors like temperature and rainfall, and biotic (living) factors like species diversity – have the capacity to shape parasite transmission. The influence of biodiversity arises from biotic factors, and, more specifically, from which species live in an ecosystem and how abundant they are.

Many parasites are shared by multiple host species that together influence the degree to which a parasite can spread. Parasites like tapeworms and flukes have complex life cycles, requiring multiple species of host for their development. Generalist parasites can infect a wide range of species and can be opportunistic in which they infect. Some parasites find themselves in certain host species because they spill over from others (i.e., colonize and potentially adapt to a new host species that was not previously part of their transmission). Thinking beyond parasites of wild organisms, we recognize these same features in parasites of humans. People experience infection by complex life cycle parasites (e.g., schistosomes that cycle between aquatic snails and human or animal hosts) and generalist parasites (e.g., *Giardia* that can be harbored by most mammals), and often must contend with disease spillover from wild hosts (e.g., Nipah virus that spills over from bats). Indeed, most human emerging infectious diseases are zoonotic (moving between animals and people, Jones et al. 2008), and even infections that are unique to humans today were likely zoonotic at an earlier point in evolutionary history (Wolfe et al., 2007). Taken together, the fact that parasites come from ecosystems, and move among species that reside within ecosystems, means that we can expect parasites to be regulated by ecological interactions.

One topic that has garnered considerable attention within the context of ecosystem health is the relationship between biological diversity and the prevalence of infectious disease. Through containing a rich suite of ecological interactions, biological diversity can impose limits on the transmission of parasites and pathogens in ecological communities. When biological diversity is mechanistically linked to a reduction in disease prevalence, such an outcome is termed a "dilution effect" because the presence of high diversity dilutes transmission of an infectious agent (Keesing & Ostfeld, 2021a). The generality of dilution effects and other diversity-disease relationships (Ostfeld & Keesing, 2000; Randolph & Dobson, 2012; Wood & Lafferty, 2013), the mechanisms responsible (Keesing et al., 2006; Keesing et al., 2010), and the scales at which they operate (Keesing & Ostfeld, 2021a, 2021b) have been the subject of debate within disease ecology. A majority of studies indicate a generalizable pattern that with increasing species diversity there is a decrease in the prevalence of specific infectious agents (Cardinale et al., 2012; Ostfeld & Keesing, 2012; Civitello et al., 2015), and the mechanisms by which these effects occur could shed light on potential interventions for maintaining ecosystem services that benefit human health. For example, predator species can consume infected hosts, either preferentially or randomly, thereby removing infected individuals from a population and reducing overall infection prevalence (Ostfeld & Holt, 2004). When the capacity to allow an infectious agent to replicate and spread varies among host species (i.e., host reservoir competence varies), competitor species that reduce the density of a reservoir host species may reduce the number of contacts between susceptible and infected individuals (Mills, 2005). Additionally, competition over resources can alter a host's susceptibility to infection, shaping the likelihood of their infection and subsequent transmission of the pathogen (Strauss et al., 2015). These potential pathways have been formalized into a set of mechanisms for when diversity is predicted to reduce pathogen prevalence (Keesing et al., 2006), understanding of which could be incorporated into principles of ecosystem management to enhance the ecosystem service of disease regulation.

Lyme disease, as well as other tick-borne illnesses, offers a particularly compelling narrative on how diversity can reduce disease exposure risk for humans. Lyme disease is caused by the bacterium *Borrelia burgdorferi*, which in the eastern United States is transmitted among hosts by the black-legged tick (*Ixodes scapularis*). These ticks have three parasitic life stages (i.e., larva, nymph, and adult), but are born free from infection with the Lyme pathogen, and so can only become infected by feeding on an infected animal host (Ostfeld, 2011). Wildlife host species vary considerably in their suitability as tick hosts and pathogen reservoirs, with some small mammal species, including the white-footed mouse (*Peromyscus leucopus*) and eastern chipmunk (*Tamias striatus*), serving as especially reservoir-competent hosts for *B. burgdorferi* and suitable hosts for *I. scapularis*. These two host species, and others that play an important role in the tick life cycle such as white-tailed deer (*Odocoileus virginianus*), have been shown to thrive in human-dominated landscapes – indeed, extirpation risk appears to be inversely correlated with host reservoir competence (Kremen & Ostfeld, 2005). When forest habitats are fragmented by human development, biodiversity loss occurs non-randomly and the host species that remain and even increase in abundance tend to be the most important contributors to the Lyme disease transmission cycle (Rosenblatt et al., 1999; Nupp & Swihart, 2000). As a result, forest fragmentation causes an increase in the density of nymph life stage ticks infected with *B. burgdorferi*, increasing human risk of exposure to Lyme disease as shown in Figure 9.1 (Allan et al., 2003; Brownstein et al., 2005), whereas more intact forested habitats contain a greater diversity of wildlife host species that deflect tick blood meals away from more competent reservoirs (LoGiudice et al., 2003). Additionally, other sources of ecosystem change such as invasions by non-native plant species (e.g., barberry, *Berberis thunbergii*) appear to further increase prevalence of the Lyme pathogen in nature and human exposure risk (Williams et al., 2009). This extensive research into the ecology of Lyme disease demonstrates that, ultimately, healthy environments often contain many species that absorb parasites but are dead-end hosts; by diluting these parasites, greater diversity can reduce disease.

While more diverse ecosystems may reduce the spread of a particular infection, they are also likely to be sources of diverse parasites and pathogens (Dunn et al., 2010; Wood et al., 2017). Consider this dichotomy: your risk of acquiring Lyme disease from a tick may be lower in an intact forest containing a species-rich mammal community than it is in a fragmented forest containing primarily white-footed mice (as proposed by "the dilution effect" above). However, your probability of experiencing spillover of a novel disease, or acquiring a zoonotic disease transmitted by wild animals, is likely higher in the depths of a forest than it is on the streets of an urbanized city. This latter case stems, again, from the fact that most parasites can or must infect multiple host species. Therefore, if an ecosystem has high host diversity, it is also likely to contain more species of parasites (Hechinger & Lafferty, 2005; Wood & Johnson, 2015). The concepts of hazard and risk have been elegantly used to describe this dichotomy (Hosseini et al., 2017). "Hazard" is a *potential* source of harm and may represent the cumulative threat posed by all parasites. Arguably, high hazard could indicate a healthy ecosystem if it means the ecosystem contains the host diversity we aim to preserve. Conversely, parasite "risk" represents the *likelihood* that an adverse event will result from a particular parasite or pathogen. When risk becomes high (beyond natural or historical levels), that suggests that a parasite is out of control (i.e., the factors that typically regulate the parasite are no longer operating effectively). In this case, it is often critical that we find interventions for preventing transmission so that we can mitigate outbreaks that damage human, plant, animal, or ecosystem health. A large body of observational studies and small-scale experimental manipulations indicate that biodiversity may help regulate the transmission of specific parasites and reduce risk (Civitello et al., 2015). But looking to the future, the One Health arena still critically needs intervention studies that measure the protective capacity of biodiversity and evaluate whether real-world conservation policies have resulted in noticeable reductions to disease risk.

The often-devastating consequences of disease have led to negative perspectives on parasites and pathogens in clinical settings. In these settings, disease is often thought of as a problem that must be solved. But as we attempt to maintain and restore ecosystems – ideally keeping them in a "healthier" state – it is important to recognize that not all parasites and pathogens are problematic. Parasitic organisms are members of natural ecosystems (Kuris et al., 2008) and are likely

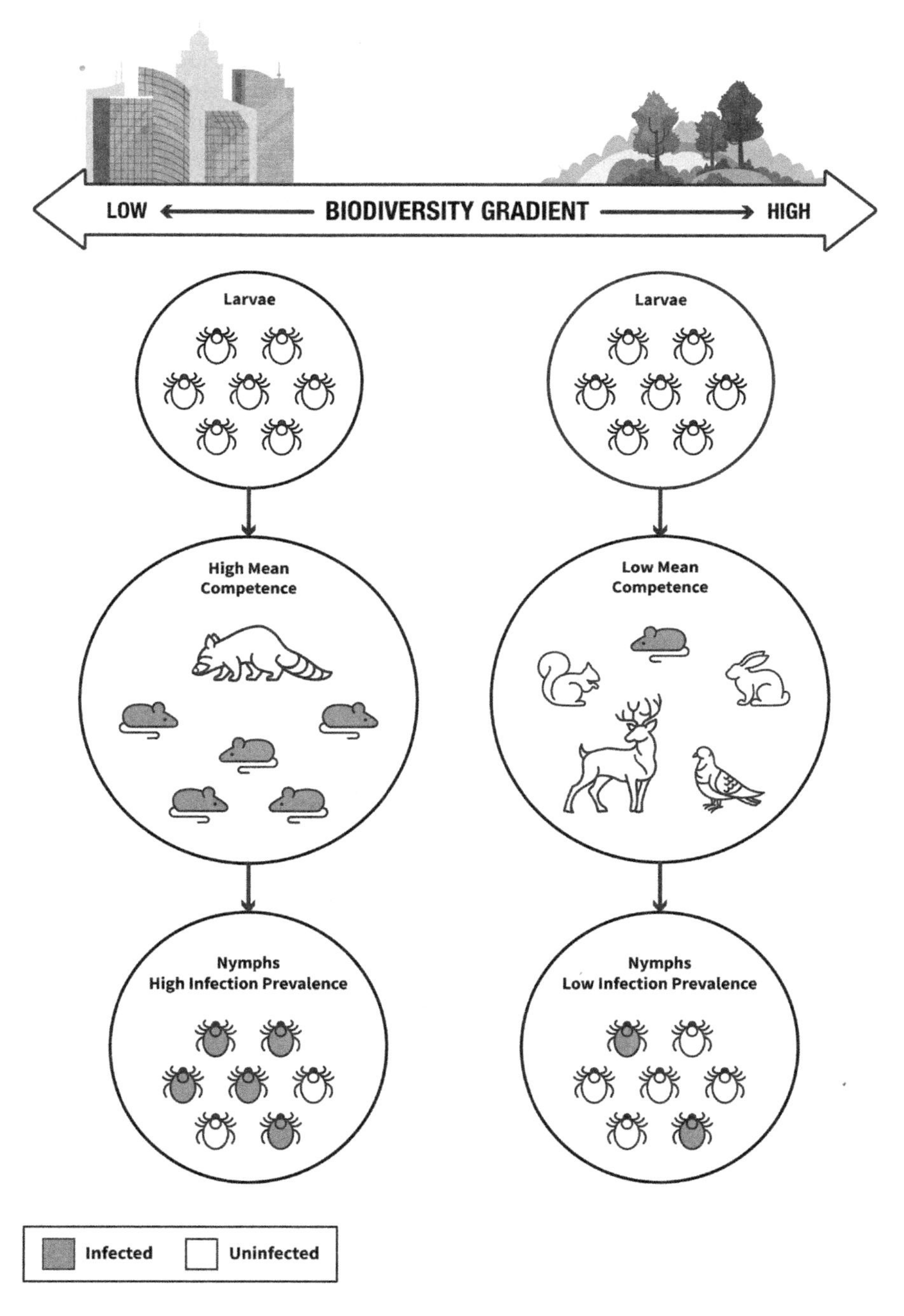

FIGURE 9.1 Example of how biodiversity can impact the spread of Lyme disease (*Borrelia burgdorferi*).

as valuable to ecosystem health as free-living species (Hudson et al., 2006). As a point of comparison, predators like wolves, bears, lions, and leopards (among others) have been viewed through a negative lens throughout human history, because they preyed upon (or were believed to prey upon) livestock and people. These perspectives led to predator eradication campaigns that were successful insofar as they reduced predator numbers; however, hindsight has also demonstrated the failures of these campaigns. Following the loss of large predators from multiple ecosystems, ungulate populations tend to explode in abundance, increasing herbivory pressure and driving declines in woody plants (Beschta & Ripple, 2009). When it comes to predators, recent conservation goals aim to restore or maintain predators' roles in ecosystems (keeping prey populations in check) while also managing human-predator conflict. The same such balance may be needed for parasites and pathogens: we may strive to keep diverse ecosystems that maintain the regulatory roles of parasites (regulating host populations) while preventing deadly outbreaks. To achieve this balance, we need careful investigations that examine the costs and benefits of disease.

In our **first case study**, we consider an example where the reintroduction of the Channel Island fox (*Urocyon littoralis*), a terrestrial predator native to the Channel Islands of southern California, created an opportunity to evaluate whether conservation efforts to protect and restore the diversity of predator species reduces the prevalence of Sin Nombre virus (SNV), a dangerous human pathogen carried by rodents.

CASE STUDY 1. PREDATOR RICHNESS PROTECTS AGAINST SNV ON THE CHANNEL ISLANDS

The Channel Islands are an eight-island archipelago in southern California where a remarkable success story in wild carnivore conservation played out that links the concepts of ecosystem health and regulation of infectious disease. Due to a history of anthropogenic disturbance on the Channel Islands, a native predator, the island fox (*U. littoralis*), was threatened with extinction. Historical use of dichlorodiphenyltrichloroethane (DDT) caused a shift in the composition of the apex predator on the islands from bald eagles (*Haliaeetus leucocephalus*) to golden eagles (*Aquila chrysaetos* (Sharpe & Garcelon, 2005)). Meanwhile, human-introduced feral swine were prey supporting a large golden eagle population, and the susceptibility of island foxes to predation by golden eagles threatened their wild populations with extinction (Coonan et al., 2010). In an effort to conserve this iconic predator endemic only to these islands, the U.S. National Park Service intervened, establishing a captive breeding program for island foxes and undertaking a campaign to remove golden eagles and eradicate feral pigs from the islands. The program was successful, and the reintroduction of the island fox to the Channel Islands capped one of the most successful carnivore conservation programs ever undertaken (Coonan et al., 2010). It also presented an extraordinary research opportunity, because the Channel Islands support a high prevalence of a dangerous human pathogen. Deer mice (*Peromyscus maniculatus*), a wild rodent species commonly found on the Channel Islands, are the known natural reservoir hosts for SNV, the causative agent of Hantavirus Pulmonary Syndrome in humans, a rare but often fatal disease. The highest prevalence of SNV ever reported from deer mice is on the Channel Islands (Orrock & Allan 2008), and island foxes are important predators of deer mice (Schwemm & Coonan, 2001). Thus, the reintroduction of the island fox presented the opportunity to evaluate whether an ecosystem restoration program, in this case the reintroduction of a lost predator species, provided additional benefits in reducing infectious disease risk to humans.

The island fox is one of the most effective predators of deer mice reported from the Channel Islands, and it is important to consider whether the conservation of predator species diversity in general affects regulation of the deer mouse population. Because SNV is transmitted among deer mice through aggressive encounters, predators may reduce SNV prevalence through directly regulating mouse population size, or by altering mouse behaviors to avoid predation that also reduce opportunities for pathogen transmission (Keesing et al., 2006). The opportunity to evaluate these hypotheses was possible because the eight Channel Islands vary considerably in several biogeographic properties, including island area and precipitation. As might be expected from the Theory of Island Biogeography (MacArthur & Wilson, 2001), larger islands generally support greater species diversity, and so there is a gradient of predator species diversity across which to compare SNV prevalence in deer mice. In an analysis performed by Orrock et al. (2011), after controlling for the effects of island area and precipitation, a significant negative relationship was detected between predator species diversity and SNV prevalence in deer mice on the Channel Islands (Orrock et al., 2011). This indicates that as the species diversity of predators increases, SNV prevalence in deer mice decreases. Possible ecological mechanisms behind this finding may include that with greater predator species diversity, there is greater ecological redundancy in the predator species available to regulate the prey population. And with the reintroduction of the island fox, one more effective deer mouse predator has been returned to the ecosystem. Thus, a conservation program that was created to restore a number of ecosystem properties to a state prior to recent human disturbance (removal of golden eagles and feral swine, reintroduction of the native island fox) may offer the additional benefit of reducing infectious disease risk to humans.

Healthy Environments

All living organisms reside in environments, and the structure and integrity of environments provide an additional pathway through which ecosystems can benefit organismal health. Whether it is a forest, reef, lake, or savannah, there are several ways that a habitat can become unhealthy. Many forms of such habitat decline, as well as their symptoms, are familiar to us. The bleaching of coral reefs and seas of plastic pollution are highly publicized forms of decline; likewise, clouds of smog emitted from vehicles and industrial plants are another dramatic and well-known example. But habitat decline can also be more subtle in natural environments. For instance, excess algae in a lake may signal that the lake is experiencing nutrient pollution, which can result in a harmful algal bloom. And temperature shifts from ongoing global climate change often occur in small increments that seem minor or gradual but result in large alterations to weather patterns and environments. Of the varied ways in which habitats can become unhealthy, they generally coalesce into habitat loss and habitat degradation. Both processes can have complex relationships with the spread of pathogens.

Just as biodiversity loss can lead to increased risk of transmission, so can habitat loss enhance pathogen spread. Natural habitat is often lost via destruction (complete removal) or fragmentation (partial removal). These two forms of habitat loss can alter disease outcomes by changing rates of contact between infectious and susceptible hosts. For example, in eastern Australia, populations of *Pteropus* fruit bats, also known as flying foxes, are becoming increasingly habituated to urban landscapes. This phenomenon is partly attributable to ongoing fragmentation of forests in eastern Australia, which increases the number of flying foxes potentially coming into contact with humans and domestic animals. This is especially problematic because flying foxes are the endemic reservoir host for Hendra virus in Australia, a pathogen with high case fatality rates in humans and horses. Plowright et al. (2011) demonstrate that timing and intensity of outbreaks of Hendra virus in flying fox populations in urban areas are affected by habitat fragmentation. While outbreaks in populations of bats concentrated in urban habitat fragments occur less frequently, when they do occur, the outbreaks are of greater intensity, likely due to the isolation of urban bat populations reducing exposure frequency which over time increases the number of individuals susceptible to infection. Spillover events of Hendra virus from bats to horse and human populations have been increasing (Plowright et al., 2011), and this pathway of pathogen spillover due to landscape change is concerning due to its potential to unleash future human pandemics (Plowright et al., 2021).

In addition to habitat loss, habitats can become unhealthy when they experience declines in quality, often due to the introduction of exogenous contaminants. Natural habitats are vulnerable to a suite of human-introduced contaminants that can cause degradation and, ultimately, influence infectious disease. Notable contaminants include oil from rigs, road salt from thoroughfares, and fertilizers from agricultural fields. Such contaminants have well-known lethal effects on organisms, and we are now beginning to learn about their sublethal consequences for organisms' immune function (Galloway & Depledge, 2001; Desforges et al., 2016). For example, mussels of the species *Mytilus edulis* experienced declines in cellular immunity following a largescale oil spill, with the innate response rebounding in the population after environmental hydrocarbon levels dropped (Dyrynda et al., 1997). Similarly, antibody responses were impaired in female glaucous gulls (*Larus hyperboreus*) that had high blood concentrations of organochlorine pollutants (Bustnes et al., 2004). Environmental pollution not only affects hosts but can also impact free-living stages of parasites and pathogens. In some cases, this may result in the death of a parasite or pathogen (Pietrock & Marcogliese, 2003), where in other cases, pollution and additional types of environmental degradation may make infective agents more abundant, transmissible, or virulent. *Vibrio cholerae* is an environmentally transmitted pathogen that replicates rapidly when water temperatures are warm and when nutrient levels are high (Lipp et al., 2002). Hence, by increasing this pathogen's abundance, factors like climate change and fertilizer runoff can amplify the risk of *V. cholerae* infection in aquatic systems. The accidental release of antibiotics into water bodies (often due to antibiotic overuse in agricultural and aquacultural settings) may also select for antibiotic-resistant bacteria, increasing the potential for more virulent – and less treatable – disease (Cabello et al., 2013). Finally, pathogens themselves can represent a form of pollution, producing a new source of disease within a degraded habitat. This frequently occurs when human and animal waste moves along waterways. Along the Pacific coast of North America, pathogens like *Giardia* spp., *Cryptosporidium* spp., and *Toxoplasma gondii* travel with effluent from terrestrial to nearshore marine environments (Fayer et al., 2004). Once they reach marine ecosystems, their infective cysts can be ingested by shellfish and other organisms, which can then serve as reservoirs of infection for people and marine mammals (Robertson, 2007; Miller et al., 2002). Indeed, this transmission pathway has been implicated in the decline of California sea otters (Enhydra lutis nereis), as infections with *T. gondii* are associated with otter morbidity and mortality (Shapiro et al., 2012).

The movement of pathogens from source to sea is a particularly good example of how healthy habitats can lead to less disease. Estuaries, or large swaths of wetland that occur where land meets ocean, are thought to filter out pathogens through mechanical processes (e.g., settlement of pathogen cysts in slow-moving channels) as well as biological processes (e.g., entrapment of pathogen cysts in biofilms) (Shapiro et al., 2010). Hence, restoring natural habitat to an intact state may provide cleaner water and reduce the movement of pathogens, a win-win that can benefit the health of people and wildlife (Hopkins et al., 2022). In our **second case study**, we consider an example where the alteration of river habitat via damming has led to an increase in human schistosomiasis.

CASE STUDY 2. A DAMMED RIVER BREEDS SCHISTOSOMIASIS IN SENEGAL

The damming of the Senegal river was an important infrastructural development for Senegal. The Diama dam reduced saltwater intrusion, thereby increasing irrigation and enhancing agriculture (Sokolow et al., 2017; Lund et al., 2021). But as with many anthropogenic changes, the dam had unintended consequences. Shortly following its construction, human infections with the helminth parasites, *Schistosoma mansoni* and *S. haematobium*, began to rise and both rapidly achieved endemicity (Sow et al., 2002). In prior decades, Senegal had not experienced *S. mansoni* infections, and *S. haematobium* occurred at only low levels (Chaine & Malek, 1983). Schistosome parasites are transmitted to humans by aquatic snails when they release infective larvae into bodies of water. The rise of both infections therefore suggested an ecological change with epidemiological consequences.

Speculated to be at the heart of this continuing outbreak was the extirpation of a prawn, *Macrobrachium vollenhoveni* (Sokolow et al., 2017). This particular prawn is a voracious predator of *Bulinus* spp. snails, the first intermediate hosts of schistosomes, and was eradicated from the Senegal River Basin after the Diama dam was constructed. This eradication likely occurred because the dam blocked the prawn's annual migration from brackish (semi-salty) habitats to inland freshwater habitats (Savaya Alkalay et al., 2014). With prawn predators absent from freshwater bodies, snails increased in their abundance, supporting the reproduction and high densities of schistosome parasites. Consequently, the people who work and recreate in flooded fields and water bodies experienced an increase in the prevalence and intensity of schistosome infection. This phenomenon marked a clear example where an environmental change – the blocking of a river – could trigger an ecological series of events – release of prey from predators – cascading to an increase in human disease (Sokolow et al., 2017).

One of the predominant challenges with schistosomiasis control is that medical treatment is only a short-term solution: if a person is medically cleared of infection, they can rapidly reacquire infection as soon as they enter a body of water containing infective larvae. Reducing transmission therefore requires a combination of both treatment and prevention of exposure. As long as snail population sizes remained high in the Senegal River Basin, there was the potential for a continual source of exposure for human communities.

The saga of schistosomes and snails in Senegal has been studied and addressed by many scientists and public health workers in Senegal and around the world. More recently, a collective called "The Upstream Alliance" has addressed the issue with a focus on undoing the ecological damage caused by the Diama dam. Specifically, this collaborative group is restoring prawns in their native freshwater habitats to limit snail numbers and reduce human infections. Given that the dam still prevents prawn migration, these restoration efforts involve stocking prawns from wild or cultivated populations. This strategy has so far seen great success. As prawn numbers have increased, human infections have subsequently declined by almost 20% (Sokolow et al., 2017). And the benefits of restoring prawns have extended beyond disease reduction. Small prawns are the most effective at consuming snails (per gram biomass; Sokolow et al., 2015), so at large sizes, the prawns can be collected and used as a food source. This case therefore represents a win-win where prawns may reduce disease, help fight hunger, and potentially alleviate poverty of the local communities (Hopkins et al., 2022). In the face of environmental changes, like dam construction, creative interventions grounded in ecology can be employed to restore ecosystem health as well as the benefits derived from healthy environments.

Effects of Health on Ecosystems

One of the advantages of the One Health approach is that it looks at systems and recognizes how positive influences can benefit other components. While we have so far mainly discussed how healthy ecosystems can reduce disease, we can also use disease management practices and interventions to benefit ecosystems and maintain their services. Implementation of classic disease control measures may not only benefit the target species of those actions but may help protect other species and have much broader positive ecological impacts. Such interventions thereby sustain or improve the overall health of the ecosystem. In essence, any means that helps sustain species vital to an ecosystem should benefit the overall health of the system by enabling that species to play their natural role.

This principle is most easily illustrated through health interventions targeting ecologically important species within an ecosystem. In the context of infectious disease, veterinary interventions, like mass vaccination to protect certain wildlife species, are relatively rare (Yabsley, 2019; Barnett & Civitello, 2020) but can have this multi-level benefit. For example, vaccination to protect black footed ferrets (*Mustela nigripes*) against the invasive sylvatic plague bacterium *Yersinia pestis* has been vital to their still tenuous reintroduction into the prairie grasslands of the American west (Salkeld, 2017). In that system, ferrets have been missing in recent history but have a critical ecological role to play as a top predator balancing prairie dog populations. While restoration of these ecosystems is a complex, ongoing process, the simplified aspect of healthy ferret populations keeping prairie dog populations in check leads to an overall healthier ecosystem (Miller et al., 1994). Vaccination to protect other top predators like Ethiopian wolves (*Canis simensis*) and African wild dogs (*Lycaon pictus*) against diseases like rabies helps maintain their similarly important ecological roles in their ecosystems (Knobel et al., 2008; Canning et al., 2019).

Because a given infectious disease can usually affect multiple host species, vaccination interventions like those described above can have protective benefits to additional species. For example, both prairie dogs and ferrets can be infected with *Y. pestis*, so vaccinating ferrets can protect prairie dogs and vice versa (Salkeld, 2017). However, the question of which species to vaccinate in these multi-host systems is not always straightforward. Basic disease ecology dictates that disease is best controlled in the reservoir species, which is how rinderpest was eradicated by vaccinating only cattle and not the often-devastated wild hoofstock species (Roeder, 2011). On the other hand, attempts to protect certain species including lions, tigers, and Ethiopian wolves from diseases like rabies and distemper by controlling those diseases in domestic dogs may be ineffective and require targeted vaccination of those wild carnivores (Viana et al., 2014; Gilbert et al., 2020; Marino et al., 2017). Mass vaccination can have further One Health benefits as well with the classic example of rabies control benefiting humans in addition to the wild or domestic species that might be the target of such campaigns (Maki et al., 2017). In fact, calls for broader application of veterinary vaccinology are increasingly made with a One Health emphasis (Entrican & Francis, 2022).

While vaccination may be the most straightforward, other wildlife health practices can benefit broader ecosystems by maintaining the health of key species in the same fashion. Animals, as well as plants, can be directly treated for certain infectious diseases with antimicrobial agents and other chemotherapeutics. Such interventions are generally very limited in scope because of the challenges of drug delivery compounded by valid concerns for fueling antimicrobial resistance in systems that are challenging to control (Wobeser, 2002). Instead, wildlife disease management strives to start with disease prevention including measures such as avoiding the introduction of exotic or feral species, health screening of translocated animals, strict hygiene protocols, and controlling disease vectors (Foufopoulos et al., 2022). Such measures, along with disease surveillance, are often the only tools available or practical, especially in the absence of vaccines or suitable treatments.

These same disease prevention strategies are widely integrated into animal and plant agriculture (including aquaculture) practices under the heading of biosecurity. Oysters are one example where extensive disease prevention efforts including movement restrictions, water treatment, and husbandry practices are vital for safeguarding a valuable industry that can ultimately have broader ecosystem benefits as outlined at the beginning of this chapter (Rodgers et al., 2019). In some cases, aquaculture itself can even serve as a health measure. For instance, addition of farmed oysters to coastal ecosystems may reduce dermo disease in vulnerable shellfish, because the farmed oysters filter and remove infective stages of the causative agent *Perkinsus marinus* from the water column (Ben-Horin et al., 2018). Importantly, this strategy will only be effective if farmed oysters are harvested before they can transmit infection but highlights how aquaculture practices grounded in knowledge on transmission can create solutions that maintain ecosystem health. Our **third case study** will further detail how ensuring the health of an individual species, mountain gorillas, can have positive feedback to the broader ecosystem.

CASE STUDY 3. GORILLA HEALTH PROMOTES ECOSYSTEM HEALTH

Because of their size, charisma, and relatedness to humans, gorillas garner much attention and are a valued wildlife resource in many range countries. Mountain gorillas (*Gorilla beringei beringei*) in particular are a unique conservation success story in which a suite of conservation measures, including individual and population level health care, have helped boost the species from critically endangered to endangered status (Robbins et al., 2011). While every species in an ecosystem may have an important role to play, maintaining the health of keystone species, such as gorillas, has broader benefits in supporting the overall health of the entire ecosystem.

Being primarily herbivores, gorillas feed on hundreds of different leaves, fruits, shoots, vines, and other vegetation with adults consuming up to 20 kg per day. Like other large herbivories, they play a vital role in seed dispersal, plant fertilization, along with vegetation regeneration and distribution (Watts, 1987; Haurez et al., 2018; Rogers et al., 2021) – all vital to sustaining healthy forests and plant communities. Their wider role in transforming the overall landscape and regulating important plant species is also recognized and is beginning to be studied further (Pinon & Butler, 2022; Kayitete et al., 2019). Healthy and diverse plant communities in turn are the primary energy producers in ecosystems helping to sustain animals while also providing ecosystem services such as capturing carbon and purifying air, water, and soil (Hector, 2022).

Gorillas have additional effects on helping maintain balanced ecosystems through interactions with other animals. For example, gorillas sometimes feed on ants and termites, which may help to control the populations of these insects and help prevent them from causing damage to the forest (Auger et al., 2023). Similarly, gorillas are often sympatric with other large mammals like forest elephants, buffalo, and chimpanzees with whom they may compete, coexist, exclude, or be excluded by depending on different factors such as habitat type or season (Head et al., 2012). Those complex relationships may not be well understood but any imbalance, such as fewer healthy gorillas, would likely affect those other species and have ecosystem altering consequences as seen when there are too many or too few other large animals like elephants or chimpanzees (Beaune, 2015; Feleha, 2018).

That same principle applies to disease ecology where protecting the health of gorillas has important implications across the ecosystem as infectious agents can affect multiple species. The most striking example may be human Ebola virus disease outbreaks that often originate from consumption or handling of diseased wildlife like gorillas and chimpanzees (Stephens et al., 2022). More subtle interspecies disease transmission interactions involving gorillas are also known or suspected and could have their transmission cycles disrupted (to the benefit of all species) by treating gorillas (Kalema-Zikusoka et al., 2018). Recent research also suggests that mountain gorillas are not exempt from the well-known risk of density-dependent increased disease transmission with their growing populations in a finite protected area, which further underscores the importance of actively safeguarding their health (Petrželková et al., 2022).

In addition to their clear keystone role in their environment, it should be noted that gorillas are exemplar umbrella species, whose targeted conservation affords protection to all the plants and animals in their ecosystem. This may even manifest beyond their environment where local human and livestock community healthcare has improved because of the concern for protecting gorillas (Kalema-Zikusoka et al., 2021; Mountain Gorilla Veterinary Project (MGVP, Inc.) & Wildlife Conservation Society (WCS), 2008). Ensuring their ongoing health, thereby, protects the health of their ecosystem both directly and indirectly.

Nature-Based Solutions

Until recently, much of the research on ecology and health has focused on characterizing problems versus finding solutions. As defined by the International Union for Conservation of Nature (IUCN), NbS are actions to protect, sustainably manage, and restore natural or modified ecosystems that address societal challenges effectively and adaptively, simultaneously providing human well-being and biodiversity benefits. NbS include sustainable planning, design, environmental management, public health measures, surveillance, and monitoring and engineering practices that weave natural features or processes into the built environment to promote adaptation and resilience (Federal Emergency Management Agency, 2023). These solutions are intended to combat climate change; prevent emerging infectious diseases; reduce flood risk; improve water quality; protect coastal property; restore and protect wetlands; stabilize shorelines; reduce urban heat; and add recreational space, to name a few benefits.

Like One Health, NbS encourage and facilitate collaborations across different sectors for the implementation of the Post-2020 Global Biodiversity Framework (GBF) and move toward achieving the Convention on Biological Diversity (CBD) 2050 Vision of "Living in harmony with nature." NbS are a potential pathway for synergies among several multilateral environmental agreements, including for biological diversity (Center for Biological Diversity), climate change (United Nations Framework Convention on Climate Change), disaster risk reduction (Sendai Framework), desertification (United Nations Convention to Combat Desertification), and the wider Sustainable Development Goals (SDGs) – and for mainstreaming nature conservation into sectoral decision-making processes (Geneva Environment Network, n.d.). NbS have become critical to face our triple planetary challenge: climate change, biodiversity loss, and pollution.

This approach aligns to One Health, and both approaches will need to be followed to influence global change in our human societies.

One of the aspects of NbS is to reconcile the blue economy (a sustainable use of ocean resources for economic growth, improved livelihoods and jobs, while protecting the health of marine and coastal ecosystems (Vierros & De Fontaubert, 2017)) with different sectors of activity and find the best solutions. For example, NbS should be considered for seafood production, especially with the negative environmental and social impacts associated with aquaculture. Promoting the One Health approach and best practices, including synergies with marine protection and coastal communities, should be essential to turn marine systems around within the next decade. Many of these solutions can solve food security and local economies and, in addition, we can solve issues of climate mitigation and social issues related to diversity, equity, and inclusion. For these frameworks to work, we need the transformation of science into policy.

We have the methodologies, infrastructure, and knowledge to change current environmental trends. We now know vastly more about bioassessment and ecological monitoring, and how to craft ecosystem management and policy responses that can be tuned to societal values, needs, and aspirations. However, the One Health paradigm that explicitly addresses links between oceans, wildlife, and human health still largely overlooks the web of linkages. The decline of ocean-based livelihoods provides a critical reminder of the linkages between oceans and human health that extend beyond infectious disease to the already growing rate of chronic illnesses associated with socioeconomic stress. We must keep in mind that the current threats to human and wildlife health may eventually seem trivial if we fail to grasp the fundamental contribution of oceans to health and global sustainability (Wilcox & Aguirre, 2004; Aguirre & Weber, 2012). Our **fourth case study** demonstrates the connections of ocean adjacent peoples and the communities of sea turtles they interact with.

Moving forward we need to bridge the barriers that drive disciplines of knowledge; empower individuals and institutions to act collaboratively; and take risks at all levels and engage in translational ecological health science. We need to embrace a transdisciplinary approach to solve the ecological and health problems we are facing. "Transdisciplinarity" or Transdisciplinary Research (TD) requires team members to share roles and systematically cross disciplinary boundaries. Role differentiation between disciplines is defined by the needs of the situation rather than by discipline-specific characteristics. Assessment, intervention, and evaluation are carried out jointly. The goal is knowledge and skills integration to allow more efficient and comprehensive assessment and interventions to address a societal problem. TD brings together academic experts, field practitioners, community members, research scientists, political leaders, students, and business owners among others to solve some of the pressing problems facing the world today and essential for the application of NbS. It is important to understand that TD and complexity or complex systems go together. Complex systems are not just typical systems, involving conventional systems thinking or approach, and these require community participation (Aguirre & Wilcox, 2008; Wilcox et al., 2019).

In addition, the idea of "resilience" is seen as an emerging property of natural resource management systems (e.g., forests, fisheries, wildlife, biodiversity), viewed as coupled human-natural systems to explain adaptive management and sustainability in One Health. Tying together humans and nature, the economics and ecology of biodiversity represent a critical breakthrough in our understanding of how and why ecosystems (as human-natural systems) exhibit complex adaptive behaviors and more explicitly justify adopting complexity-focused conceptual and methodological orientations.

Of promise is One Health's application to improving understanding of the complex synergies involved in infectious disease transmission at the wildlife, domestic animal, and human interface. The objectives of One Health correspond with the UN SDGs and as such understanding their areas of overlap will help strengthen a common operational framework as well as enable productive sharing of resources, funding pathways, and capacity building following NbS (Wilcox et al., 2019).

CASE STUDY 4. ONE HEALTH AS A POWERFUL MOTIVATOR TO REDUCE THE ILLEGAL TAKE AND TRADE OF SEA TURTLES IN MEXICO

Despite the bans and international laws on sea turtle fishing and derived products that have been in place for several decades, sea turtle consumption remains prevalent throughout Mexico. Between 75% and 100% of the sea turtle mortality rate for various sea turtle species is due to human consumption and more than 30,000 turtles are consumed each year in Baja California peninsula, Mexico (Mancini & Koch, 2009; Mancini et al., 2011). At least one-quarter of the residents in this area consume sea turtles monthly. A survey of dumps found hundreds of loggerhead turtle carcasses, the majority of which had been eaten. Most of these dead turtles were juveniles – a critical life stage for population health. This suggests that continued human consumption will devastate sea turtle populations in coming years. While consumption of sea turtle meat has decreased in some regions, turtle trafficking and illegal harvesting are continuing threats to environmental security of the region. But poaching is not only a threat to animal and human health; it is also a threat to ecological health (Aguirre & Nichols, 2020).

Sea turtle meat and other products are frequently contaminated with bacteria, parasites, organochlorine compounds, and heavy metals, including cadmium and mercury. These chemicals can cause kidney and liver failure, neurological disorders, and other forms of severe human disease when they accumulate in the body. The consumption of sea turtles may have adverse human health effects due to the presence of bacteria, parasites, biotoxins, and environmental contaminants. Reported consumption of sea turtles, or their eggs infected with zoonotic infectious agents, cause diarrhea, vomiting, and extreme dehydration. Consumption has resulted in hospitalization and death on several occasions. Moreover, outbreaks of gastroenteritis caused by *Salmonella chester* and *Vibrio mimicus* have been reported. Other zoonotic agents found in sea turtles include *Escherichia coli* and *Cryptosporidium* spp. Human fatalities and illness induced by poisoning from eating marine turtles have been reported throughout the Indo-Pacific region. To the best of our knowledge, no studies have been performed correlating sea turtle organic contaminant or heavy metal levels, consumption of meat or eggs, and risks to human health. Although contaminant levels vary according to sea turtle species and location, recent research suggests that turtles in Baja California may have elevated contaminant levels, and their consumption is cause for concern (Aguirre et al., 2006).

In 2017, we conducted a knowledge, attitudes, and practices (KAP) survey with a convenience sample of more than 200 residents of 14 fishing communities in northwestern Mexico. We asked participants about their nutritional and health status, dietary and risk behaviors, and perceptions of local ecological issues. We also collected hair samples that could be tested for heavy metals. About one-quarter of the participants reported consuming sea turtles in the past month. Laboratory tests showed a high prevalence of elevated levels of arsenic, lead, and mercury. Preliminary analyses show that people who eat sea turtle meat are more likely than their neighbors to have high levels of mercury in their hair. Conservation efforts may be more successful when they appeal to people's self-interest rather than merely focusing on ecological benefits. Concerns about toxins in sharks, tuna, and other types of deep-sea fish have reduced human consumption of some species. Sea turtle conservation efforts may benefit from awareness campaigns that emphasize the adverse health outcomes associated with eating turtle meat while continuing to affirm the economic benefits of healthy ecosystems (Aguirre et al., 2023).

The One Health case study presented herein provides a compelling argument for the reduction of human consumption of sea turtles, which in turn promotes a continuing culture of poaching and bribery to avoid legal penalties. Dissemination of this information may improve public, animal, and ecosystem health and simultaneously result in enhanced conservation of these endangered species. To date, attempts to halt sea turtle consumption have largely failed. We realize that we must appeal to people's self-interests to save the turtles.

Transdisciplinary research that draws on diverse fields such as ecology, epidemiology, toxicology, environmental law, and public policy provides a valuable foundation for solving One Health issues. Creative reframing of biodiversity concerns in countries around the globe will be necessary for promoting One Health in a time of accelerating environmental change.

CONCLUSION

Our world is experiencing many changes that put the benefits of healthy ecosystems at risk. Species around the globe and across diverse taxa are being lost to extinction at an accelerating rate, resulting in dramatically reduced biodiversity that some have called the "sixth mass extinction" event (Dirzo et al., 2014; Wake & Vredenburg, 2008; Barnosky et al., 2011). Concurrently, the ecosystems and habitats upon which species rely are being destroyed, converted away from their natural state, or degraded by pollutants and contaminants. As these forms of loss occur, rapid human population growth and development means there is an ever-growing demand for resources, both harvested and cultivated. At the intersection of these challenges is disease, which we are beginning to experience in new and, at times, disastrous ways.

The 2019 emergence of COVID-19 has been a particularly potent example of the devastation that can occur from a novel pathogen. The etiological agent of COVID-19 (severe acute respiratory syndrome coronavirus (SARS-CoV)-2) likely spilled over to humans from a wild animal, due to environmental degradation that put humans in close contact with reservoirs and/or traded wild animal hosts together at markets (Lytras et al., 2022; Jiang & Wang, 2022). In either case, high human densities and international travel networks allowed the rapid spread of SARS-CoV-2 and underscores a core lesson: in a time when the health of our global population is intimately connected, we are primed for the explosive spread of novel pathogens. We must be prepared for these events and work to prevent them. Recognizing the connections between people, animals, plants, environments, and disease, we can harness the knowledge of epidemiologists, health practitioners (both medical and veterinary), and ecologists to understand and attempt to mitigate the consequences of infection. Communication and collaboration that unite these areas are needed to generate powerful strategies and solutions for our changing world.

The challenge to achieve ecological health in One Health is in building bridges between theories, disciplines, methods, practices, and mandates that require breaking down disciplinary silos. Thinking ecologically about health and equally approaching health as a social-ecological issue is at the intersection of new work and emerging collaborations; and at the same time, we must simultaneously address issues of environmental justice, ecological integrity, and social, physical, and mental health and well-being (Aguirre et al. 2023) to ensure that the benefits of ecosystem health are experienced by all.

REFERENCES

Aguirre, A. A., Gardner, S. C., Marsh, J. C., Delgado, S. G., Limpus, C. J., & Nichols, W. J. (2006). Hazards associated with the consumption of sea turtle meat and eggs: A review for health care workers and the general public. *EcoHealth*, *3*(3), 141–153. https://doi.org/10.1007/s10393-006-0032-x

Aguirre, A.A., Fleming, L.C., Sandoval-Lugo, A.G., Leal-Moreno, R., Ley-Quiñónez, C.P., Zavala-Norzagaray, A.A., &Jacobsen, K.H. 2023. Conservation and health policy implications linked to the human consumption of sea turtles in northwestern Mexico. *World Medical & Health Policy*, 16, 57–69. doi: 10.1002/wmh3.596.

Aguirre, A. A., & Nichols, W. J. (2020). *The conservation mosaic approach to reduce corruption and the illicit sea turtle take and trade.* Targeting Natural Resource Corruption. https://www.worldwildlife.org/pages/tnrc-practice-note-the-conservation-mosaic-approach-to-reduce-corruption-and-the-illicit-sea-turtle-take-and-trade

Aguirre, A. A., Ostfeld, R. S., Tabor, G. M., House, C., & Pearl, M. C. (Eds.). (2002). *Conservation medicine: Ecological health in practice*. Oxford University Press.

Aguirre, A. A., & Weber, E. S. (2012). Living ocean, an evolving oxymoron. In R. A. Meyers (Ed.), *Encyclopedia of sustainability science and technology* (pp. 6178–6201). Springer New York. https://doi.org/10.1007/978-1-4419-0851-3_910

Aguirre, A. A, & Wilcox, B. A. (2008). EcoHealth: Envisioning and creating a truly global transdiscipline. *EcoHealth*, *5*(3), 238–239. https://doi.org/10.1007/s10393-008-0197-6

Aguirre, A.A., Basu, N., Kahn, L. H., Morin, X. K., Echaubard, P., Wilcox, B. A., & Beasley, V. R. (2019). Transdisciplinary and social-ecological health frameworks— Novel approaches to emerging parasitic and vector-borne diseases. *Parasite Epidemiology and Control*, 4, e00084. https://doi.org/10.1016/j.parepi.2019.e00084

Allan, B. F., Keesing, F., & Ostfeld, R. S. (2003). Effect of forest fragmentation on Lyme disease risk. *Conservation Biology*, *17*(1), 267–272. https://doi.org/10.1046/j.1523-1739.2003.01260.x

Auger, C., Cipolletta, C., Todd, A., Fuh, T., Sotto-Mayor, A., Pouydebat, E., & Masi, S. (2023). Feeling a bit peckish: Seasonal and opportunistic insectivory for wild gorillas. *American Journal of Biological Anthropology*, *182*(2), 210–223. https://doi.org/10.1002/ajpa.24811

Barnett, K. M., & Civitello, D. J. (2020). Ecological and evolutionary challenges for wildlife vaccination. *Trends in Parasitology*, *36*(12), 970–978. https://doi.org/10.1016/j.pt.2020.08.006

Barnosky, A. D., Matzke, N., Tomiya, S., Wogan, G. O. U., Swartz, B., Quental, T. B., Marshall, C., McGuire, J. L., Lindsey, E. L., Maguire, K. C., Mersey, B., & Ferrer, E. A. (2011). Has the Earth's sixth mass extinction already arrived? *Nature*, *471*(7336), 51–57. https://doi.org/10.1038/nature09678

Beaune, D. (2015). What would happen to the trees and lianas if apes disappeared? *Oryx*, *49*(3), 442–446. https://doi.org/10.1017/S0030605314000878

Ben-Horin, T., Burge, C., Bushek, D., Groner, M., Proestou, D., Huey, L., Bidegain, G., & Carnegie, R. (2018). Intensive oyster aquaculture can reduce disease impacts on sympatric wild oysters. *Aquaculture Environment Interactions*, *10*, 557–567. https://doi.org/10.3354/aei00290

Beschta, R. L., & Ripple, W. J. (2009). Large predators and trophic cascades in terrestrial ecosystems of the western United States. *Biological Conservation*, *142*(11), 2401–2414. https://doi.org/10.1016/j.biocon.2009.06.015

Bonds, M. H., Keenan, D. C., Rohani, P., & Sachs, J. D. (2010). Poverty trap formed by the ecology of infectious diseases. *Proceedings of the Royal Society B: Biological Sciences*, *277*(1685), 1185–1192. https://doi.org/10.1098/rspb.2009.1778

Bratman, G. N., Hamilton, J. P., & Daily, G. C. (2012). The impacts of nature experience on human cognitive function and mental health. *Annals of the New York Academy of Sciences*, *1249*(1), 118–136. https://doi.org/10.1111/j.1749-6632.2011.06400.x

Brownstein, J. S., Skelly, D. K., Holford, T. R., & Fish, D. (2005). Forest fragmentation predicts local scale heterogeneity of Lyme disease risk. *Oecologia*, *146*(3), 469–475. https://doi.org/10.1007/s00442-005-0251-9

Bustnes, J. O., Hanssen, S. A., Folstad, I., Erikstad, K. E., Hasselquist, D., & Skaare, J. U. (2004). Immune function and organochlorine pollutants in arctic breeding glaucous gulls. *Archives of Environmental Contamination and Toxicology*, *47*(4), 530–541. https://doi.org/10.1007/s00244-003-3203-6

Cabello, F. C., Godfrey, H. P., Tomova, A., Ivanova, L., Dölz, H., Millanao, A., & Buschmann, A. H. (2013). Antimicrobial use in aquaculture re-examined: Its relevance to antimicrobial resistance and to animal and human health. *Environmental Microbiology*, *15*(7), 1917–1942. https://doi.org/10.1111/1462-2920.12134

Canning, G., Camphor, H., & Schroder, B. (2019). Rabies outbreak in African Wild Dogs (*Lycaon pictus*) in the Tuli region, Botswana: Interventions and management mitigation recommendations. *Journal for Nature Conservation*, *48*, 71–76. https://doi.org/10.1016/j.jnc.2019.02.001

Cardinale, B. J., Duffy, J. E., Gonzalez, A., Hooper, D. U., Perrings, C., Venail, P., Narwani, A., Mace, G. M., Tilman, D., Wardle, D. A., Kinzig, A. P., Daily, G. C., Loreau, M., Grace, J. B., Larigauderie, A., Srivastava, D. S., & Naeem, S. (2012). Biodiversity loss and its impact on humanity. *Nature*, *486*(7401), 59–67. https://doi.org/10.1038/nature11148

Chaine, J. P., & Malek, E. A. (1983). Urinary schistosomiasis in the Sahelian region of the Senegal River Basin. *Tropical and Geographical Medicine*, *35*(3), 249–256.

Civitello, D. J., Cohen, J., Fatima, H., Halstead, N. T., Liriano, J., McMahon, T. A., Ortega, C. N., Sauer, E. L., Sehgal, T., Young, S., & Rohr, J. R. (2015). Biodiversity inhibits parasites: Broad evidence for the dilution effect. *Proceedings of the National Academy of Sciences*, *112*(28), 8667–8671. https://doi.org/10.1073/pnas.1506279112

Coonan, T. J., Schwemm, C. A., & Garcelon, D. K. (2010). *Decline and recovery of the island fox: A case study for population recovery*. Cambridge University Press.

Costanza, R., d'Arge, R., De Groot, R., Farber, S., Grasso, M., Hannon, B., Limburg, K., Naeem, S., O'Neill, R. V., Paruelo, J., Raskin, R. G., Sutton, P., & Van Den Belt, M. (1997). The value of the world's ecosystem services and natural capital. *Nature*, *387*(6630), 253–260. https://doi.org/10.1038/387253a0

Desforges, J.-P. W., Sonne, C., Levin, M., Siebert, U., De Guise, S., & Dietz, R. (2016). Immunotoxic effects of environmental pollutants in marine mammals. *Environment International*, *86*, 126–139. https://doi.org/10.1016/j.envint.2015.10.007

Dirzo, R., Young, H. S., Galetti, M., Ceballos, G., Isaac, N. J. B., & Collen, B. (2014). Defaunation in the Anthropocene. *Science*, *345*(6195), 401–406. https://doi.org/10.1126/science.1251817

Dobson, A., Cattadori, I., Holt, R. D., Ostfeld, R. S., Keesing, F., Krichbaum, K., Rohr, J. R., Perkins, S. E., & Hudson, P. J. (2006). Sacred cows and sympathetic squirrels: The importance of biological diversity to human health. *PLoS Medicine*, *3*(6), e231. https://doi.org/10.1371/journal.pmed.0030231

Dunn, R. R., Davies, T. J., Harris, N. C., & Gavin, M. C. (2010). Global drivers of human pathogen richness and prevalence. *Proceedings of the Royal Society B: Biological Sciences*, *277*(1694), 2587–2595. https://doi.org/10.1098/rspb.2010.0340

Dyrynda, E. A., Law, R. J., Dyrynda, P. E. J., Kelly, C. A., Pipe, R. K., Graham, K. L., & Ratcliffe, N. A. (1997). Modulations in cell-mediated immunity of *Mytilus Edulis* following the 'Sea Empress' oil spill. *Journal of the Marine Biological Association of the United Kingdom*, *77*(1), 281–284. https://doi.org/10.1017/S0025315400033993

Entrican, G., & Francis, M. J. (2022). Applications of platform technologies in veterinary vaccinology and the benefits for one health. *Vaccine*, *40*(20), 2833–2840. https://doi.org/10.1016/j.vaccine.2022.03.059

Fayer, R., Dubey, J. P., & Lindsay, D. S. (2004). Zoonotic protozoa: From land to sea. *Trends in Parasitology*, *20*(11), 531–536. https://doi.org/10.1016/j.pt.2004.08.008

Federal Emergency Management Agency. (2023, October 13). *Nature-based solutions*. https://www.fema.gov/emergency-managers/risk-management/climate-resilience/nature-based-solutions

Feleha, D. D. (2018). Impact of african elephant (*Loxodonta africana*) on flora and fauna community and options for reducing the undesirable ecological impacts. *Journal of Biology, Agriculture and Healthcare*, *8*. https://core.ac.uk/download/pdf/234662563.pdf

Foufopoulos, J., Wobeser, G. A., & McCallum, H. (2022). *Infectious disease ecology and conservation* (1st ed.). Oxford University Press. https://doi.org/10.1093/oso/9780199583508.001.0001

Galloway, T. S., & Depledge, M. H. (2001). Immunotoxicity in invertebrates: Measurement and ecotoxicological relevance. *Ecotoxicology*, *10*(1), 5–23. https://doi.org/10.1023/A:1008939520263

Geneva Environment Network. (n.d.). *Nature-based solutions*. UN Environment Programme. Retrieved June 5, 2024, from https://www.genevaenvironmentnetwork.org/resources/updates/nature-based-solutions/

Gilbert, M., Sulikhan, N., Uphyrkina, O., Goncharuk, M., Kerley, L., Castro, E. H., Reeve, R., Seimon, T., McAloose, D., Seryodkin, I. V., Naidenko, S. V., Davis, C. A., Wilkie, G. S., Vattipally, S. B., Adamson, W. E., Hinds, C., Thomson, E. C., Willett, B. J., Hosie, M. J., … Cleaveland, S. (2020). Distemper, extinction, and vaccination of the Amur tiger. *Proceedings of the National Academy of Sciences*, *117*(50), 31954–31962. https://doi.org/10.1073/pnas.2000153117

Haurez, B., Tagg, N., Petre, C., Brostaux, Y., Boubady, A., & Doucet, J. (2018). Seed dispersal effectiveness of the western lowland gorilla (*Gorilla gorilla gorilla*) in Gabon. *African Journal of Ecology*, *56*(2), 185–193. https://doi.org/10.1111/aje.12449

Head, J. S., Robbins, M. M., Mundry, R., Makaga, L., & Boesch, C. (2012). Remote video-camera traps measure habitat use and competitive exclusion among sympatric chimpanzee, gorilla and elephant in Loango National Park, Gabon. *Journal of Tropical Ecology*, *28*(6), 571–583. https://doi.org/10.1017/S0266467412000612

Hechinger, R. F., & Lafferty, K. D. (2005). Host diversity begets parasite diversity: Bird final hosts and trematodes in snail intermediate hosts. *Proceedings of the Royal Society B: Biological Sciences*, *272*(1567), 1059–1066. https://doi.org/10.1098/rspb.2005.3070

Hector, A. (2022). The importance of biodiverse plant communities for healthy soils. *Proceedings of the National Academy of Sciences*, *119*(1), e2119953118. https://doi.org/10.1073/pnas.2119953118

Hocking, D. J., & Babbitt, K. J. (2014). Amphibian contributions to ecosystem services. *Herpetological Conservation and Biology*, *9*(1), 1–17.

Hopkins, S. R., Lafferty, K. D., Wood, C. L., Olson, S. H., Buck, J. C., De Leo, G. A., Fiorella, K. J., Fornberg, J. L., Garchitorena, A., Jones, I. J., Kuris, A. M., Kwong, L. H., LeBoa, C., Leon, A. E., Lund, A. J., MacDonald, A. J., Metz, D. C. G., Nova, N., Peel, A. J., … Sokolow, S. H. (2022). Evidence gaps and diversity among potential win–win solutions for conservation and human infectious disease control. *The Lancet Planetary Health*, *6*(8), e694–e705. https://doi.org/10.1016/S2542-5196(22)00148-6

Hosseini, P. R., Mills, J. N., Prieur-Richard, A.-H., Ezenwa, V. O., Bailly, X., Rizzoli, A., Suzán, G., Vittecoq, M., García-Peña, G. E., Daszak, P., Guégan, J.-F., & Roche, B. (2017). Does the impact of biodiversity differ between emerging and endemic pathogens? The need to separate the concepts of hazard and risk. *Philosophical Transactions of the Royal Society B: Biological Sciences*, *372*(1722), 20160129. https://doi.org/10.1098/rstb.2016.0129

Hudson, P. J., Dobson, A. P., & Lafferty, K. D. (2006). Is a healthy ecosystem one that is rich in parasites? *Trends in Ecology & Evolution*, *21*(7), 381–385. https://doi.org/10.1016/j.tree.2006.04.007

Jackson, J. B. C. (2001). What was natural in the coastal oceans? *Proceedings of the National Academy of Sciences*, *98*(10), 5411–5418. https://doi.org/10.1073/pnas.091092898

Jiang, X., & Wang, R. (2022). Wildlife trade is likely the source of SARS-CoV-2. *Science*, *377*(6609), 925–926. https://doi.org/10.1126/science.add8384

Jones, K. E. Levy, M. A., Storeygard, A., Balk, D., Gittleman, J.L. & Daszak, P. (2008). Global trends in emerging infectious diseases. *Nature 451*, 990–993

Kalema-Zikusoka, G., Ngabirano, A., & Rubanga, S. (2021). Gorilla conservation and One Health. In S. C. Underkoffler & H. R. Adams (Eds.), *Wildlife biodiversity conservation* (pp. 371–381). Springer International Publishing. https://doi.org/10.1007/978-3-030-64682-0_13

Kalema-Zikusoka, G., Rubanga, S., Mutahunga, B., & Sadler, R. (2018). Prevention of cryptosporidium and giardia at the human/gorilla/livestock interface. *Frontiers in Public Health*, *6*, 364. https://doi.org/10.3389/fpubh.2018.00364

Kayitete, L., Van Der Hoek, Y., Nyirambangutse, B., & Derhé, M. A. (2019). Observations on regeneration of the keystone plant species *Hagenia abyssinica* in Volcanoes National Park, Rwanda. *African Journal of Ecology*, *57*(2), 274–278. https://doi.org/10.1111/aje.12585

Keesing, F., Belden, L. K., Daszak, P., Dobson, A., Harvell, C. D., Holt, R. D., Hudson, P., Jolles, A., Jones, K. E., Mitchell, C. E., Myers, S. S., Bogich, T., & Ostfeld, R. S. (2010). Impacts of biodiversity on the emergence and transmission of infectious diseases. *Nature*, *468*(7324), 647–652. https://doi.org/10.1038/nature09575

Keesing, F., Holt, R. D., & Ostfeld, R. S. (2006). Effects of species diversity on disease risk. *Ecology Letters*, *9*(4), 485–498. https://doi.org/10.1111/j.1461-0248.2006.00885.x

Keesing, F., & Ostfeld, R. S. (2021a). Dilution effects in disease ecology. *Ecology Letters*, *24*(11), 2490–2505. https://doi.org/10.1111/ele.13875

Keesing, F., & Ostfeld, R. S. (2021b). Impacts of biodiversity and biodiversity loss on zoonotic diseases. *Proceedings of the National Academy of Sciences*, *118*(17), e2023540118. https://doi.org/10.1073/pnas.2023540118

Knobel, D. L., Fooks, A. R., Brookes, S. M., Randall, D. A., Williams, S. D., Argaw, K., Shiferaw, F., Tallents, L. A., & Laurenson, M. K. (2008). Trapping and vaccination of endangered Ethiopian wolves to control an outbreak of rabies. *Journal of Applied Ecology*, *45*(1), 109–116. https://doi.org/10.1111/j.1365-2664.2007.01387.x

Kremen, C., & Ostfeld, R. S. (2005). A call to ecologists: Measuring, analyzing, and managing ecosystem services. *Frontiers in Ecology and the Environment*, *3*(10), 540–548. https://doi.org/10.1890/1540-9295(2005)003[0540:ACTEMA]2.0.CO;2

Kuris, A. M., Hechinger, R. F., Shaw, J. C., Whitney, K. L., Aguirre-Macedo, L., Boch, C. A., Dobson, A. P., Dunham, E. J., Fredensborg, B. L., Huspeni, T. C., Lorda, J., Mababa, L., Mancini, F. T., Mora, A. B., Pickering, M., Talhouk, N. L., Torchin, M. E., & Lafferty, K. D. (2008). Ecosystem energetic implications of parasite and free-living biomass in three estuaries. *Nature*, *454*(7203), 515–518. https://doi.org/10.1038/nature06970

Lafferty, K. D., Harvell, C. D., Conrad, J. M., Friedman, C. S., Kent, M. L., Kuris, A. M., Powell, E. N., Rondeau, D., & Saksida, S. M. (2015). Infectious diseases affect marine fisheries and aquaculture economics. *Annual Review of Marine Science*, *7*(1), 471–496. https://doi.org/10.1146/annurev-marine-010814-015646

Lafferty, K. D., & Kuris, A. M. (2002). Trophic strategies, animal diversity and body size. *Trends in Ecology & Evolution*, *17*(11), 507–513. https://doi.org/10.1016/S0169-5347(02)02615-0

Lipp, E. K., Huq, A., & Colwell, R. R. (2002). Effects of global climate on infectious disease: The cholera model. *Clinical Microbiology Reviews*, *15*(4), 757–770. https://doi.org/10.1128/CMR.15.4.757-770.2002

LoGiudice, K., Ostfeld, R. S., Schmidt, K. A., & Keesing, F. (2003). The ecology of infectious disease: Effects of host diversity and community composition on Lyme disease risk. *Proceedings of the National Academy of Sciences*, *100*(2), 567–571. https://doi.org/10.1073/pnas.0233733100

Lund, A. J., Rehkopf, D. H., Sokolow, S. H., Sam, M. M., Jouanard, N., Schacht, A.-M., Senghor, S., Fall, A., Riveau, G., De Leo, G. A., & Lopez-Carr, D. (2021). Land use impacts on parasitic infection: A cross-sectional epidemiological study on the role of irrigated agriculture in schistosome infection in a dammed landscape. *Infectious Diseases of Poverty*, *10*(1), 35. https://doi.org/10.1186/s40249-021-00816-5

Lytras, S., Hughes, J., Martin, D., Swanepoel, P., De Klerk, A., Lourens, R., Kosakovsky Pond, S. L., Xia, W., Jiang, X., & Robertson, D. L. (2022). Exploring the natural origins of SARS-CoV-2 in the light of recombination. *Genome Biology and Evolution*, *14*(2), evac018. https://doi.org/10.1093/gbe/evac018

MacArthur, R. H., & Wilson, E. O. (2001). *The theory of island biogeography*. Princeton University Press.

Maki, J., Guiot, A.-L., Aubert, M., Brochier, B., Cliquet, F., Hanlon, C. A., King, R., Oertli, E. H., Rupprecht, C. E., Schumacher, C., Slate, D., Yakobson, B., Wohlers, A., & Lankau, E. W. (2017). Oral vaccination of wildlife using a vaccinia–rabies-glycoprotein recombinant virus vaccine (RABORAL V-RG®): A global review. *Veterinary Research*, *48*(1), 57. https://doi.org/10.1186/s13567-017-0459-9

Mancini, A., & Koch, V. (2009). Sea turtle consumption and black market trade in Baja California Sur, Mexico. *Endangered Species Research*, *7*, 1–10. https://doi.org/10.3354/esr00165

Mancini, A., Senko, J., Borquez-Reyes, R., Póo, J. G., Seminoff, J. A., & Koch, V. (2011). To poach or not to poach an endangered species: Elucidating the economic and social drivers behind illegal sea turtle hunting in Baja California Sur, Mexico. *Human Ecology*, *39*(6), 743–756. https://doi.org/10.1007/s10745-011-9425-8

Marino, J., Sillero-Zubiri, C., Deressa, A., Bedin, E., Bitewa, A., Lema, F., Rskay, G., Banyard, A., & Fooks, A. R. (2017). Rabies and distemper outbreaks in smallest ethiopian wolf population. *Emerging Infectious Diseases*, *23*(12), 2102–2104. https://doi.org/10.3201/eid2312.170893

Messenger, A. M., Barnes, A. N., & Gray, G. C. (2014). Reverse zoonotic disease transmission (Zooanthroponosis): A systematic review of Seldom-documented human biological threats to animals. *PLoS One*, *9*(2), e89055. https://doi.org/10.1371/journal.pone.0089055

Miller, B., Ceballos, G., & Reading, R. (1994). The prairie dog and biotic diversity. *Conservation Biology*, *8*(3), 677–681. https://doi.org/10.1046/j.1523-1739.1994.08030677.x

Miller, M. A., Gardner, I. A., Kreuder, C., Paradies, D. M., Worcester, K. R., Jessup, D. A., Dodd, E., Harris, M. D., Ames, J. A., Packham, A. E., & Conrad, P. A. (2002). Coastal freshwater runoff is a risk factor for *Toxoplasma gondii* infection of southern sea otters (*Enhydra lutris nereis*). *International Journal for Parasitology*, *32*(8), 997–1006. https://doi.org/10.1016/S0020-7519(02)00069-3

Mills, J. N. (2005). Regulation of rodent-borne viruses in the natural host: Implications for human disease. In C. J. Peters & C. H. Calisher (Eds.), *Infectious diseases from nature: Mechanisms of viral emergence and persistence* (pp. 45–57). Springer-Verlag. https://doi.org/10.1007/3-211-29981-5_5

Mountain Gorilla Veterinary Project (MGVP, Inc.) & Wildlife Conservation Society (WCS). (2008). Conservation medicine for gorilla conservation. In T. S. Stoinski, H. D. Steklis, & P. T. Mehlman (Eds.), *Conservation in the 21st century: Gorillas as a case study* (pp. 57–78). Springer US. https://doi.org/10.1007/978-0-387-70721-1_2

Murray, C. J., Ikuta, K. S., Sharara, F., Swetschinski, L., Robles Aguilar, G., Gray, A., Han, C., Bisignano, C., Rao, P., Wool, E., Johnson, S. C., Browne, A. J., Chipeta, M. G., Fell, F., Hackett, S., Haines-Woodhouse, G., Kashef Hamadani, B. H., Kumaran, E. A. P., McManigal, B., … Naghavi, M. (2022). Global burden of bacterial antimicrobial resistance in 2019: A systematic analysis. *The Lancet*, S0140673621027240. https://doi.org/10.1016/S0140-6736(21)02724-0

Nupp, T. E., & Swihart, R. K. (2000). Landscape-level correlates of small-mammal assemblages in forest fragments of farmland. *Journal of Mammalogy*, *81*(2), 512–526. https://doi.org/10.1644/1545-1542(2000)081<0512:LLCOSM>2.0.CO;2

Orrock, J. L. & Allan, B. F. (2008). Sin Nombre Virus Infection in Deer Mice, Channel Islands, California, *14*.

Orrock, J. L., Allan, B. F., & Drost, C. A. (2011). Biogeographic and ecological regulation of disease: Prevalence of Sin Nombre virus in island mice is related to island area, precipitation, and predator richness. *The American Naturalist*, *177*(5), 691–697. https://doi.org/10.1086/659632

Ostfeld, R. S. (2011). *Lyme disease: The ecology of a complex system*. Oxford University Press.

Ostfeld, R. S., & Holt, R. D. (2004). Are predators good for your health? Evaluating evidence for top-down regulation of zoonotic disease reservoirs. *Frontiers in Ecology and the Environment*, *2*(1), 13–20. https://doi.org/10.1890/1540-9295(2004)002[0013:APGFYH]2.0.CO;2

Ostfeld, R. S., & Keesing, F. (2000). Biodiversity series: The function of biodiversity in the ecology of vector-borne zoonotic diseases. *Canadian Journal of Zoology*, *78*(12), 2061–2078. https://doi.org/10.1139/z00-172

Ostfeld, R. S., & Keesing, F. (2012). Effects of host diversity on infectious disease. *Annual Review of Ecology, Evolution, and Systematics*, *43*(1), 157–182. https://doi.org/10.1146/annurev-ecolsys-102710-145022

Petrželková, K. J., Samaš, P., Romportl, D., Uwamahoro, C., Červená, B., Pafčo, B., Prokopová, T., Cameira, R., Granjon, A. C., Shapiro, A., Bahizi, M., Nziza, J., Noheri, J. B., Syaluha, E. K., Eckardt, W., Ndagijimana, F., Šlapeta, J., Modrý, D., Gilardi, K., … Cranfield, M. (2022). Ecological drivers of helminth infection patterns in the Virunga Massif mountain gorilla population. *International Journal for Parasitology: Parasites and Wildlife*, *17*, 174–184. https://doi.org/10.1016/j.ijppaw.2022.01.007

Pietrock, M., & Marcogliese, D. J. (2003). Free-living endohelminth stages: At the mercy of environmental conditions. *Trends in Parasitology*, *19*(7), 293–299. https://doi.org/10.1016/S1471-4922(03)00117-X

Pinon, A., & Butler, D. R. (2022). An analysis of gorillas as zoogeomorphic agents. *Revista de Geomorfologie*, *24*(2), 5–15.

Plowright, R. K., Foley, P., Field, H. E., Dobson, A. P., Foley, J. E., Eby, P., & Daszak, P. (2011). Urban habituation, ecological connectivity and epidemic dampening: The emergence of Hendra virus from flying foxes (*Pteropus* spp.). *Proceedings of the Royal Society B: Biological Sciences*, *278*(1725), 3703–3712. https://doi.org/10.1098/rspb.2011.0522

Plowright, R. K., Reaser, J. K., Locke, H., Woodley, S. J., Patz, J. A., Becker, D. J., Oppler, G., Hudson, P. J., & Tabor, G. M. (2021). Land use-induced spillover: A call to action to safeguard environmental, animal, and human health. *The Lancet Planetary Health*, *5*(4), e237–e245. https://doi.org/10.1016/S2542-5196(21)00031-0

Randolph, S. E., & Dobson, A. D. M. (2012). Pangloss revisited: A critique of the dilution effect and the biodiversity-buffers-disease paradigm. *Parasitology*, *139*(7), 847–863. https://doi.org/10.1017/S0031182012000200

Rapport, D. J., Regier, H. A., & Hutchinson, T. C. (1985). Ecosystem behavior under stress. *The American Naturalist*, *125*(5), 617–640. https://doi.org/10.1086/284368

Robbins, M. M., Gray, M., Fawcett, K. A., Nutter, F. B., Uwingeli, P., Mburanumwe, I., Kagoda, E., Basabose, A., Stoinski, T. S., Cranfield, M. R., Byamukama, J., Spelman, L. H., & Robbins, A. M. (2011). Extreme conservation leads to recovery of the Virunga mountain gorillas. *PLoS One*, *6*(6), e19788. https://doi.org/10.1371/journal.pone.0019788

Robertson, L. J. (2007). The potential for marine bivalve shellfish to act as transmission vehicles for outbreaks of protozoan infections in humans: A review. *International Journal of Food Microbiology*, *120*(3), 201–216. https://doi.org/10.1016/j.ijfoodmicro.2007.07.058

Rodgers, C., Arzul, I., Carrasco, N., & Furones Nozal, D. (2019). A literature review as an aid to identify strategies for mitigating ostreid herpesvirus 1 in *Crassostrea gigas* hatchery and nursery systems. *Reviews in Aquaculture*, *11*(3), 565–585. https://doi.org/10.1111/raq.12246

Roeder, P. L. (2011). Rinderpest: The end of cattle plague. *Preventive Veterinary Medicine*, *102*(2), 98–106. https://doi.org/10.1016/j.prevetmed.2011.04.004

Rogers, H. S., Donoso, I., Traveset, A., & Fricke, E. C. (2021). Cascading impacts of seed disperser loss on plant communities and ecosystems. *Annual Review of Ecology, Evolution, and Systematics*, *52*(1), 641–666. https://doi.org/10.1146/annurev-ecolsys-012221-111742

Rosenblatt, D. L., Heske, E. J., Nelson, S. L., Barber, D. M., Miller, M. A., & MacAllister, B. (1999). Forest fragments in east-central Illinois: Islands or habitat patches for mammals? *The American Midland Naturalist*, *141*(1), 115–123. https://doi.org/10.1674/0003-0031(1999)141[0115:FFIECI]2.0.CO;2

Salkeld, D. J. (2017). Vaccines for conservation: Plague, prairie dogs & black-footed ferrets as a case study. *EcoHealth*, *14*(3), 432–437. https://doi.org/10.1007/s10393-017-1273-6

Savaya Alkalay, A., Rosen, O., Sokolow, S. H., Faye, Y. P. W., Faye, D. S., Aflalo, E. D., Jouanard, N., Zilberg, D., Huttinger, E., & Sagi, A. (2014). The prawn *Macrobrachium vollenhovenii* in the Senegal River Basin: Towards sustainable restocking of all-male populations for biological control of schistosomiasis. *PLoS Neglected Tropical Diseases*, *8*(8), e3060. https://doi.org/10.1371/journal.pntd.0003060

Scheele, B. C., Pasmans, F., Skerratt, L. F., Berger, L., Martel, A., Beukema, W., Acevedo, A. A., Burrowes, P. A., Carvalho, T., Catenazzi, A., De La Riva, I., Fisher, M. C., Flechas, S. V., Foster, C. N., Frías-Álvarez, P., Garner, T. W. J., Gratwicke, B.,

Guayasamin, J. M., Hirschfeld, M., … Canessa, S. (2019). Amphibian fungal panzootic causes catastrophic and ongoing loss of biodiversity. *Science*, *363*(6434), 1459–1463. https://doi.org/10.1126/science.aav0379

Schwemm, C. A., & Coonan, T. J. (2001). *Status and ecology of deer mice (Peromyscus maniculatus subsp.) on Anacapa, Santa Barbara, and San Miguel Islands, California: Summary of monitoring 1992–2000*. United States Department of Interior, National Park Service, Channel Islands National Park.

Shapiro, K., Conrad, P. A., Mazet, J. A. K., Wallender, W. W., Miller, W. A., & Largier, J. L. (2010). Effect of estuarine wetland degradation on transport of *Toxoplasma gondii* surrogates from land to sea. *Applied and Environmental Microbiology*, *76*(20), 6821–6828. https://doi.org/10.1128/AEM.01435-10

Shapiro, K., Miller, M., & Mazet, J. (2012). Temporal association between land-based runoff events and California sea otter (*Enhydra lutris nereis*) protozoal mortalities. *Journal of Wildlife Diseases*, *48*(2), 394–404. https://doi.org/10.7589/0090-3558-48.2.394

Sharpe, P. B., & Garcelon, D. K. (2005). Restoring and monitoring bald eagles in southern California: The legacy of DDT. *Proceedings of the Sixth California Islands Symposium*, Ventura, CA. https://sbbotanicgarden.org/wp-content/uploads/2022/09/Sharpe-Garcelon-2005-bald-eagle-restor-monitor.pdf

Sokolow, S. H., Huttinger, E., Jouanard, N., Hsieh, M. H., Lafferty, K. D., Kuris, A. M., Riveau, G., Senghor, S., Thiam, C., N'Diaye, A., Faye, D. S., & De Leo, G. A. (2015). Reduced transmission of human schistosomiasis after restoration of a native river prawn that preys on the snail intermediate host. *Proceedings of the National Academy of Sciences*, *112*(31), 9650–9655. https://doi.org/10.1073/pnas.1502651112

Sokolow, S. H., Jones, I. J., Jocque, M., La, D., Cords, O., Knight, A., Lund, A., Wood, C. L., Lafferty, K. D., Hoover, C. M., Collender, P. A., Remais, J. V., Lopez-Carr, D., Fisk, J., Kuris, A. M., & De Leo, G. A. (2017). Nearly 400 million people are at higher risk of schistosomiasis because dams block the migration of snail-eating river prawns. *Philosophical Transactions of the Royal Society B: Biological Sciences*, *372*(1722), 20160127. https://doi.org/10.1098/rstb.2016.0127

Sow, S., De Vlas, S. J., Engels, D., & Gryseels, B. (2002). Water-related disease patterns before and after the construction of the Diama dam in northern Senegal. *Annals of Tropical Medicine & Parasitology*, *96*(6), 575–586. https://doi.org/10.1179/000349802125001636

Stephens, P. R., Sundaram, M., Ferreira, S., Gottdenker, N., Nipa, K. F., Schatz, A. M., Schmidt, J. P., & Drake, J. M. (2022). Drivers of African filovirus (Ebola and Marburg) outbreaks. *Vector-Borne and Zoonotic Diseases*, *22*(9), 478–490. https://doi.org/10.1089/vbz.2022.0020

Stewart, T. E., & Schnitzer, S. A. (2017). Blurred lines between competition and parasitism. *Biotropica*, *49*(4), 433–438. https://doi.org/10.1111/btp.12444

Strauss, A. T., Civitello, D. J., Cáceres, C. E., & Hall, S. R. (2015). Success, failure and ambiguity of the dilution effect among competitors. *Ecology Letters*, *18*(9), 916–926. https://doi.org/10.1111/ele.12468

The World Bank. (2010). *People, pathogens, and our planet: Volume one: Towards a one health approach for controlling Zoonotic diseases (English)*. World Bank Group. https://documents.worldbank.org/curated/en/214701468338937565/Volume-one-towards-a-one-health-approach-for-controlling-zoonotic-diseases

Vaughn, C. C., & Hoellein, T. J. (2018). Bivalve impacts in freshwater and marine ecosystems. *Annual Review of Ecology, Evolution, and Systematics*, *49*(1), 183–208. https://doi.org/10.1146/annurev-ecolsys-110617-062703

Viana, M., Mancy, R., Biek, R., Cleaveland, S., Cross, P. C., Lloyd-Smith, J. O., & Haydon, D. T. (2014). Assembling evidence for identifying reservoirs of infection. *Trends in Ecology & Evolution*, *29*(5), 270–279. https://doi.org/10.1016/j.tree.2014.03.002

Vierros, M., & De Fontaubert, C. (2017). *The potential of the blue economy: Increasing long-term benefits of the sustainable use of marine resources for small island developing states and coastal least developed countries (English)* [Working paper]. World Bank Group. https://documents.worldbank.org/curated/en/523151496389684076/The-potential-of-the-blue-economy-increasing-long-term-benefits-of-the-sustainable-use-of-marine-resources-for-small-island-developing-states-and-coastal-least-developed-countries

Wake, D. B., & Vredenburg, V. T. (2008). Are we in the midst of the sixth mass extinction? A view from the world of amphibians. *Proceedings of the National Academy of Sciences*, *105*(supplement_1), 11466–11473. https://doi.org/10.1073/pnas.0801921105

Watts, D. P. (1987). Effects of mountain gorilla foraging activities on the productivity of their food plant species. *African Journal of Ecology*, *25*(3), 155–163. https://doi.org/10.1111/j.1365-2028.1987.tb01102.x

Wilcox, B. A., Aguirre, A. A., De Paula, N., Siriaroonrat, B., & Echaubard, P. (2019). Operationalizing One Health employing social-ecological systems theory: Lessons from the greater Mekong sub-region. *Frontiers in Public Health*, *7*, 85. https://doi.org/10.3389/fpubh.2019.00085

Wilcox, Bruce A., & Aguirre, A. A. (2004). One Ocean, One Health. *EcoHealth*, *1*(3). https://doi.org/10.1007/s10393-004-0122-6

Williams, S. C., Ward, J. S., Worthley, T. E., & Stafford, K. C. (2009). Managing Japanese barberry (Ranunculales: Berberidaceae) infestations reduces blacklegged tick (Acari: Ixodidae) abundance and infection prevalence with *Borrelia burgdorferi* (Spirochaetales: Spirochaetaceae). *Environmental Entomology*, *38*(4), 977–984. https://doi.org/10.1603/022.038.0404

Wobeser, G. (2002). Disease management strategies for wildlife. *Revue Scientifique et Technique de l'OIE*, *21*(1), 159–178. https://doi.org/10.20506/rst.21.1.1326

Wolfe, N. D., Dunavan, C. P., & Diamond, J. (2007). Origins of major human infectious diseases. *Nature*, *447*(7142), 279–283. https://doi.org/10.1038/nature05775

Wood, C. L., & Johnson, P. T. (2015). A world without parasites: Exploring the hidden ecology of infection. *Frontiers in Ecology and the Environment*, *13*(8), 425–434. https://doi.org/10.1890/140368

Wood, C. L., & Lafferty, K. D. (2013). Biodiversity and disease: A synthesis of ecological perspectives on Lyme disease transmission. *Trends in Ecology & Evolution*, *28*(4), 239–247. https://doi.org/10.1016/j.tree.2012.10.011

Wood, C. L., McInturff, A., Young, H. S., Kim, D., & Lafferty, K. D. (2017). Human infectious disease burdens decrease with urbanization but not with biodiversity. *Philosophical Transactions of the Royal Society B: Biological Sciences*, *372*(1722), 20160122. https://doi.org/10.1098/rstb.2016.0122

Yabsley, M. J. (2019). Vaccination of wildlife species. In S. M. Hernandez, H. W. Barron, E. A. Miller, R. F. Aguilar, & M. J. Yabsley (Eds.), *Medical management of wildlife species* (1st ed., pp. 85–96). Wiley. https://doi.org/10.1002/9781119036708.ch7

10 Economics of One Health

Susan Horton and Jonathan Rushton

INTRODUCTION

Zoonotic diseases (diseases which affect both humans and animals and can be transmitted between them) were making headlines well before the coronavirus disease (COVID-19) pandemic. One earlier estimate is that zoonotic diseases account for two-thirds of pathogens affecting humans, and three-quarters of the emerging and re-emerging ones (Coker et al., 2011). These diseases can have enormous economic impacts: Baum et al. (2017) cite previous estimates of the cost of severe acute respiratory syndrome (SARS) in 2003 of $54 bn; the loss of 12% of GDP to Guinea, Liberia, and Sierra Leone over the period 2014–2016 due to Ebola (Baum et al., 2017); and the loss of hundreds of millions of dollars due to the effect of Nipah virus in 1998–1999 affecting swine in Southeast Asia. The World Bank (2012) estimates that over the period 1997 to 2009, there were losses of $80 bn in total due to six outbreaks of zoonoses (Nipah virus in Malaysia, West Nile virus in the United States, SARS in Asia and Canada, highly pathogenic avian influenza in Asia and Europe, bovine spongiform encephalopathy (BSE) in the United States and UK, and Rift Valley Fever in Tanzania, Kenya, and Somalia) (The World Bank, 2012).

All these estimates pale by comparison to the estimated losses attributable to COVID-19, which caused the greatest economic declines since the Great Depression. World GDP fell by 4.3% ($2.9 trillion) in 2020 (by 9.6% in India, 9.5% in the UK, and 8.0% in Latin America and the Caribbean) (Department of Economic and Social Affairs, United Nations, 2021), there were huge healthcare expenditure costs, and an estimated 14.9 million excess deaths as of the end of 2021 (World Health Organization, 2022).

The evidence presented points to the importance of zoonotic diseases, particularly those that are capable of spilling over into human populations and being maintained and spread independently of the original animal source. The range of the estimates also indicates a need to understand how these large numbers have been derived, as to date there has been little standardization in this area. Therefore, economics, and in particular the economics of One Health, is important to understanding the potential consequences of zoonoses and the ongoing zoonotic threats. Of course, One Health economics is not only about zoonotic diseases, and conversely not all zoonotic diseases require a full One Health economic analysis. This chapter aims to provide an introduction intelligible to

DOI: 10.1201/9781003232223-10

a broad audience, and to briefly outline where One Health economics can and should be used.

Economic analysis has proven very useful in the study of global health. For human health, economic analysis is widely used as a tool for assisting in priority setting (e.g., Jamison et al., 2018) and in making budget estimates for health interventions (e.g., Stenberg et al., 2017). Economic analysis has grown in importance in national health priority setting (e.g., the National Institute for Clinical Excellence – NICE – in the UK, which was set up in 1999). Economic analysis has also played an important role in the global health interventions associated with the Millennium Development Goals and now the Sustainable Development Goals. The World Health Organization (WHO) developed a set of economic tools initially called WHO-CHOICE (World Health Organization, 2014); and international organizations such as the Global Fund, Gavi, and UNITAID make extensive use of economics methods. Economic analysis can also inform broader policy analysis of human health interventions, with impacts on equity being particularly salient (for example, by socio-economic status, gender, rural/urban location, and across countries, discussed further below).

Similar developments have been made in parallel in animal health. Economic analysis initially became important for examining national public investments in animal health in the 1960s (Rushton et al., 2007), reducing the risks associated with the intensification of livestock production. As high-income countries succeeded in exerting more control over outbreaks of domestic origin, they became more concerned about re-introducing infection through transboundary diseases, and hence analyzing the international policy environment (Rushton et al., 2007). Despite these advances animal health is still behind the human health field on burden estimations and has only just begun attempts at looking at the burden of animal diseases (Rushton et al., 2018; Rushton, Huntington, et al., 2021).

Emerging disease threats have led to a need to integrate the study of the economics of human and animal health and the link to the environment. One example is food-borne threats to human health associated with intensification of livestock farming, such as BSE, salmonella, campylobacteriosis, *E. coli*, and listeria. Another example is zoonotic diseases where both humans and animals are susceptible to pathogens, and there are possibilities of transmission from animals to humans and vice versa. A third example is the growing risk of antimicrobial resistance (AMR) based on the use of the same antibiotics in both animals and humans.

It is already challenging enough to combine an understanding of health and of economics to analyze health economics; so, bringing together economics of both human and animal health into economics of One Health involves even greater complexity. Some work and thought have been applied to bring burden estimates together from human and animal health (Torgerson et al., 2018; Shaw et al., 2017). What is essential is to figure out those cases where economics of One Health is required, and those cases where either economics of human health or economics of animal health will suffice, albeit acknowledging the broader context.

We first present some basic concepts of the economics of human health, and then the same for animal health, focusing on key methodologies (costing, cost-benefit analysis, and cost-effectiveness analysis), and discuss how these may be combined in a One Health approach. We also mention how economics may be used to examine equity aspects of health issues, and how economics may be incorporated into more complex problems requiring a One Health approach.

We then present case studies to exemplify the use of some of these techniques in the One Health approach. In the final discussion, we briefly outline how economics can contribute to One Health analysis and some of the limitations, and how the current food system is leading to increased risks. While One Health is generally concerned with the environment, due to space considerations, we do not cover environmental economics.

The chapter is directed to a diverse audience of those from various backgrounds with a common interest in One Health. Those familiar with methods of economics of human health and those familiar with economics of animal health may wish to skip the corresponding sections on methods. Non-economists interested in pursuing more technical details on economic methods for One Health are referred to the chapter by Canali et al. (2018) in the *Network for the Evaluation of One Health* handbook.

BASIC ECONOMIC TOOLS

Analyzing Costs

Economics seeks to examine how scarce resources can best be directed to address desires and needs, and health is an area where needs are considerable. The country (excluding one very small island state) spending the most on health as a share of national income (the United States, 16.77% of Gross Domestic Product in 2019) (The World Bank, 2023)) did not even make it into the top 20 list of countries ranked according to life expectancy at birth. Hence, allocation of health spending matters.

Economists have various precise definitions of **costs**, which are not always intuitively obvious. **Opportunity cost** is the value of a resource in its next best alternative use. Community health volunteers are not a free good: their time could be used elsewhere and receive a salary. The time taken by a farmer to bring their animals for a vaccination could be spent managing their crops or doing some off-farm work. Time is never free, hence, to quote an old adage, there is no such thing as a free lunch. A related concept is the distinction between economic costs and financial costs, the former taking into account the opportunity costs of resources and services, whereas the latter reflects the market price of these items.

Costs and benefits can occur in different time periods, yet, there is a difference in perceived value of a current and future cost or benefit. Future benefits are less valuable than current benefits because individuals prefer to receive benefits immediately, and costs also have a different value with people preferring to postpone paying for purchases (or working) into the future. Therefore, economists

discount future costs and benefits in order to determine present values and allow comparisons across time.

Economists always ask the question about costs and benefits, and they will want to know the **incremental or marginal change** in these costs and benefits in order to understand if an optimal point of allocation of a resource or service has been reached. This optimization (or profit maximizing) point is important, yet so is answering the question who benefits? When analyzing health, the **perspective** of an individual person, the health sector of the economy, the public sector as a whole, and society as a whole are all different, and encompass an increasingly broad accounting of both costs and benefits.

Economists distinguish carefully between **private** and **public** costs and benefits. Individuals are assumed to be typically motivated by private costs and benefits in economic models (profit or utility maximizers). This causes problems where there are what economists term **externalities**. For example, if individuals with an infectious disease do not self-isolate because it is expensive and inconvenient, they may transmit the infection to other people (a negative externality). Hence public health policymakers may mandate quarantine. Similarly, pollution due to a firm or farm's activities may not harm their own business, but can have negative impacts on others, and environmental policymakers may choose to tax pollution, or to impose environmental restrictions and to penalize those who do not comply.

Analyzing Benefits: Human Health

Just as costs of health interventions have to be calculated using an agreed-upon set of rules, of which some key ones are mentioned above, benefits of health interventions equally have to be put into a common metric. Some human health interventions reduce disability or improve individuals' feeling of well-being, while others may avert mortality at a range of different ages. Saving the life of an infant is not the same as saving the life of a senior. Although evaluating health outcomes in a common metric is not easy and may involve making value judgments, it is essential in prioritizing health spending.

There are two main approaches which have been used to evaluate benefits of human health improvements. One is to use the economists' concept of **utility** which relies on trying to understand people's preferences, and the other is to try to value all benefits in monetary terms. The former approach, when combined with information on costs, leads to **cost-utility analysis**, of which **cost-effectiveness analysis** is one example. The latter approach can lead to **cost-benefit analysis** where the outcome(s) can be converted into monetary benefit streams.

A commonly used measure of utility of health is the **QALY** (quality-adjusted life year), and has been used in high-income countries. A similar metric which has been widely used for cost-effectiveness analysis in global health is the **DALY** (disability-adjusted life year). Both measures seek to be able to incorporate comparisons between interventions which save lives, and those which improve the quality of life.

The QALY relies on survey information about self-reported quality of life. Individuals are asked to evaluate and rank different health states. This may sound difficult to do, but individuals are often put in the situation that they have to choose. An individual suffering from a heart condition may be offered by the doctor the opportunity to undergo surgery in which case the outcome might be an 80% chance of surviving with improved mobility and quality of life, but a 20% chance of not surviving the surgery. Forgoing the surgery might entail surviving, but with a reduced quality of life (being bedridden, or with very restricted mobility).

Interventions with a lower cost per QALY saved or DALY averted are preferable to those with a higher cost. Countries have developed rules of thumb (threshold values) based on their current health budgets in order to evaluate newly proposed health interventions. Those interventions costing less than the threshold value are more likely to be adopted. The United States has used a threshold value of $50,000 over an extended time period (Neumann et al., 2014), and NICE in the UK has a similar but lower threshold, reflecting the lower per capita income of this country. The WHO has in the past provided guidance on thresholds for low- and middle-income countries based on their much lower health budgets (Bertram et al., 2016).

Obviously, there are value judgments involved in rating quality of life. For example, how do we evaluate the health preferences of those who cannot readily speak for themselves, such as small children, or individuals living with mental illness or reduced cognitive ability? Another topic for debate is the practice of discounting life years saved further in the future, which has been more recently challenged.

Using cost-effectiveness (or cost-utility) methods obviates putting a monetary value on lives saved, which some health economists find difficult. It makes it possible to prioritize among health interventions and to undertake comprehensive comparisons across many diverse areas of health (e.g., the Disease Control Priorities series (Jamison et al., 2018)). One disadvantage is that although it allows prioritization within the health sector, it does not allow comparisons with investments in other sectors of the economy.

Private businesses use profit as a tool to decide on which activities to undertake. Cost-benefit analysis is a similar tool widely used to evaluate government spending priorities and has been used in the United States for close to a century. In this methodology, benefits have to be measured in monetary terms, in order to compare them with costs. One valuation method is the human capital approach, where health interventions which improve the ability to work (as measured by higher wages), or prevent premature mortality, can be evaluated. Another alternative is to use the Value of Statistical Life (VSL), which measures peoples' willingness to pay for lower mortality risks. The VSL method has been used widely for public policy decisions on environmental projects and on road safety. This method tends to lead to estimates which are considerably higher than those generated by the human capital approach. For more detailed discussion of VSL see

Robinson et al. (2019). For a more detailed introduction to economic methods as used in One Health, see Canali et al. (2018).

Analyzing Benefits: Animal Health

A majority of the domesticated animals are kept for producing meat, milk, eggs, wool, traction power, and manure. The role in most societies is economic with variations in terms of culture and social norms, so for example cattle in India are used to produce milk, provide traction power to plow fields and pull carts, and produce manure, yet in many parts of the country there is a religious and legislative ban on consuming beef. The importance of these animals in the agricultural system is such that it would not be sensible to eat them; it would be like a farm in the Midwest of the United States dismantling and selling their tractor components at the end of the harvest. Despite such cultural issues, these animals have an economic role and are often referred to as livestock which, as the word implies, are live animals that have stock value. In addition, the products and services these livestock generate have markets with prices. Some other domesticated animals (e.g., dogs) have a role as companion animals, and it is more difficult to assign a monetary value to their services.

It is also difficult to assign a monetary value to wild animals. A good discussion and exploration of methods similar to QALY and DALY can be found in Teng et al. (2018). While in some countries it is possible to make estimates of the tourism revenues generated associated with wild animals, this probably only captures a small share of the benefits received by humans. Capturing the benefits that are not reflected in market prices requires methods such as willingness to pay and contingent valuation (Fox-Rushby & Cairns, 2009). We focus here on livestock, where most of the economic analysis is concentrated and where markets exist and give a good indication of the value of both the animals and the products they produce. We do recognize that such a focused analysis can ignore wider implications on public health and the environment, and for a true economic analysis these aspects need to be included.

The economic value of livestock and livestock products allows economists to look at animal health in a way where the ethical and moral issues of valuing an animal life or an animal product are less complicated. The value can be obtained by what people are willing to pay for the animals and their products, and hence if there is a death due to a disease or a loss of production due to a disease being present, it is possible to convert the change in animal numbers and production into a benefit stream as a monetary value. The estimation of the burden of animal disease is generated as a monetary value (see Knight-Jones & Rushton, 2013 for an estimation of the foot and mouth disease burdens). When assessing an intervention such as a vaccination (Perry et al., 2020) or a change in farm-level animal health management (Damaso & Rushton, 2017), cost-benefit analysis or partial budget analysis is more commonly used. These methods compare the additional costs and benefits of a change with partial budget analysis looking at short periods of time, whereas cost-benefit analysis looks at costs and benefits that are spread over years that requires the cost and benefit streams to be discounted.

There have been calls to use cost-effectiveness analysis more widely in animal health (Babo Martins & Rushton, 2014) and the need to learn lessons from the way that human health economics applies these approaches (Thomas et al., 2019).

Other Applications of Economic Methods

We have focused so far on economic methods used for prioritization and costing of health interventions. Economic methodologies have also been used in other areas to examine socio-economic outcomes of health events and policies. A large number of studies have examined the impact of out-of-pocket expenditures on healthcare and their impact on poverty and inequality (e.g., Essue et al., 2017). A recent economic study with comprehensive global data shows that the rapid global spread of COVID-19 in 2020 was associated with the largest increase in global inequity and poverty since at least 1990 and possibly much earlier (comparable data are not available prior to 1990). One hundred and forty-three million people fell into extreme poverty using the World Bank definition, undoing all the progress that had been made over the previous five years (Yonzan et al., 2022). These analyses use standard economic methods.

One Health models often examine interactions involving several disciplinary groups (environmental, health, and economic, among others), aiming to understand highly complex systems with substantial amounts of interaction and feedback. Canali et al. (2018) provide an overview of how economic tools fit into these more complex models, mentioning socio-ecological systems models, agrarian systems analysis, which can encompass farm household models used in economics, dynamic transmission models, and bio-economic models, among others (Canali et al., 2018). They pick out dynamic transmission models (e.g., Zinsstag et al., 2009) as the backbone of economic analyses of One Health. The study of Canali et al. (2018) is a good place to start to investigate this topic further.

CASE STUDIES

A FRAMEWORK FOR ECONOMIC ANALYSES OF ZOONOSES USING ONE HEALTH

As discussed above, analyzing zoonoses is one of the key applications for a One Health approach (although not the only one). Figure 10.1 and Table 10.1 provide some suggestions for analyzing costs for zoonoses. One key takeaway from the figure and table is that costs are likely to increase the further away from initial transmission of the disease in animals. Costs also vary by the type of animal host, the transmissibility of the disease, the mode of transmission (e.g., respiratory; consumption of products of infected animals; skin-to-skin contact; insect vector biting affected people

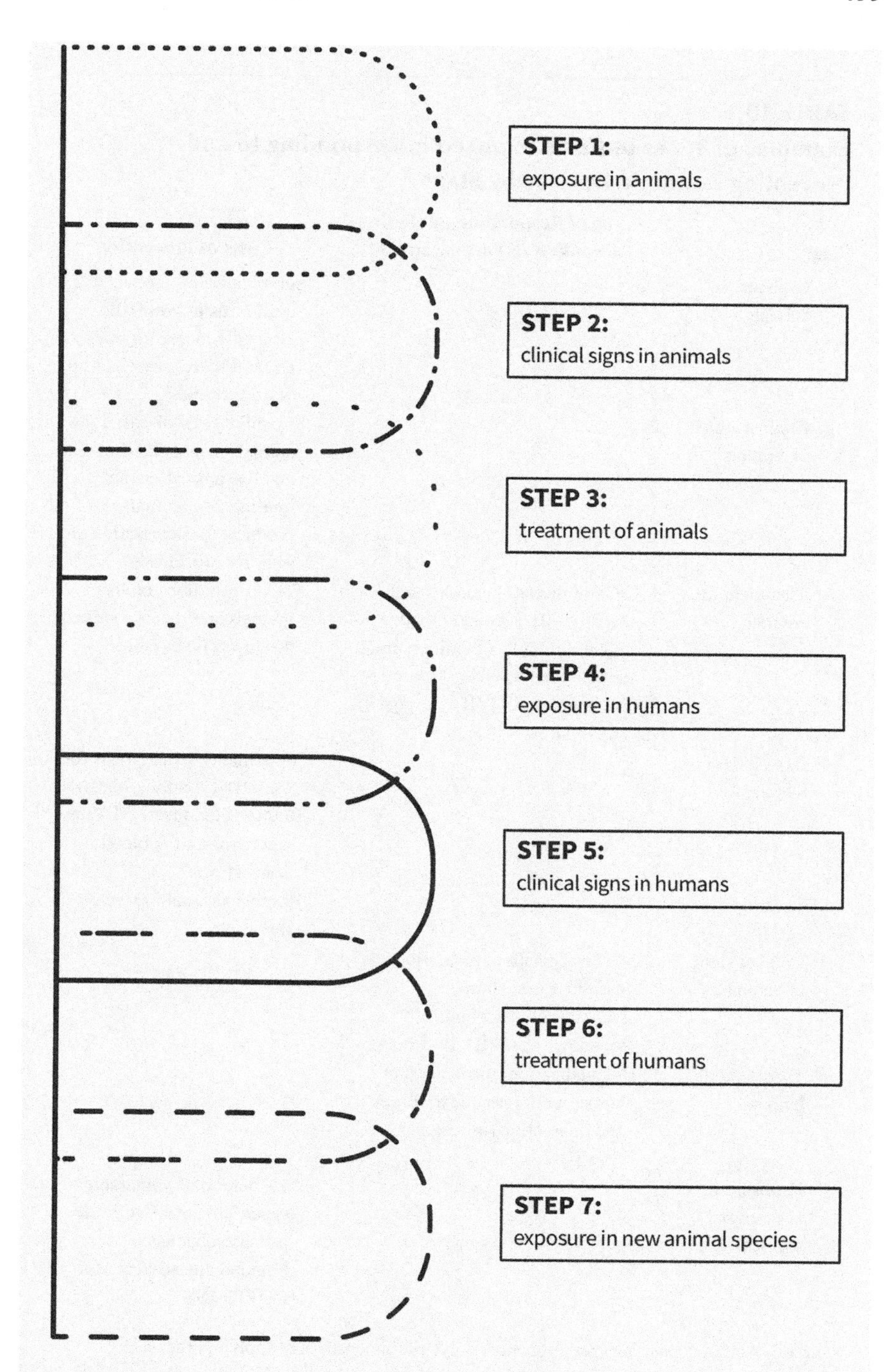

FIGURE 10.1 Stage of intervention in One Health disease pathway and impacts on costs.

TABLE 10.1
Examples of Types of Costs Incurred in Responding to and Preventing Zoonotic Disease by Stage

Stage	Costs of Responding, Including Prevention of Further Spread	Costs of Prevention
1. Exposure in animals		Surveillance of animals (e.g., dead birds in West Nile virus; blood test for dogs in Lyme disease); vaccination of dogs (rabies)
2. Clinical signs in animals		Surveillance of diseased animals in slaughterhouses and live animal markets; treatment of animal products (pasteurization of milk for brucellosis)
3. Treatment of animals	Culls of diseased animals and herds (BSE, foot and mouth, avian flu); culls of entire animal populations whether diseased or not (mink in COVID-19; pigs in Nipah virus)	Trade restrictions on live animals and specific animal products (BSE, rabies)
4. Exposure in humans		Surveillance (wastewater for COVID-19, polio, others) Routine blood tests (HIV in newborns and in blood donors) Routine vaccination (many diseases)
5. Clinical signs in humans	Mass vaccination; restrictions on mobility; quarantine/self-isolation; economic lockdown (COVID-19; Ebola)	
6. Treatment of humans	Hospitalization; medication at home; work loss due to illness and quarantine (influenza; COVID-19)	
7. Exposure in new animal species		Vaccination of vulnerable species in zoos (large cats, primates, hyenas, hippopotamuses, etc., for COVID-19)

Source: Authors. Examples are indicative only: similar costs may apply in other zoonoses not listed.

and animals), and the type of animal(s) involved (livestock, poultry, larger wild animals, companion animals, rodents, etc.). A more detailed framework is suggested in a very helpful figure (Figure 10.2) from Babo Martins et al. (2016).

One Health analyses can be used to address key policy questions surrounding zoonoses. Some examples of these questions include: Should some (or all) animals be vaccinated against the disease? Should some or all humans be vaccinated against the disease? Is either eradication or elimination a possibility? Is there a possibility of surveillance in an animal population to identify an outbreak before the disease reaches the human population? and Can environmental changes (e.g., reduction of deforestation) be used to reduce disease?

SELECTED CASE STUDIES OF ZOONOSES

Existing studies of zoonoses using a One Health approach provide guidance as to how broad the economic evaluation needs to be. There seem to be only a few such studies prior to 2015. Canali et al. (2018) identify one study of the cost of an integrated vaccination system for humans and animals in Chad (Schelling et al., 2007) and one of the cost-effectiveness of vaccination for rabies in dogs in Chad (Zinsstag et al., 2009). A systematic survey

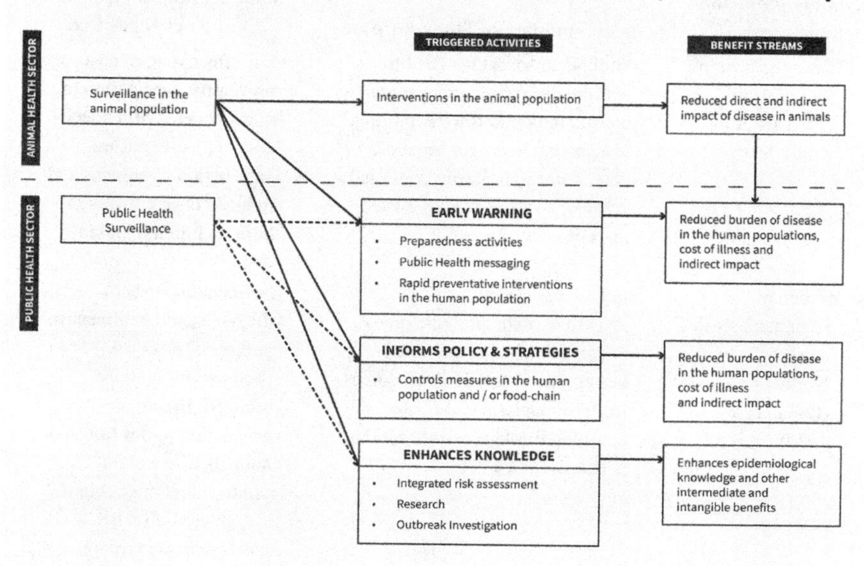

FIGURE 10.2 Conceptual framework of the links between surveillance of zoonoses in the animal population, the wider public health disease mitigation system, and benefit components associated.

from 2003 to 2015 undertaken by Baum et al. (2017) searching for quantitative evaluations of One Health interventions identified only one study with economic costs and outcomes ((Häsler et al., 2014), on rabies in Colombo).

Table 10.2 summarizes information on costs and benefits of five studies published from 2015 onward, of which three stated that they incorporated a One Health approach. This is not a systematic survey, rather a selection of various studies to illustrate the variations in approach, with an effort to include some of the more costly zoonoses.

The five studies in Table 10.2 encompass a range of diseases: affecting companion animals (rabies), poultry (avian influenza), livestock (leishmaniasis and brucellosis), and both wild and companion animals (West Nile). Two studies primarily focused on costs in farmed animals (avian influenza and brucellosis), while the other three also included health outcomes or costs associated with human illness (rabies, leishmaniasis, and West Nile

TABLE 10.2
Main Cost Drivers in Five Examples of Recent Economic Studies of Zoonoses

Study Topic and Reference	Brief Summary of Study Purpose	Main Economic Costs and Outcomes Measured
Rabies control in Tamil Nadu (Fitzpatrick et al., 2016)	Tamil Nadu has a One Health coordination committee; study considered options for capturing, vaccinating and – for female dogs – sterilizing stray dogs; and voluntary vaccination of owned dogs	Cost-effectiveness: costs of preventive vaccination in dogs; benefits of reduced cost of post-exposure vaccination in humans and reduced DALY losses in humans bitten by rabid dogs
Burden of leishmaniasis in one region in Northern Italy (Canali et al., 2020)	Study developed an economic-epidemiologic model to identify key cost drives of disease, and identify data sources for leishmaniasis (study also considered echinococcus but this is not discussed here)	Cost-benefit: costs of diagnosis and treatment in people; cost of lost productivity in those affected; livestock production losses (affected animals lose weight, produce less milk, and if slaughtered offal cannot be consumed and requires disposal); cost of preventive treatment of exposed dogs

(*Continued*)

TABLE 10.2 (*Continued*)
Main Cost Drivers in Five Examples of Recent Economic Studies of Zoonoses

Study Topic and Reference	Brief Summary of Study Purpose	Main Economic Costs and Outcomes Measured
One Health versus human-only surveillance for West Nile virus in one region in northern Italy (Paternoster et al., 2017)	Study compared the costs of a unisectoral surveillance approach (testing all blood donations for the rest of the year and the following year, following a case detection in a human), with a One Health strategy (with surveillance of human, insect, horse, and wild bird populations which triggered testing of blood donations in any year, following a case detection)	Cost-benefit: costs of surveillance in humans only, compared to costs of surveillance in humans, horses, wild birds, and insects. Benefits: One Health surveillance approach led to a more targeted screening of blood donations, and reduced cost of infections (hospitalization cost, cost of compensation to those who were unknowingly infected via a tainted blood transfusion)
Cost-benefit analysis of three alternative strategies for prevention and control of brucellosis in cows and buffalo in India (Singh et al., 2018)	Three alternative strategies considered: annual vaccination of replacement calves for 20 years; initial vaccination of both adult and calves followed by annual vaccination for replacement calves; and annual vaccination for replacement calves for 10 years, followed by 10 years of test and slaughter	Cost-benefit: Vaccination cost; slaughter price of healthy versus diseased animals. Second option has highest benefit:cost; all strategies have benefits which exceed costs; third option is not viable since religion proscribes slaughter of cows, and no government subsidy exists to purchase diseased cows
Analysis of alternative strategies to control highly pathogenic avian influenza in the Netherlands (Backer et al., 2015)	Consider combinations of usual strategy components (culling infected flocks; pre-emptively culling nearby flocks; zoning; screening dangerous contacts; transport regulations)	Develops a complex epidemiological-economic model. Costs include value of culled poultry; costs of culling; number of farms on which infection is detected; costs of zoning and screening; depends on duration in which no production occurs

virus). The studies used different methodologies: cost-effectiveness (rabies); a livestock cost model (avian influenza); and cost-benefit (leishmaniasis, West Nile, and brucellosis). The studies of rabies and, to a lesser extent, West Nile virus focused on outcomes for humans. The study of avian influenza only considered the economic costs associated with reduced poultry production, perhaps not surprisingly, since transmission to humans until recently has been restricted mainly to those handling poultry and their byproducts. Similarly, the study of brucellosis focused on costs associated with livestock.

The studies can be categorized as follows. The study of rabies adopts a One Health perspective, but the methodology is the standard one used in studies of human health and uses cost-effectiveness. The studies of avian influenza and brucellosis do not mention a One Health perspective, and the methodologies used are standard ones in the economics of animal health. The other three studies adopt a One Health perspective and all use cost-benefit analysis in a more complete One Health economic analysis. The differences in methodology reflect different patterns of disease transmission and different effects of the diseases.

A couple of studies in Table 10.2 explicitly note that they did not examine all costs, for example, the study of leishmaniasis did not include the cost of voluntary vaccination of horses and the benefits of reduced direct and indirect health improvements on horses. This illustrates a more general point that it is more important to focus on getting the big costs correct than to do an exhaustive inventory of costs. If the big costs are all incurred in humans (rabies example) then the appropriate economic methodology might be a standard one for human health – even though the intervention involves animals. If the big costs are all in animals (avian influenza example), then a standard economics of animal health approach might be used. And if the interventions being compared are fairly narrowly focused on a single outcome (brucellosis example), a full One Health economic analysis might not be required.

A general conclusion might be that the initial analysis of a zoonosis should likely adopt a One Health perspective to identify key impacts and types of costs, but that not all ensuing economic analyses need to be full One Health ones. However, it is also important to consider the equity implications of interventions. The dollar value of benefits from an intervention is often lower if the beneficiaries have low incomes, but the intervention may nevertheless be socially valuable.

The next section examines the converse: although many examples of One Health economic analyses are of zoonoses, there are important examples with a broader ambit.

SELECTED CASE STUDIES NOT INVOLVING INDIVIDUAL ZOONOSES

In this section, we summarize another two strands of analysis using a One Health perspective, from a broader perspective than that of a single zoonosis. We first review a couple of recent papers tackling the ambitious agenda of improved surveillance systems per se – a vital step in advocating for more resources for such systems. There have been calls for such a system to detect emerging zoonoses for at least a decade (e.g., Babo Martins et al., 2016), which have garnered greater urgency following the COVID-19 experience. We secondly review the thorny issue of growing AMR where the human and animal aspects are closely intertwined and a One Health approach would seem indispensable.

INVESTING IN WIDE-RANGING SURVEILLANCE AGAINST ZOONOSES

New zoonotic pandemics in the early 21st century led to increased international concern but somewhat more limited action. The SARS outbreak of 2002–2004 highlighted the risks of international transmissibility of novel pathogens, and led to international reports (e.g., OECD, 2011). There is also the ongoing panzootic of avian influenza that began in the late 1990s and is yet to be controlled, which threatens human health indirectly through inefficiencies in food supply and directly with disease spillover. To place the latter in context, an influenza pandemic did occur in 2009 that was thought to be a spillover from pigs in North America and rapidly spread across the world. The impacts were limited by therapeutic treatments and vaccines, yet still caused an increase in human deaths. The Ebola epidemics in West Africa of 2014–2016 similarly heightened concern. One response was the setting up of the World Bank Pandemic Financing Facility in 2016, which was exhausted and phased out in 2021, following the financial needs occasioned by the COVID-19 pandemic. The unprecedented scale of the COVID-19 pandemic led to the WHO beginning work at the end of 2021 on a pandemic pact, focusing on prevention, preparedness, and response (World Health Organization, 2021).

Various studies argue the case for more resources for integrated surveillance systems to improve the capacity to anticipate emerging zoonotic epidemics and pandemics. Some ((The World Bank, 2012) and (Bernstein et al., 2022)) make quantitative estimates of the resources required, while another (Babo Martins et al., 2016) provides a conceptual framework. Table 10.3 summarizes the differences between the methods and results of The World Bank (2012) and Bernstein et al. (2022) estimates, and we provide a more detailed discussion below.

TABLE 10.3
Summary of Different Estimates of Costs and Benefits of Prevention and Control of Zoonoses

	World Bank (2012)	Bernstein et al. (2022)
Time period for data used	6 major zoonotic outbreaks, 1997–2007	29 major zoonotic outbreaks, 1918–2021
Proposed intervention	Biosecurity (bring country standards up to OIE and WHO levels)	• Biosecurity • Increased viral research • Surveillance of wildlife trade • End farming of wildlife for human consumption in China • Halve rate of deforestation
Anticipated costs averted/ benefits	• Health services costs • Cost of culled animals • Better human health (measured in DALYs, monetized as productivity gains) • Other income gains: tourism, trade, tax (some studies)	Mortality reductions, monetized using the VSL method Loss of productivity (GDP loss) of 0.6% per outbreak, occurring with probability 0.4 each year
Cost estimates (annual)	$1.86–$3.3 bn	$20 bn
Benefits estimates (annual)	$6.9 bn (if outbreaks completely prevented); more if a "severe pandemic" averted once per century, saving an additional $30 bn annually	$350 bn to $21 trillion, depending on VSL value selected
Benefit:cost ratio	2.1 to 3.71 8.1 (including "severe pandemic" event)	17.5:1 to 1050:1 (chapter authors' calculation)

The World Bank (2012) study has a narrower focus than that of Bernstein et al. (2022) and estimates the cost of improving prevention and control of zoonoses, based on the experience with six major zoonoses over the period 1997–2009. The two major components of the estimated costs are for disease surveillance (both in animals and humans) and control, measured as the cost of bringing systems for containment of disease both in animals and humans up to standards of the WOAH (World Organisation for Animal Health) and WHO respectively. They also allow for efficiencies in improved coordination of the separate animal and human surveillance and control systems. Using detailed data analysis based on experience of the six major outbreaks, they estimate the costs of prevention and control as between $1.86 bn and $3.3 bn annually. Against this, they estimate the benefits that would have ensued from averting these six epidemics (using data from previously published individual studies of each one separately). The benefits depend on the disease, and whether the main consequences were for animal or human health. They term some costs as direct, including cost of veterinary services and compensation (for animals which are culled), as well as health services or hospital costs, DALY losses, and losses to national income and tourism. They also note indirect costs such as losses to trade and to tax revenue, which some of the studies considered.

The World Bank (2012) estimates that the six zoonoses combined led to an annual average loss of $6.9 bn (i.e., the potential benefit of completely averting these epidemics would be $6.9 bn). This compares to the $1.86-$3.36 bn cost of prevention and control, which would yield a benefit-cost ratio of between 2:1 and 3.7:1, which is a good return on investment, if improved prevention and control managed to completely contain these epidemics. They also hypothesize that if prevention and control prevented a severe pandemic (causing an estimated $3 trillion in losses, happening once every 100 years or $30 bn/year) the benefit:cost ratio would be even more favorable (8.1:1, even using the higher end of the range of estimates of costs of prevention and control).

Bernstein et al. (2022) attempt a more ambitious estimate of the global cost of outbreaks of novel viral zoonoses over just over a century, from the 1918 flu pandemic COVID-19 to the end of 2021. This period also includes human immunodeficiency virus (HIV), and these three conditions have the greatest loss of human life of all the pandemics. They estimate the annual deaths from outbreaks of novel zoonoses as 3.3 m, averaged over the century, and monetize these losses using the willingness-to-pay metric (VSL). They use estimates of the VSL as ranging from $107,000 per life to $7 m per life and come up with a benefit ranging from $350 bn per year to $21 trillion. In addition, they estimate that the annual average loss of gross national income per year is 0.6% of global income per novel outbreak, and that these events occur about once every 2.5 years (i.e., there is a probability of 0.4 of one occurring in a single year). This would appear to imply some double-counting, since the VSL estimates should include labor productivity losses. This makes the expected annual economic loss of $212 bn, markedly higher than the World Bank estimate, which is over a shorter period and which does not encompass any of the three largest outbreaks.

Note that VSL estimates tend to be a factor of about seven larger than estimates of lost productivity (human capital) or DALYs, which might range from at most $16,000 (in a country with an annual per capita income of $500) to $1.6 m (in a country with an annual per capita income of $50,000). (This calculation is the authors', and assumes that maximum life expectancy at birth is about 32 DALYs, following World Bank (1993), and that the monetary value of a DALY is around the same as per capita income, the threshold value for being very cost-effective according to WHO (2001; Sachs, 2001).)

Bernstein et al. (2022) also have a more extensive set of costs. They include the costs of increased biosecurity, similar to the World Bank estimates regarding improving containment of disease, and surveillance. But they also add in costs of viral discovery (i.e., research), surveillance of wildlife trade, closing down wildlife farming for human consumption in China, and halving the rate of deforestation. The big costs here are the last two, each costing multiple billions of dollars. They assess the total costs as $20 bn per year, compared to annual losses of per capita income of $212 bn per year and losses of life evaluated as between $350 bn and $21 trillion per year (depending on whether one uses the lowest or the highest numbers for VSL, as given above). They argue that even if the prevention measures only prevent 10% of the losses, the benefit:cost is very favorable.

Bernstein et al.'s (2022) estimates are considerably more speculative than those of the World Bank. The cost of halting farming of wildlife in China for human consumption is likely less than $19 bn annual estimate they use, since $19 bn assumes that those resources cannot be moved to another use. However, almost certainly they can be put to another use, although perhaps not as valuable. The use of VSL to value lives lost is about a factor of seven larger than using estimates of lost productivity, and the overall cost of prevention is heavily dominated by the estimates of the costs of reducing deforestation, which are highly speculative. It is also extremely difficult to estimate what the exact impact of prevention is, and by what percentage it reduces the probability of another major pandemic. It is unclear that the set of measures contemplated in Bernstein et al. (2022) would have averted a pandemic similar to COVID-19 if the virus had escaped from a laboratory, as some continue to speculate. The estimates by Bernstein et al. (2022) can perhaps be regarded as a first cut at the "big picture." Getting traction on a major international effort to prevent major zoonotic viral outbreaks may require such big thinking.

ANTIMICROBIAL RESISTANCE

AMR is a large and very concerning issue. Drug-resistant tuberculosis (TB) is considerably more costly to treat than drug-sensitive TB. Patients in hospitals with already weakened immune systems are at risk for hospital-acquired infections such as *Clostridium difficile.* Dadgostar (2019) cites the concerning Centers for Disease Control and Prevention estimates of 2 million infections with antibiotic-resistant conditions annually, resulting in 23,000 deaths in the United States, $20 bn annually from increased healthcare costs, and $35 bn in lost productivity (Dadgostar,

2019). There are considerable concerns that the "last line" antibiotics will fail, and mortality from currently treatable conditions will then increase.

One cause for the increase in AMR is the externality issue discussed earlier: people with low incomes facing out-of-pocket costs (and unpleasant side effects) of medication may be tempted to stop treatment once they feel better and not take the full course of antibiotics. The negative externality is that although the individual may recover, those pathogens which survive are more resistant to the antibiotic used. Weaker regulatory regimes in low- and middle-income countries also contribute by permitting over-the-counter sales of treatments that would only be available with a prescription in high-income countries. Physicians worldwide may be tempted to prescribe antibiotics without first ascertaining that the patient has a bacterial infection rather than a viral one: this is all the more likely in countries where laboratories for doing culture are not available at the primary care level.

There is also a zoonotic component to the AMR problem, in that globally more antibiotics are consumed by animals than people: in the United States 80% of antibiotics sold are applied to animal feed (Dadgostar, 2019), much of this prophylactically, in order to save money both on animal feed and housing. This is a complicated One Health issue. Antibiotic resistance is bad for human health but potentially costs of animal products could increase if prophylactic antibiotic use becomes less effective. Higher costs of animal products will in turn have potentially adverse effects on nutrition, particularly nutrition of small children in countries with already high levels of stunting. There are, in addition, environmental implications, among other things, from intensive livestock production causing issues of disposal of animal waste potentially containing antibiotic residue. Queenan et al. (2016) list (without quantifying) the kinds of cost component estimates required for an appropriate AMR surveillance system: modeling a system with this degree of complexity would not be easy (Queenan et al., 2016).

CONCLUSION

Economics has a role to play as the One Health movement continues. It provides analytical frameworks that help to assess the burdens of diseases that impact people, animals, plants, and the environment and thereby creates a baseline that can help identify critical interventions. The assessment of the interventions will also require economic assessment tools such as cost-effectiveness and cost-benefit analysis in order to direct resources to activities and investments that will give us the greatest returns. Finally, once these interventions have been implemented, economic analyses can be used to determine their overall success. Within all these steps there is a need for some humility of the current economic methods: they need to be adapted, and in some cases there will be a need for innovation, in order to reflect the wide context in which One Health operates and the need for economists to work with other disciplines to parameterize their models.

In terms of future One Health challenges, we see the continuation of problems with emerging and re-emerging diseases as human populations grow, economies change, and resource use shifts. These challenges occur in a socio-economic setting and economics should help to identify risky behaviors. In addition to the infectious diseases, we also need to be conscious of the problems our food systems are creating in terms of public health through poor nutrition and food-related non-communicable diseases. These problems are becoming increasingly important and reflect a culture of excessive calorie intake. The costs of overconsumption are twofold – greater costs to the health system and poor health outcomes and overuse of the environment for food we do not need. These challenges from the food system are complex and require a One Health approach (Rushton, McMahon, et al., 2021) within which economics and its methods have a role.

REFERENCES

Babo Martins, S., & Rushton, J. (2014). Cost-effectiveness analysis: Adding value to assessment of animal health, welfare and production. *Revue Scientifique et Technique de l'OIE*, *33*(3), 681–689. https://doi.org/10.20506/rst.33.3.2312

Babo Martins, S., Rushton, J., & Stärk, K. D. C. (2016). Economic assessment of zoonoses surveillance in a 'One Health' context: A conceptual framework. *Zoonoses and Public Health*, *63*(5), 386–395. https://doi.org/10.1111/zph.12239

Backer, J. A., Van Roermund, H. J. W., Fischer, E. A. J., Van Asseldonk, M. A. P. M., & Bergevoet, R. H. M. (2015). Controlling highly pathogenic avian influenza outbreaks: An epidemiological and economic model analysis. *Preventive Veterinary Medicine*, *121*(1–2), 142–150. https://doi.org/10.1016/j.prevetmed.2015.06.006

Baum, S. E., Machalaba, C., Daszak, P., Salerno, R. H., & Karesh, W. B. (2017). Evaluating one health: Are we demonstrating effectiveness? *One Health*, *3*, 5–10. https://doi.org/10.1016/j.onehlt.2016.10.004

Bernstein, A. S., Ando, A. W., Loch-Temzelides, T., Vale, M. M., Li, B. V., Li, H., Busch, J., Chapman, C. A., Kinnaird, M., Nowak, K., Castro, M. C., Zambrana-Torrelio, C., Ahumada, J. A., Xiao, L., Roehrdanz, P., Kaufman, L., Hannah, L., Daszak, P., Pimm, S. L., & Dobson, A. P. (2022). The costs and benefits of primary prevention of zoonotic pandemics. *Science Advances*, *8*(5), eabl4183. https://doi.org/10.1126/sciadv.abl4183

Bertram, M. Y., Lauer, J. A., De Joncheere, K., Edejer, T., Hutubessy, R., Kieny, M.-P., & Hill, S. R. (2016). Cost–effectiveness thresholds: Pros and cons. *Bulletin of the World Health Organization*, *94*(12), 925–930. https://doi.org/10.2471/BLT.15.164418

Canali, M., Aragrande, M., Angheben, A., Capelli, G., Drigo, M., Gobbi, F., Tamarozzi, F., & Cassini, R. (2020). Epidemiologic-economic models and the One Health paradigm: Echinococcosis and leishmaniasis, case studies in Veneto region, Northeastern Italy. *One Health*, *9*, 100115. https://doi.org/10.1016/j.onehlt.2019.100115

Canali, M., Aragrande, M., Cuevas, S., Cornelsen, L., Bruce, M., Rojo-Gimeno, C., & Häsler, B. (2018). The economic evaluation of One Health. In S. R. Rüegg, B. Häsler, & J. Zinsstag (Eds.), *Integrated approaches to health: A handbook for the evaluation of One Health*. Wageningen Academic Publishers. https://doi.org/10.3920/978-90-8686-875-9

Coker, R., Rushton, J., Mounier-Jack, S., Karimuribo, E., Lutumba, P., Kambarage, D., Pfeiffer, D. U., Stärk, K., & Rweyemamu, M. (2011). Towards a conceptual framework to support one-health research for policy on emerging zoonoses. *The Lancet Infectious Diseases*, *11*(4), 326–331. https://doi.org/10.1016/S1473-3099(10)70312-1

Dadgostar, P. (2019). Antimicrobial resistance: Implications and costs. *Infection and Drug Resistance*, *12*, 3903–3910. https://doi.org/10.2147/IDR.S234610

Damaso, A. F., & Rushton, J. (2017). Economic impact of a *Salmonella* outbreak on a Welsh dairy farm and an estimation of the breakeven point for vaccination. *Veterinary Record Case Reports*, *4*(2), e000359. https://doi.org/10.1136/vetreccr-2016-000359

Department of Economic and Social Affairs, United Nations. (2021). *World Economic Situation and Prospects as of mid-2021*. https://www.un.org/development/desa/dpad/publication/world-economic-situation-and-prospects-as-of-mid-2021/

Essue, B. M., Laba, T.-L., Knaul, F., Chu, A., Van Minh, H., Nguyen, T. K. P., & Jan, S. (2017). Chapter 6: Economic burden of Chronic Ill health and injuries for households in low- and middle-income countries. In D. T. Jamison, H. Gelband, S. Horton, P. Jha, R. Laxminarayan, C. N. Mock, & R. Nugent (Eds.), *Disease control priorities, third edition (Volume 9): Improving health and reducing poverty*. The World Bank. https://doi.org/10.1596/978-1-4648-0527-1

Fitzpatrick, M. C., Shah, H. A., Pandey, A., Bilinski, A. M., Kakkar, M., Clark, A. D., Townsend, J. P., Abbas, S. S., & Galvani, A. P. (2016). One Health approach to cost-effective rabies control in India. *Proceedings of the National Academy of Sciences*, *113*(51), 14574–14581. https://doi.org/10.1073/pnas.1604975113

Fox-Rushby, J., & Cairns, J. A. (Eds.). (2009). *Economic evaluation* (Reprinted). Open University Press.

Häsler, B., Hiby, E., Gilbert, W., Obeyesekere, N., Bennani, H., & Rushton, J. (2014). A One Health framework for the evaluation of Rabies Control Programmes: A case study from Colombo City, Sri Lanka. *PLoS Neglected Tropical Diseases*, *8*(10), e3270. https://doi.org/10.1371/journal.pntd.0003270

Jamison, D. T., Gelband, H., Horton, S., Jha, P., Laxminarayan, R., Mock, C. N., & Nugent, R. A. (Eds.). (2018). *Improving health and reducing poverty* (3rd ed.). World Bank Group. https://doi.org/10.1596/978-1-4648-0527-1

Neumann, P. J., Cohen, J. T., & Weinstein, M. C. (2014). Updating cost-effectiveness—The curious resilience of the $50,000-per-QALY threshold. *New England Journal of Medicine*, *371*(9), 796–797. https://doi.org/10.1056/NEJMp1405158

OECD. (2011). *Future global shocks: Improving risk governance*. OECD. https://doi.org/10.1787/9789264114586-en

Paternoster, G., Babo Martins, S., Mattivi, A., Cagarelli, R., Angelini, P., Bellini, R., Santi, A., Galletti, G., Pupella, S., Marano, G., Copello, F., Rushton, J., Stärk, K. D. C., & Tamba, M. (2017). Economics of One Health: Costs and benefits of integrated West Nile virus surveillance in Emilia-Romagna. *PLoS One*, *12*(11), e0188156. https://doi.org/10.1371/journal.pone.0188156

Perry, B., Rich, K. M., Rojas, H., Romero, J., Adamson, D., Bervejillo, J. E., Fernandez, F., Pereira, A., Pérez, L., Reich, F., Sarno, R., Vitale, E., Stanham, F., & Rushton, J. (2020). Integrating the technical, risk management and economic implications of animal disease control to advise policy change: The example of foot-and-mouth disease control in Uruguay. *EcoHealth*, *17*(3), 381–387. https://doi.org/10.1007/s10393-020-01489-6

Queenan, K., Häsler, B., & Rushton, J. (2016). A One Health approach to antimicrobial resistance surveillance: Is there a business case for it? *International Journal of Antimicrobial Agents*, *48*(4), 422–427. https://doi.org/10.1016/j.ijantimicag.2016.06.014

Robinson, L. A., Hammitt, J. K., & O'Keeffe, L. (2019). Valuing mortality risk reductions in global benefit-cost analysis. *Journal of Benefit-Cost Analysis*, *10*(S1), 15–50. https://doi.org/10.1017/bca.2018.26

Rushton, J., Bruce, M., Bellet, C., Torgerson, P., Shaw, A., Marsh, T., Pigott, D., Stone, M., Pinto, J., Mesenhowski, S., & Wood, P. (2018). Initiation of global burden of animal diseases programme. *The Lancet*, *392*(10147), 538–540. https://doi.org/10.1016/S0140-6736(18)31472-7

Rushton, J., Huntington, B., Gilbert, W., Herrero, M., Torgerson, P. R., Shaw, A. P. M., Bruce, M., Marsh, T. L., Pendell, D. L., Bernardo, T. M., Stacey, D., Grace, D., Watkins, K., Bondad-Reantaso, M., Devleesschauwer, B., Pigott, D. M., Stone, M., & Mesenhowski, S. (2021). Roll-out of the global burden of animal diseases programme. *The Lancet*, *397*(10279), 1045–1046. https://doi.org/10.1016/S0140-6736(21)00189-6

Rushton, J., McMahon, B. J., Wilson, M. E., Mazet, J. A. K., & Shankar, B. (2021). A food system paradigm shift: From cheap food at any cost to food within a One Health framework. *NAM Perspectives*, *11*. https://doi.org/10.31478/202111b

Rushton, J., Viscarra, R., Otte, J., McLeod, A., & Taylor, N. (2007). Animal health economics where have we come from and where do we go next? *CABI Reviews*. https://doi.org/10.1079/PAVSNNR20072031

Sachs, J. (Ed.). (2001). *Macroeconomics and health: Investing in health for economic development; report of the Commission on Macroeconomics and Health*. World Health Organization.

Schelling, E., Bechir, M., Ahmed, M. A., Wyss, K., Randolph, T. F., & Zinsstag, J. (2007). Human and animal vaccination delivery to remote nomadic families, Chad. *Emerging Infectious Diseases*, *13*(3), 373–379. https://doi.org/10.3201/eid1303.060391

Shaw, A. P. M., Rushton, J., Roth, F., & Torgerson, P. R. (2017). DALYs, dollars and dogs: How best to analyse the economics of controlling zoonoses: *Revue Scientifique et Technique de l'OIE*, *36*(1), 147–161. https://doi.org/10.20506/rst.36.1.2618

Singh, B. B., Kostoulas, P., Gill, J. P. S., & Dhand, N. K. (2018). Cost-benefit analysis of intervention policies for prevention and control of brucellosis in India. *PLoS Neglected Tropical Diseases*, *12*(5), e0006488. https://doi.org/10.1371/journal.pntd.0006488

Stenberg, K., Hanssen, O., Edejer, T. T.-T., Bertram, M., Brindley, C., Meshreky, A., Rosen, J. E., Stover, J., Verboom, P., Sanders, R., & Soucat, A. (2017). Financing transformative health systems towards achievement of the health Sustainable Development Goals: A model for projected resource needs in 67 low-income and middle-income countries. *The Lancet Global Health*, *5*(9), e875–e887. https://doi.org/10.1016/S2214-109X(17)30263-2

Teng, K. T.-Y., Devleesschauwer, B., Maertens De Noordhout, C., Bennett, P., McGreevy, P. D., Chiu, P.-Y., Toribio, J.-A. L. M. L., & Dhand, N. K. (2018). Welfare-Adjusted Life Years (WALY): A novel metric of animal welfare that combines the impacts of impaired welfare and abbreviated lifespan. *PLoS One*, *13*(9), e0202580. https://doi.org/10.1371/journal.pone.0202580

The World Bank. (2012). *People, pathogens and our planet: Volume two: The economics of One Health*. https://hdl.handle.net/10986/11892

The World Bank. (2023, April 7). World Bank Open Data. https://data.worldbank.org/indicator/SH.XPD.CHEX.GD.ZS

Thomas, L. F., Bellet, C., & Rushton, J. (2019). Using economic and social data to improve veterinary vaccine development: Learning lessons from human vaccinology. *Vaccine, 37*(30), 3974–3980. https://doi.org/10.1016/j.vaccine.2018.10.044

Torgerson, P. R., Rüegg, S., Devleesschauwer, B., Abela-Ridder, B., Havelaar, A. H., Shaw, A. P. M., Rushton, J., & Speybroeck, N. (2018). zDALY: An adjusted indicator to estimate the burden of zoonotic diseases. *One Health, 5*, 40–45. https://doi.org/10.1016/j.onehlt.2017.11.003

World Health Organization. (2014, January 1). *WHO-CHOICE*. https://www.who.int/news-room/questions-and-answers/item/who-choice-frequently-asked-questions

World Health Organization. (2021, December 1). *World Health Assembly agrees to launch process to develop historic global accord on pandemic prevention, preparedness and response*. https://www.who.int/news/item/01-12-2021-world-health-assembly-agrees-to-launch-process-to-develop-historic-global-accord-on-pandemic-prevention-preparedness-and-response

World Health Organization. (2022, May 5). *14.9 million excess deaths associated with the COVID-19 pandemic in 2020 and 2021*. https://www.who.int/news/item/05-05-2022-14.9-million-excess-deaths-were-associated-with-the-covid-19-pandemic-in-2020-and-2021

Yonzan, N., Mahler, D. G., & Lakner, C. (2022, August 22). The impact of COVID-19 on global inequality and poverty. *Session 4A-2, New Developments in Poverty Measurement with a Focus on National Statistical Office Efforts II*. 37th IARIW General Conference. https://iariw.org/wp-content/uploads/2022/08/Mahler-Yonzan-Lakner-IARIW-2022.pdf

Zinsstag, J., Dürr, S., Penny, M. A., Mindekem, R., Roth, F., Gonzalez, S. M., Naissengar, S., & Hattendorf, J. (2009). Transmission dynamics and economics of rabies control in dogs and humans in an African city. *Proceedings of the National Academy of Sciences, 106*(35), 14996–15001. https://doi.org/10.1073/pnas.0904740106

11 Cultural Competency – A Cornerstone for Effective Collaborations

Mariam Reda and Sulagna Chakraborty

INTRODUCTION

The need for cultural competency arose due to existing barriers in various settings, be it in healthcare settings (Ahmann, 2002), policymaking (SenGupta et al., 2004), international aid (Southall et al., 2017), business/commerce (Brett, 2000), and public affairs (Carrizales, 2010). The basic premise of cultural competency is to train individuals about distinct cultures, traditions, norms, languages, beliefs, and behaviors so that they can reduce biases, address inequalities, form stronger collaborations, and provide effective solutions that account for these cultural, behavioral, and linguistic differences. From a healthcare standpoint, Betancourt et al. (2002) defined cultural competence as the ability of systems to provide care to patients with diverse values, beliefs, and behaviors, including tailoring delivery to meet patients' social, cultural, and linguistic needs. Cultural competency has to be part of the One Health framework because only when we recognize the array of cultural, ethnic, linguistic, and behavioral differences can we learn to solve problems that affect humans, animals, plants, and the environment globally.

WHAT IS CULTURAL COMPETENCY?

The concept of culture has evolved over time. In 2003, Betancourt et al. defined culture as being linked to patterns of learned beliefs, values, and behaviors within a group including language, styles of communication, practices, customs, and views on roles and relationships (Betancourt et al., 2003). In 2007, Carpenter-Song et al. highlighted the dynamic nature of culture by defining it as a dynamic, relational process of shared meaning originating from interactions between individuals (Carpenter-Song et al., 2007). In 2010, Schein underscored the population's perception that their behaviors, attitudes, and perspectives are the correct way to think, act, and feel through the shared beliefs, values, and assumptions that they learn from one another and teach to others (Schein, 2010). The U.S. Centers for Disease Control and Prevention (U.S. CDC) described culture as encompassing the different patterns of human behavior demonstrated in language, thoughts, communications, actions, institutions of race, ethnicity, customs, values, beliefs,

DOI: 10.1201/9781003232223-11

traditions, religion, or social groups (Centers for Disease Control and Prevention, 2024). A person's culture must be considered in the historical, social, political, and economic contexts the group this person belongs to shares the environment with (Gregory et al., 2010). Determinants of culture and cultural dynamics impact the individual's sensemaking mechanisms, subsequently affecting their behavioral response to the environment around them. Hence, people from different cultural backgrounds have varying perceptions, interpretations, explanations, and predictions of behaviors (Muzychenko, 2008).

While cultural intelligence emphasizes one's ability to interpret a stranger's behaviors (Muzychenko, 2008), cultural competence focuses on one's ability to understand different cultures and tailor behaviors and interactions accordingly. Several definitions exist for cultural competence. Williams defined cultural competence as the ability of individuals and systems to work or respond effectively across cultures in a way that acknowledges and respects the culture of the person or organization being served (Williams, 2001). Then, Muzychenko focused the description of cultural competence of an individual's behavior in a foreign environment specifying the need for appropriateness and effectiveness of one's behavior (Muzychenko, 2008). In 2013, Wilson et al. linked the definition of cultural competence to the acquisition and maintenance of culture-specific skills for very practical reasons such as functioning effectively in new cultures and interacting with people from varying cultural backgrounds (Wilson et al., 2013). Finally, the U.S. CDC focused the definition of cultural competence at the structural level as a set of congruent behaviors, attitudes, and policies that come together in a system, agency, or among professionals that enables effective work in cross-cultural situations (Centers for Disease Control and Prevention, 2024). This absence of consistency, harmonization, and understanding in the way cultural competence is defined reflects the difficulty of the concept but also the challenges with standardization of intercultural competency training in the fields of education, public health, and clinical practice for interdisciplinary teams.

The world currently lives in a post-globalization era comprised of extensive societal and cultural interactions and cross-fertilization between groups, populations, and countries. It is more important now than ever in human society to understand and embrace the notion of cultural competence. Globally, culture and group norms are constantly changing. It can be easy to stay within one's cultural comfort zone, frame of reference, and perspective. However, to better build relations, educators, clinicians, global health actors, and interdisciplinary practitioners must think more holistically and consider strategies to foster inclusivity in the education sector, clinical setting, and global health programming. The notion of "competence" requires acquiring the capacity to effectively work with individuals, communities, or organizations with consideration of their cultural beliefs, behaviors, and needs (Centers for Disease Control and Prevention, 2024). Building cultural competence means having awareness to the cross-cultural variations; acquiring the ability to adapt in new cultures and environments; and constantly reflecting on how one's thoughts and behaviors might have effects on others and the work being done (Hansen et al., 2000; Chao et al., 2011).

WHY IS CULTURAL COMPETENCY IMPORTANT IN GLOBAL HEALTH?

This chapter focuses on cultural competence in global health and, more specifically, One Health. One Health challenges include a wide spectrum of issues including emerging and re-emerging zoonotic diseases such as coronavirus disease (COVID)-19, Zika, and Mpox; antimicrobial resistance; food security; climate change; etc. The concept of One Health appreciates the close interconnectedness and mutual impact of people, animals, and ecosystems. The One Health approach requires the collaboration of professionals from human health (doctors, nurses, public health practitioners, epidemiologists), animal health (veterinarians, paraprofessionals, agricultural workers), environment (ecologists, wildlife experts), and other areas of expertise (Centers for Disease Control and Prevention, 2022). This collaboration and coordination represent the interdisciplinary practitioners' teams that this chapter will be referring to. Interdisciplinary practitioners including health educators, public health experts, and medical staff all influence population health and outcomes across various sectors. From the training of clinical professionals and public health experts to the design and implementation of health programming to the delivery of health services, these practitioners are all vital in ensuring healthy populations, health security, and better health for all (World Health Organization, 2023).

Interdisciplinary practitioners must develop their cultural competence skills for them to relate, connect, and work with individuals and communities from varying backgrounds and cultures. More importantly, these skills prevent these practitioners from bringing their own cultural frame of reference to apply it in different contexts. Cultural competency skills are important for health practitioners to adapt and tailor their support based on the cultural background of the individuals, groups, and communities they are assisting (Hofstede et al., 2002). With less barriers between countries, diversity and intercultural connectedness have become the norm of our current context. This setting mandates that health practitioners have the ability to overcome the barriers posed by cultural and societal differences in norms, habits, behaviors, and customs. Overcoming those hurdles allows for the creation of interactions, relations, and linkages that strengthen health programming and clinical services toward achievement of high-quality health outcomes (Garneau & Pepin, 2015). The absence of cultural competence skills can cause practitioners to inadvertently deliver low-quality care due to misinterpretation of patient behavior and language barriers causing inaccurate description of patient signs and symptoms leading to inappropriate treatment.

Brunett and Shingles compiled evidence on the effects of having culturally competent providers and staff on patients' experiences and level of satisfaction (Brunett & Shingles, 2018). The evidence revealed: (i) a significant correlation ($r=0.193$) between nurse practitioners' cultural competence and Latina patients' level of satisfaction, underscoring the importance of health systems to engage practitioners with cultural competence skills, higher education, and ability to speak the same language as the primary population (Castro & Ruiz, 2009); (ii) patients are less likely to follow medical directives if they perceive significant cultural differences with practitioners due to lack of culturally competent communication of treatment plans (Ohana & Mash, 2015); (iii) patients have higher level of satisfaction when served by practitioners with culturally competent attitudes and actions evident through their increased willingness and openness to learn about other cultures (Paez et al., 2009); (iv) the importance for nonphysician staff to be polite to patients and aware of existence of implicit biases to avoid perceptions of implicit discrimination and ensure more patient-centered care (Tajeu et al., 2015); and (v) behavior change suggestions by providers with higher cultural competence are better received by patients (Thom & Tirado, 2006).

In a conceptual line of thinking, cultural competency merges cultural awareness, knowledge, sensitivity, skill, proficiency, and dynamicity (Sharifi et al., 2019). This thinking implies that a health practitioner who can portray those elements will have the ability to deliver culturally sensitive services. However, the pragmatic application of this is much more complicated and complex as there are many other factors that impact global health competency (Purnell, 2016). Global health competency requires an understanding of disease morbidity; acknowledgment of the social determinants of health (SDH); health determinants; mastery of the core public health skills; soft skills; and cultural competence (Ferorelli et al., 2020). Furthermore, cultural competency proved to be the main and unique factor influencing global health competency. This reinforces the importance of cultural competence skills in building culturally sensitive health workforce and delivering culturally tailored clinical services and global health programming (Campinha-Bacote, 2002).

In the next section, we will explore the importance and level of application of cultural competence in the education sector, clinical setting, and public health programming. The final section of this chapter will provide recommendations for strengthening the application of cultural competence in these three sectors.

APPLICATION OF CULTURAL COMPETENCY ACROSS SECTORS

Application of Cultural Competency in the Education Sector

In 2014, the Lancet Commission on Culture and Health stressed the importance of understanding culture in medicine strengthening the need for cultural competence skills in medical education (Holmes, 2014). Furthermore, the American Association of Colleges of Nursing (Calvillo et al., 2009), the Association of American Medical Colleges (2005), the National Association of Social Workers (2001), and the Association of School and Programs of Public Health (Expert Panel on Cultural Competence Education for Students in Medicine and Public Health, 2012) have issued reports including the recommendation for the adoption of cultural competence capacity strengthening in academic institutions.

Researchers in South Korea conducted, in 2022, eight years after the Lancet commission recommendations, a cross-sectional descriptive study to assess the factors affecting global health and cultural competencies of nursing students.

They proved that global health competence has a positive correlation with cultural competency ($r=0.49$), cultural nursing confidence ($r=0.26$), and metacognition (awareness of own thinking process; $r=0.22$) (Cho & Kim, 2022). Additionally, Lynn et al. conducted self-administered surveys of attendees at the National League for Nursing Education Summit to assess the level of global health competencies (knowledge and skills) of nursing faculty in U.S. prelicensure programs. This study demonstrated cultural competence as a significant factor associated with nursing educators' global health competency (Lynn et al., 2021).

Despite these findings highlighting the importance of cultural competence in building the future workforce of practitioners, the topic is still addressed sporadically and inconsistently at academic institutions and in training curricula. There is great heterogeneity in cultural competence definitions, teaching strategies, teaching format, and faculty qualifications. For example, while German and Swiss universities' undergraduate programs cover some aspects of the topic, Anglo-American universities have decades of experience and expertise addressing cultural competence in education (Mews et al., 2018). In many universities, courses related to this topic are still optional, un-integrated, and without a unified structure of content (Kaffes et al., 2016).

Finally, public health professionals deliver a substantial portion of public health services to populations through health programs. The field of public health has greatly evolved over the years especially in the wake of zoonotic diseases such as the COVID-19 pandemic. For public health practitioners to contribute effectively in improving population health, they need to develop the skills required to continuously self-reflect and acknowledge the impact of their cultural background on their interaction with differing cultures (Lachance & Oxendine, 2015). There is a dearth of evidence from public health institutions on the level of integration of cultural competence into training curricula and subsequent impact. However, we hypothesize that, similar to medical professionals' education, if public health professionals are equipped with cultural competence skills, they will be more effective and successful at delivering culturally sensitive and tailored programs and services.

Application of Cultural Competency in Clinical Setting

The American Hospital Association (AHA) defines a culturally competent healthcare system as one with "the ability of systems to provide care to patients with diverse values, beliefs and behaviors, including the tailoring of health care delivery to meet patients' social, cultural and linguistic needs". Further, the AHA definition "acknowledges the importance of culture, incorporates the assessment of cross-cultural relations, recognizes the potential impact of cultural differences, expands cultural knowledge, and adapts services to meet culturally unique needs." The goal of cultural competence in healthcare settings is to reduce racial, economic, ethnic, and social disparities when meeting a community's healthcare needs (Health Research & Educational Trust, 2013). In delivering healthcare services, cultural competence is defined as the ability to understand cultural influences and to provide care in harmony with a patient's culture (Campinha-Bacote, 2002).

Medical practitioners are typically the first line of contact with a patient in the health system (Tzeng & Yin, 2006). There is growing recognition of the mandate on healthcare systems to recognize and adapt health services to consider the diverse cultural background of the patients it serves as well as the dynamics between stakeholders to meet the varying needs of communities, especially the minorities and marginalized (Betancourt et al., 2003). When clinical practitioners work in a culturally competent environment, they can adapt their approach to patient care with a consideration to the patient's background and experiences but more importantly mitigate the impact of their and other different cultures' practices and experiences in informing patient care and treatment. Clinicians should tailor medical care to the patients' own cultural practices and experiences. We assume that when nurses deliver services to patients from various cultural backgrounds, they must tailor their care based on their understanding of that background. But in reality, a large portion of medical professionals provide care to patients with different cultural backgrounds in the same way they provide care to patients from their own culture. Kim et al. conducted in-depth interviews with Korean nurses showing common themes of deep cultural difference, difficulty in providing care in multicultural settings, sticking to care in our own way, and not being ready to accept culturally different women (Kim et al., 2014).

Further, the COVID-19 pandemic has had a significant impact on the global dynamics including health labor markets, individual health workers, and populations' social and economic well-being (Kim et al., 2014). Global health is constantly evolving with new challenges which has a direct impact on the population well-being and countries' abilities to achieve health for all (Lynn et al., 2021). Estevan and Solano Ruíz showed that strengthening nurses' cultural competencies increases patient's trust, facilitates disease tracking and management, and improves provision and safety of appropriate and equitable health services (Estevan & Solano Ruíz, 2017). Furthermore, Jeong et al. shared evidence that cultural competence in nursing care is multilayered, comprising of cognitive, affective, and behavior domains with emphasis on the critical attributes of sensitivity, equality, awareness, openness, and coherence. Hence, this is a complex skill and must be purposefully taught to decrease health disparities, improve quality of care, and increase client satisfaction (Jeong et al., 2016). Finally, a study conducted by Hwang and Jang examining immigrant patients from foreign countries, ethnicities, and cultural backgrounds has noted inequalities in service delivery and limited access to health services (Hwang & Jang, 2017).

This evidence shows that it is crucial to prepare the next generation of medical practitioners through the establishment of education programs that continuously apply and strengthen the clinical professionals' cultural competencies. More importantly, these efforts must be linked to medical providers' improvements and performance data to allow for tracking and continuous learning throughout their career.

Application of Cultural Competency in Public Health Programming

According to the World Health Organization (WHO) definition, health is a "state of complete physical, mental and social well-being and not merely the absence of disease or infirmity." The main question is, which part of the health system contributes directly to the health of populations? Whether people are healthy or not is not only determined by the health system but by the contextual and environmental circumstances in which they are born, grow, work, live, and age. The WHO defines the determinants of health as: (i) income and social status, (ii) education, (iii) physical environment, (iv) social support networks, (v) genetics, (vi) health services, and (vii) gender (World Health Organization, 2024). This is where the role of the SDH becomes important to understand. There is increased acknowledgment of the impact of SDH on population health. SDH "are the non-medical factors that influence health outcomes." These include: (i) the social and economic environment, (ii) the physical environment, and (iii) the person's individual characteristic behaviors. These factors must be taken into consideration as organizations design and implement public health programs to ensure culturally sensitive programming (World Health Organization, 2017) (Figure 11.1).

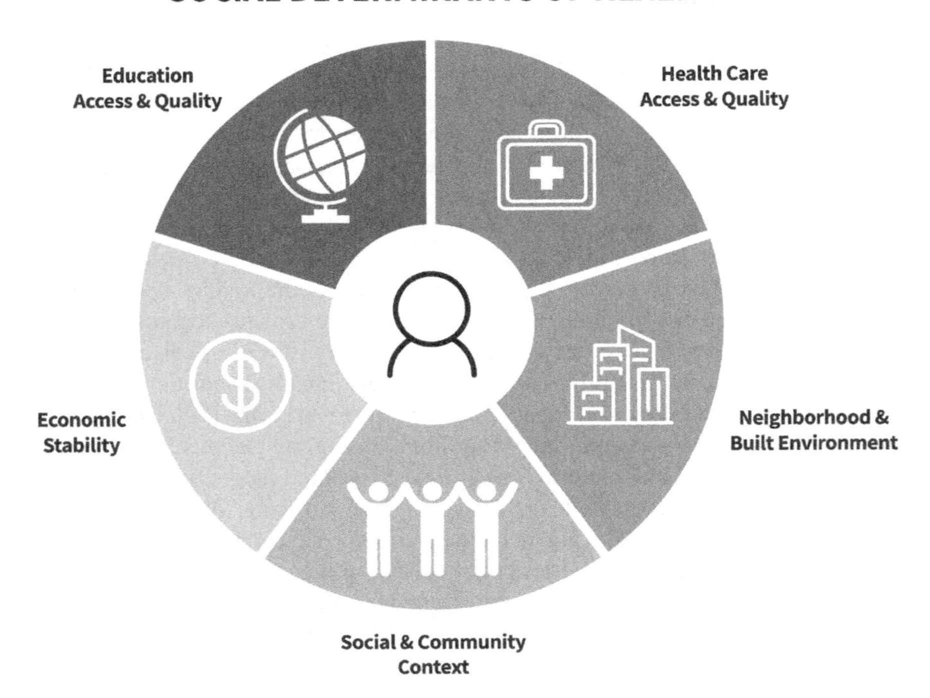

FIGURE 11.1 Social determinants of health.

In global public health specifically, cultural competence is defined as "the ability of providers and organizations to effectively deliver health care services that meet the social, cultural, and linguistic needs of patients" (Betancourt, 2004). Hence, public health interventions must acknowledge the importance of culture as imperative for effective communication while aiming to improve availability, accessibility, affordability, and quality of health services for people from different ethnic and cultural backgrounds without exception and with intentional inclusion of minority groups (Truong et al., 2014). In the field of global public health, cultural competence has a direct impact on the performance of a program, its acceptability by the community, and its potential to elicit sustainable impact. Hence, public health organizations must embrace the cultural and societal factors impacting the health of communities prior to any health behavior interventions (Centers for Disease Control and Prevention, 2024). As an example, McKinsey and company showed that companies in the top quartile for ethnic and cultural diversity on executive teams were 33% more likely to have industry-leading profitability. Companies in the bottom 25% for both gender and ethnic/cultural diversity were 29% less likely to experience profitability above the industry average (Hunt et al., 2018).

One successful example of the application of cultural competence in public health programming comes from the Ebola Virus Disease response. This response supported women and men based on their role in the community. Hence, women were trained for example on treating family members, quarantine procedures, and personal protection to limit transmission. This use of gendered roles in a high-risk community allowed for the creation of effective strategies for community-based surveillance systems and treatment in areas with poor infrastructure and limited the strain on poorly resourced health clinics (Southall et al., 2017). In Thailand and Malaysia, migrants and refugees reported challenges with accessing health systems and discrimination by service providers. To address these challenges, culturally sensitive interventions included: (i) ensuring the availability of interpreter services to mitigate language errors and misunderstandings, (ii) migrant peer educators especially in outpatient settings where interpreters are not available, (iii) capacity strengthening of health personnel on providing culturally appropriate care which is especially relevant for mental health services, and (iv) migrant education programs tailored to their cultural background to encourage health-seeking behaviors (Pocock et al., 2020).

CULTURAL COMPETENCY AND GENDER

A critical component of both cultural competency and One Health is to first determine the existing disparities, the areas of miscommunication, and identify underserved and underrepresented groups and populations so that we can then devise effective solutions. Utilizing a gender equity and social inclusion lens, we look at some of the critical problems and barriers plaguing women and minorities and how these can be addressed.

Healthcare Barriers to Gender Equality

Gender inequality continues to be a major challenge in healthcare worldwide. Women in both developed and developing countries suffer from biases in healthcare such as physicians' perception and treatment of women's symptoms and diagnoses differently from those of men (Safran et al., 1997; Samulowitz et al., 2018), lack of access to safe and timely healthcare options (Azad et al., 2020; Bonita & Beaglehole, 2014), a continued need to develop global understanding of how chronic and infectious diseases impact men and women differently, low dissemination of this information to the public (Vlassoff, 2007; Regitz-Zagrosek, 2012), and the myth that female healthcare issues of importance are mostly those related to reproductive health (Bonita & Beaglehole, 2014; Bustreo et al., 2012). On the contrary, two-thirds of women significantly suffer from cardiovascular problems, stroke, chronic diseases, and violence (Bonita & Beaglehole, 2014). Due to the social designation of women in society as caregivers as opposed to men who have more decision-making power and are considered as the providers, women's health has suffered globally (Barros et al., 2012; Langer et al., 2015). In addition to women, other sexual and gender minorities also face discrimination, inequalities, and prejudice in medical treatment (Davison et al., 2021; Lau et al., 2020). To protect these vulnerable populations, it is important to highlight the various barriers and difficulties that they face at local, state, and federal levels while seeking care and develop measures that will systematically address these challenges and increase a culturally competent workforce in healthcare. For example, people of color are more likely to trust their physician if they share racial, gender, or cultural backgrounds with that of their physician (Torres, 2018). Physicians of color are more likely to serve underserved and underrepresented people (Torres, 2018). Thus, creating pathways that allow for culturally diverse individuals to become healthcare professionals is essential in providing care to culturally diverse patients. Additionally, healthcare clinics and organizations should regularly evaluate their geographic area and their client base in order to understand the demographics of their patients and how they can best provide care and support to their patients.

Culturally, racially, ethnically, and gender-diverse clientele need specific tailor-made interventions, and in these cases, care cannot operate on a simplified method. Smedley et al. (2003) discuss three conceptual approaches (i.e., (i) The Cultural Sensitivity/Awareness Approach, (ii) The Multicultural/Categorical Approach, and (iii) The Cross-Cultural Approach) that have to be implemented together in order to reduce patient dissatisfaction, increase communication and trust between provider and patient, and provide the best care to patients. Increasing self-reflection among trainees and senior healthcare providers, learning-focused didactics, strategies to collaborate with the patient and family for optimal patient care, learner community immersion, and the use of vignettes, problem-based learning cases, medical encounter videos, and individual case-based discussion are several strategies by which healthcare professionals can be culturally competent (Smedley et al., 2003; Brooks et al., 2019; Jongen et al., 2018). There are other ways in which healthcare professionals lack the tools to provide help. For example, a study found that health professionals in Afghanistan expressed inadequate medical knowledge and interpersonal skills to address sensitive issues, such as domestic, physical, and sexual violence, that women and minorities often face (Samar et al., 2014). Schachter et al. (2004) spoke to survivors of child abuse and asked them the crucial factors they want to see when they interact with their physician/physical therapist/healthcare provider (Schachter et al., 2004). The participants responded that patients needed to feel a sense of safety with their healthcare provider, disclosure about past abuse should be left to patients (if they feel safe and if their provider believes them, then participants were more likely to disclose), physicians should practice more sensitivity and respect for their patients, physicians should share treatment and care information with patients, and involve coordinated care through interdisciplinary treatment teams. Similar results were found in another study by McGregor et al. (2010) that focused on specific steps that physicians need to undertake and be aware of before providing care to female survivors of child abuse (McGregor et al., 2010). In order to provide optimal care to women and minorities, an array of educational, organizational, and cultural changes have to take place in schools, clinics, healthcare institutions, and health insurance settings.

Institutional and Workplace Barriers to Gender Equality

Cultural competency and One Health also need to focus on social barriers and institutional challenges that cause gender inequity. In both private and public settings, these barriers include occupational segregation; lack of access to land, capital, financial resources, and technology; sexism; insufficient representation of women in leadership roles and higher positions within organizations; gender stereotypes surrounding higher positions that are discriminating; as well as gender-based violence (UN Women, 2022; Barreto et al., 2009). The terms "glass ceiling" and "labyrinth" refer to invisible yet complicated and exhausting challenges that women have to face in the workforce especially as they try to move to executive positions which include prejudice, sexual harassment, tokenism, and isolation (Barreto et al., 2009; Schwanke, 2013). In addition, due to the cultural expectation of women as caregivers for children and the elderly, women tend to lose out on experiences and opportunities at their workplace that could have given them the skills necessary to reach executive positions in the company (Schwanke, 2013; Guerrero, 2011).

Given these challenges in the workplace, companies must not only work on protecting women and minorities from discrimination but also train employees to adapt their worldview to that of a global mindset. When companies realize that cultural competency at the workplace can increase creativity, help develop and strengthen ties in international business, improve communication and working styles of employees, increase productivity, and have a healthy environment in the workplace (Goodman, 2012), they can invest in cultural competency and proactively increase gender diversity. Including women in business and in leadership positions leads to stronger company performance; women challenge gender norms and structure in business; women increase conversations around societal issues; and women brainstorm ways as to how businesses can help to address these issues (Loop & DeNicola, 2019). Gender diversity can be elevated in the corporate sector by expanding the size of executive boards, change recruiting strategies to incorporate meritorious candidates from all genders and backgrounds, hire more women and minorities, and pay attention to what women and minorities have to say in the workplace (Loop & DeNicola, 2019). Another way to address the cultural shortcomings in the workforce is by creating interdisciplinary courses that teach students elements of business, psychology, social sciences, race and gender, and cultural communication to enable them to check their biases and form the next generation of forward-thinking leaders and businesspeople (Egan & Bendick, 2008). Further measures that need to be undertaken to increase cultural competency in organizations are regularly assessing the cultural awareness of employees, identifying gaps in knowledge and attitudes, developing training, workshop, and communication materials to fill these gaps, increasing linguistic literacy in multiple languages, reducing ethnocentrism among employees and executives, increasing cultural flexibility, soliciting feedback and comments from team members, addressing the grievances or issues raised, and practicing effective cross-cultural team building (Caligiuri & Tarique, 2012; Lamson, 2018; Goodman, 2012).

In order to increase cultural competency and gender inclusion in organizations and workplaces, one has to also understand the systemic, cultural, structural, organizational, and institutional barriers that hinder women and minorities. For example, Orfan et al. (2021) found that instructional materials used in schools in Afghanistan are one of the areas in which gender inequality is very easily institutionalized (Orfan, 2023). The findings showed that women were substantially underrepresented, whereas men were significantly overrepresented in text and illustrations. Due to attitudes, access issues, organizational structures, and culture, there are not enough girls and women who get into science and medicine. The National Academies of Science, Engineering, and Medicine published a report stating that the pervasive, persistent, and damaging culture of harassment limits participation and advancement of women in science, technology, engineering, and mathematics (STEM) (Committee on the Impacts of Sexual Harassment in Academia et al., 2018). Coe et al. (2019) have outlined measures that should be taken to systematically reduce institutional barriers for women and minorities, particularly in science (Coe et al., 2019). These measures include changing structures and organizational climates toward a more inclusive academic and research setting, highlighting the role of academic societies in advocating for gender equality, visibly emphasizing the contributions of women in science, medicine, research, and innovation, increasing inclusive leadership in organizations, and increasing allyship at various levels to support women and minorities (Coe et al., 2019). In addition, Oh et al. (2021) discussed that increasing mentorship opportunities for women at medical institutions that will lead to career promotions, funding and investing in research by women from various socio-economic backgrounds, understanding the balancing act played by women as caregivers and working women, mitigating the "minority tax," and increasing equity in decision-making, promotion, mentoring, and hiring processes are some ways female faculty, researchers, and clinicians can advance (Oh et al., 2021). Conscious efforts to recruit, promote, retain, and celebrate women leaders, scientists, and researchers, understanding their barriers and grievances, and amplifying their concerns are some of the best ways to reduce institutional barriers. On the policy level, the International Labor Organization drafted a report that focuses on five key areas where government policies should be made to reduce gender inequality. These include achieving equal pay for work of equal value; preventing and eliminating violence and harassment in the world of work; creating a harmonious work-life balance for both women and men; promoting women's equal representation in leadership in the world of work; and investing in a future of work that works for women (International Labour Organization, 2021).

Social Barriers to Gender Equality

Women face a plethora of societal barriers that prevent them and other minorities from participating in the workforce and having more autonomy. These barriers include poor education, poverty, burden of unpaid work, gender wage gap, lack of proper sex education and access to contraceptives, unwanted child and teenage pregnancies, and child marriage (Samar et al., 2014; Economic Commission for Latin America and the Caribbean, 2019; Hof & Richters, 1999). There are several pathways and measures that must be implemented in order to achieve an inclusive and culturally competent society. Issues that plague women and other gender minorities often lack effective justice. To amend this, UN Women partnered with the Pathfinders for Peaceful, Just and Inclusive Societies; the International Development Law Organization; the World Bank; and the Task Force on Justice to form the High-Level Group (HLG) on Justice for Women (UN Women, IDLO, The World Bank, Task Force on Justice, 2019). The HLG consisted of the governments of Argentina, Canada, The Gambia, the Commonwealth of Australia, the Association of Women in Development, and other organizations (UN Women, IDLO, The World Bank, Task Force on Justice, 2019). This group recommend a collection of approaches to achieving justice for women. These include eliminating legal discrimination against women, preventing and responding to intimate partner violence, overcoming disadvantages for poor and marginalized women, empowering women economically and as rights holders, and including women as decision-makers (UN Women, IDLO, The World Bank, Task Force on Justice, 2019). These approaches have to be combined with countries making significant legal reform to protect women and minorities; increasing legal and cultural literacy among justices, para-legal firms, advocates/lawyers, police forces, and emergency and support services; increasing access to legal aid; eradicating patriarchal biases in law; enabling legal identity; equal representation of women in decision-making at all levels in the justice sector; and making investments in data and monitoring and evidence-based policies (UN Women, IDLO, The World Bank, Task Force on Justice, 2019).

GENDER EQUALITY AND LAND OWNERSHIP

Land ownership is important for an individual to secure their livelihood and for acquiring the necessary resources to thrive in society. Despite legal protections in place, women own less than 20% of land globally (World Economic Forum, 2017). Often, women and minorities lose out on their rights to the land due to legal and social barriers, cultural norms, gender-based violence, and underrepresentation in decision-making (IUCN, 2020). Land rights tend to be held by men or kinship groups controlled by men, and women have access mainly through a male relative, usually a father or husband (Kimani, 2008). Women face additional barriers when HIV is involved; if the woman's husband dies of HIV, the family tends to blame the woman as the cause of death. They not only lose the land and resources but are left to fend for themselves and depend on squatting, begging, and, in many cases, relegated to sex work and face violence (Kimani, 2008; Odeny, 2013). Ironically, research has shown that when women manage land, they contribute a greater proportion of their agricultural and land-based income to their household than men, which improves food security, children's health, and education as a result (Salcedo-La Viña, 2021). Additionally, when women manage land, their management leads to positive environmental impacts as opposed to men (Cook et al., 2019; Ding et al., 2016; Salcedo-La Viña, 2020). Involvement of women in the decision-making process, such as providing them secure rights to land and local forests, empowers women, gives them agency, helps promote biodiversity, protects forests and natural areas, and can even lead to implementation of key policies such as in the case of the Chipko movement in India that prompted the Indian government to amend the Indian Forest Act, 1927, and introduced the Forest Conservation Act 1980 (Shiva & Bandyopadhyay, 1986; Singh & Mishra, 2019) as well as the massive success of the Green Belt Movement in Kenya (The Green Belt Movement, 2024; U.S. Agency for International Development, 2011).

Concerted efforts between local, state, national, and international agencies and organizations have to be conducted in order to increase women's land ownership, women's land rights, and empowerment. Odeny (2013) discusses some initiatives that have been taken in African countries to increase women's land rights and ownership (Odeny, 2013). These include affirmative government policies (setting quotas of seats for women in various levels of government), decentralized and democratized customary land administration, traditional leaders becoming drivers for women's land rights, land registration and certification, and grassroots organizations that can increase awareness and result in information dissemination. It is essential for governments and international organizations to ensure that:

- Laws protecting women's land rights and ownership are upheld and implemented;
- Gender-responsive land administration systems are established;
- Capacity building at all levels of government is conducted to encourage participation of women and minorities;
- Land tenure and titling security reforms are undertaken;

- Punitive actions when women are denied access to their land are enforced;
- Further research is conducted into measures that will aid in women's land ownership, food security, and health;
- Continuous advocacy work is conducted to raise awareness that will lead to enactment of laws and policies;
- Training, technology, and resources are provided to women so they can effectively use their land; and
- Regular tracking of progress of these interventions, and legal policies are enacted (Odeny, 2013) (Swedish International Development Cooperation Agency (SIDA), 2009) (Salcedo-La Viña, 2020; Asiimwe, 2001; Quisumbing & Pandolfelli, 2010).

LIVESTOCK OWNERSHIP AND GENDER EQUALITY

In addition to land ownership, women play a key role in the livestock industry in several nations, but their work is often underestimated and not highlighted. Shafiq (2008) discusses women's roles in Baluchistan where they are responsible for feeding the animals, cleaning them and their shelters, converting manure into fuel, processing wool and hair of the animals thereby significantly contributing to the livestock industry, yet their contributions are often underreported (Shafiq, 2008). Zahoor et al. (2013) recommended that rural women in Pakistan who engage in livestock and crop production should be provided with credit facilities, extension services and agricultural education facilities, and training to increase their agency and empowerment (Zahoor et al., 2013). Problems associated with women's access to livestock are similar to those surrounding women's access to land and hence require similar laws, policies, governmental and non-governmental aid, and changes in societal norms and cultural beliefs (Swanepoel et al., 2010; Kechero, 2008; Kristjanson et al., 2014).

RECOMMENDATIONS FOR APPLICATION OF CULTURAL COMPETENCE IN INTERDISCIPLINARY TEAMS

Recommendations for the Application of Cultural Competency in Education Sector

The following recommendations are aimed at academic institutions and educators to facilitate the discussion of strategies to integrate cultural competence in curricula aimed at building the capacity of interdisciplinary practitioners:

- Embedding systematically topics of cultural competency in health professionals' education to guarantee they are receiving training where they can ensure high-quality, individualized care for all patients.
- Synergizing and linking the cultural competence skills and curricula with the medical content to allow educators to address them as complementary rather than separate.

- Connecting students with global health perspectives and experience with medical/nursing students to allow for cross-fertilization of cultural competence teaching and skills.
- Making cultural competence courses pre-requisites and mandatory as part of health professionals' educational curricula with courses that offer in-depth deep dives in the topic.
- Updating faculty qualifications, skills, and teaching methodologies to ensure didactic and practical sessions related to cultural competence training are up to date, relevant, and suitable to prepare students for the real-world application.
- Encouraging academics to contribute to the global knowledge base and learning related to the impact of cultural competence on health as well as the best practices that optimize its application, especially in the context of interdisciplinary teams.
- Accounting in the design of cultural competence as a multidisciplinary topic comprising cultural competence, which includes culture, society, law, politics, religion, economics, medical ethics, and other disciplines.
- Focusing on building students' cultural knowledge and skills but also go beyond that through embracing new frameworks such as the intercultural competence (ICC) framework (Deardorff, 2006), especially for public health training institutions, which focus on the mutual and dual need for cultural understanding and interaction between health professionals and patients (Figure 11.2).
- Conducting continuous evaluation of cultural competence training curricula through self-assessments and structured research, especially in the context of a rapidly evolving global health field and cross-border movements impacting population behaviors.

Recommendations for the Application of Cultural Competency in Clinical Settings

Culturally competent healthcare systems focus on decreasing the disparities in quality of care provided through delivering clinical services that are adapted to the patient's culture, gender, ethnicity, social, and economic factors. We share the following recommendations for consideration by actors in clinic settings to support culturally competent health workforce and clinical services:

- Adapting to the evolving needs of the populations served by the healthcare services to ensure continuous adaptation and responsiveness.
- Including cultural competence assessments and training in clinical practitioners' onboarding.
- Building continuously the knowledge and skills of clinical practitioners on cultural competency and its application in the clinical setting.

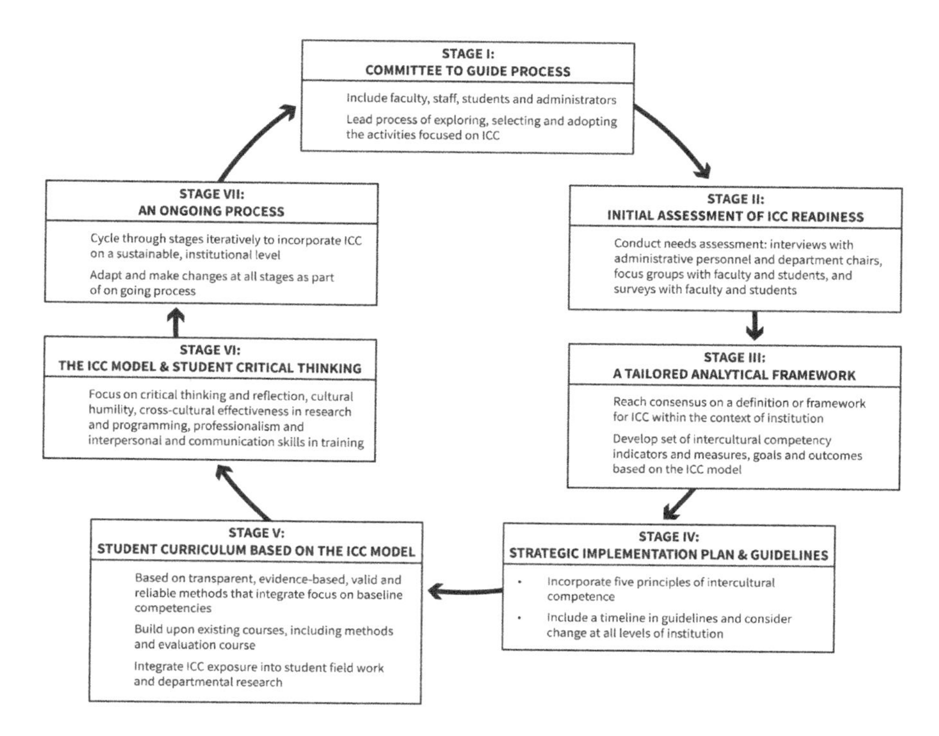

FIGURE 11.2 Intercultural competence framework.

- Ensuring clinical practitioners tailor treatment plans and behavior change recommendations to the patients' cultural, social, and economic backgrounds.
- Strengthening clinical practitioners' communication skills and equipping them with strategies to tailor their communication to the diverse patient population they are serving.
- Ensuring healthcare centers are accessible to patients from diverse cultural backgrounds and mitigating any accessibility challenges.
- Instituting systems to allow for the recruitment of a diverse workforce at health facilities with the competencies required to serve the target population.
- Providing interpreters and language support services to overcome the language barriers between patients and practitioners.

Recommendations for the Application of Cultural Competency in Global Health Programming

These recommendations serve to guide global health programming to be responsive to the evolving public health challenges especially in the context of increasing zoonotic diseases and the growing need for interdisciplinary teams of practitioners:

- Making sure cultural norms and frame of reference of individuals working in organizations are not influencing programming and imposing inappropriate interventions to foreign populations.
- Valuing the perspective of the country targeted by health programming through intentionally taking into account the country's cultural and social norms in the design and implementation of interventions.
- Encouraging public health professionals to conduct continuous self-assessments to increase awareness of their bias, actively embrace change, and be more open toward new ideas and strategies more suitable for foreign countries.
- Leading from behind, meaning that public health professionals and organizations must let countries lead their own agenda and vision and focus global health programming on tailored technical assistance as needed and where needed.
- Designing solutions with an eye on diversity and inclusion to ensure health programming is effective and reaching the marginalized populations who most often have their own cultural and social dynamics to account for.
- Applying systems thinking and complexity-aware theories that acknowledge the dynamic, complex, interconnected, and evolving nature of health systems, SDH, and cultural competence.
- Deepening understanding of the interrelations, interactions, and varying perspectives of health systems components which impact barriers and facilitate availability, access, quality, and reliability of health services for all.
- Looking beyond quantitative data, through the implementation of qualitative methods, to explore root causes of social and cultural determinants of health and design public health programs with inputs from minorities and marginalized populations.

Recommendation for Strategies to Ensure Cultural Competencies Are Applied in a Gender and Socially Inclusive Manner

To ensure smooth functioning of the One Health approach, societies, governments, institutions, and people have to adopt cultural competency as well as a gender and socially inclusive lens. Here are some recommendations on the same:

- Updating the outdated concepts of gender and sex and increasing knowledge and awareness on this topic
- Including women and minorities in research and policymaking

- Utilizing a socially inclusive approach to solving challenges: including women while defining the problem, asking their input, and having them be a part of the solution
- Updating and enacting institutional policies, care practices, social attitudes, and information practices required to provide safe and quality healthcare
- Working on developing curricula and educational materials that foster from an early age in children the importance of including all voices in decision-making and solution generation
- Building effective communication materials in institutions, healthcare, governments, schools, businesses, and non-governmental organizations (NGOs)
- Regularly checking for biases and stereotypes in individuals and organizations, reducing ethnocentrism, and increasing a culturally inclusive mindset
- Enforcing laws, policies, and punitive actions in a timely manner to prevent gender inequality
- Actively hiring, recruiting, and promoting women in all fields of work, business, commerce, trade, and research

REFERENCES

Ahmann, E. (2002). Developing cultural competence in health care settings. *Pediatric Nursing*, *28*(2), 133.

Asiimwe, J. (2001). Making Women's land rights a reality in Uganda: Advocacy for co-ownership by spouses. *Yale Human Rights and Development Journal*, *4*(1 (8)), 171–187.

Association of American Medical Colleges. (2005). *Cultural competence education for medical students*. Association of American Medical Colleges. https://www.aamc.org/media/20856/download

Azad, A. D., Charles, A. G., Ding, Q., Trickey, A. W., & Wren, S. M. (2020). The gender gap and healthcare: Associations between gender roles and factors affecting healthcare access in Central Malawi, June–August 2017. *Archives of Public Health*, *78*(1), 119. https://doi.org/10.1186/s13690-020-00497-w

Barreto, M., Ryan, M. K., & Schmitt, M. T. (2009). Introduction: Is the glass ceiling still relevant in the 21st century? In M. Barreto, M. K. Ryan, & M. T. Schmitt (Eds.), *The glass ceiling in the 21st century: Understanding barriers to gender equality.* (pp. 3–18). American Psychological Association. https://doi.org/10.1037/11863-001

Barros, A. J., Ronsmans, C., Axelson, H., Loaiza, E., Bertoldi, A. D., França, G. V., Bryce, J., Boerma, J. T., & Victora, C. G. (2012). Equity in maternal, newborn, and child health interventions in Countdown to 2015: A retrospective review of survey data from 54 countries. *The Lancet*, *379*(9822), 1225–1233. https://doi.org/10.1016/S0140-6736(12)60113-5

Betancourt, J. R. (2004). Cultural competence—Marginal or mainstream movement? *New England Journal of Medicine*, *351*(10), 953–955. https://doi.org/10.1056/NEJMp048033

Betancourt, J. R., Green, A. R., & Carrillo, J. E. (2002). *Cultural competence in health care: Emerging frameworks and practical approaches* (Vol. 576). Commonwealth Fund, Quality of Care for Underserved Populations New York, NY.
Betancourt, J. R., Green, A. R., Carrillo, J. E., & Ananeh-Firempong, O. (2003). Defining cultural competence: A practical framework for addressing racial/ethnic disparities in health and health care. *Public Health Reports*, *118*(4), 293–302. https://doi.org/10.1016/S0033-3549(04)50253-4
Bonita, R., & Beaglehole, R. (2014). Women and NCDs: Overcoming the neglect. *Global Health Action*, *7*(1), 23742. https://doi.org/10.3402/gha.v7.23742
Brett, P. (2000). *Developing cross-cultural competence in business through multimedia courseware*. *ReCALL*, *12*(2), 196–208. https://doi.org/10.1017/S0958344000000628
Brooks, L. A., Manias, E., & Bloomer, M. J. (2019). Culturally sensitive communication in healthcare: A concept analysis. *Collegian*, *26*(3), 383–391. https://doi.org/10.1016/j.colegn.2018.09.007
Brunett, M., & Shingles, R. R. (2018). Does having a culturally competent Health Care provider affect the patients' experience or satisfaction? A critically appraised topic. *Journal of Sport Rehabilitation*, *27*(3), 284–288. https://doi.org/10.1123/jsr.2016-0123
Bustreo, F., Knaul, F., Bhadelia, A., Beard, J., & de Carvalho, I. A. (2012). Women's health beyond reproduction: Meeting the challenges. *Bulletin of the World Health Organization*, *90*(7), 478–478. https://doi.org/10.2471/BLT.12.103549
Caligiuri, P., & Tarique, I. (2012). Dynamic cross-cultural competencies and global leadership effectiveness. *Journal of World Business*, *47*(4), 612–622. https://doi.org/10.1016/j.jwb.2012.01.014
Calvillo, E., Clark, L., Ballantyne, J. E., Pacquiao, D., Purnell, L. D., & Villarruel, A. M. (2009). Cultural competency in Baccalaureate nursing education. *Journal of Transcultural Nursing*, *20*(2), 137–145. https://doi.org/10.1177/1043659608330354
Campinha-Bacote, J. (2002). The process of cultural competence in the delivery of Healthcare Services: A model of care. *Journal of Transcultural Nursing*, *13*(3), 181–184. https://doi.org/10.1177/10459602013003003
Carpenter-Song, E. A., Schwallie, M. N., & Longhofer, J. (2007). Cultural competence reexamined: Critique and directions for the future. *Psychiatric Services*, *58*(10), 1362–1365.
Carrizales, T. (2010). Exploring cultural competency within the Public Affairs curriculum. *Journal of Public Affairs Education*, *16*(4), 593–606. https://doi.org/10.1080/15236803.2010.12001616
Castro, A., & Ruiz, E. (2009). The effects of nurse practitioner cultural competence on Latina patient satisfaction. *Journal of the American Academy of Nurse Practitioners*, *21*(5), 278–286. https://doi.org/10.1111/j.1745-7599.2009.00406.x
Centers for Disease Control and Prevention. (2022). *About One Health*. One Health. https://www.cdc.gov/one-health/about/index.html
Centers for Disease Control and Prevention. (2024, February 23). *Cultural competence in health and human services | National prevention information network*. https://npin.cdc.gov/pages/cultural-competence-health-and-human-services
Chao, M. M., Okazaki, S., & Hong, Y. (2011). The quest for multicultural competence: Challenges and lessons learned from clinical and organizational research. *Social and Personality Psychology Compass*, *5*(5), 263–274. https://doi.org/10.1111/j.1751-9004.2011.00350.x
Cho, M.-K., & Kim, M. Y. (2022). Factors affecting the global health and cultural competencies of nursing students. *International Journal of Environmental Research and Public Health*, *19*(7), 4109. https://doi.org/10.3390/ijerph19074109

Coe, I. R., Wiley, R., & Bekker, L.-G. (2019). Organisational best practices towards gender equality in science and medicine. *The Lancet*, *393*(10171), 587–593. https://doi.org/10.1016/S0140-6736(18)33188-X

Committee on the Impacts of Sexual Harassment in Academia, Committee on Women in Science, Engineering, and Medicine, Policy and Global Affairs, & National Academies of Sciences, Engineering, and Medicine. (2018). *Sexual harassment of women: Climate, culture, and consequences in academic sciences, engineering, and medicine* (P. A. Johnson, S. E. Widnall, & F. F. Benya, Eds.; p. 24994). National Academies Press. https://doi.org/10.17226/24994

Cook, N. J., Grillos, T., & Andersson, K. P. (2019). Gender quotas increase the equality and effectiveness of climate policy interventions. *Nature Climate Change*, *9*(4), 330–334. https://doi.org/10.1038/s41558-019-0438-4

Davison, K., Queen, R., Lau, F., & Antonio, M. (2021). Culturally competent gender, sex, and sexual orientation information practices and electronic health records: Rapid review. *JMIR Medical Informatics*, *9*(2), e25467. https://doi.org/10.2196/25467

Deardorff, D. K. (2006). Identification and assessment of intercultural competence as a student outcome of internationalization. *Journal of Studies in International Education*, *10*(3), 241–266. https://doi.org/10.1177/1028315306287002

Ding, H., Veit, P. G., Blackman, A., Gray, E., Reytar, K., Altamirano, J. C., & Hodgdon, B. (2016). *Climate benefits, tenure costs: The economic case for securing indigenous land rights in the Amazon*. World Resources Institute. https://www.wri.org/research/climate-benefits-tenure-costs

Economic Commission for Latin America and the Caribbean. (2019). *Women's autonomy in changing economic scenarios* (LC/CRM. 14/3). United Nations. https://repositorio.cepal.org/bitstream/handle/11362/45037/S1900722_en.pdf

Egan, M. L., & Bendick, M. (2008). Combining multicultural management and diversity into one course on cultural competence. *Academy of Management Learning & Education*, *7*(3), 387–393. https://doi.org/10.5465/amle.2008.34251675

Estevan, M. D. G., & Solano Ruíz, M. D. C. (2017). La aplicación del modelo de competencia cultural en la experiencia del cuidado en profesionales de Enfermería de Atención Primaria. *Atención Primaria*, *49*(9), 549–556. https://doi.org/10.1016/j.aprim.2016.10.013

Expert Panel on Cultural Competence Education for Students in Medicine and Public Health. (2012). *Cultural competence education for students in medicine and public health: Report of an expert panel*. Association of American Medical Colleges and Association of Schools of Public Health. https://www.pcpcc.org/sites/default/files/resources/Cultural%20Competence%20Education%20for%20Students%20in%20Medicine%20%26%20Public%20Health.pdf

Ferorelli, D., Mandarelli, G., & Solarino, B. (2020). Ethical challenges in Health Care policy during COVID-19 pandemic in Italy. *Medicina*, *56*(12), 691. https://doi.org/10.3390/medicina56120691

Garneau, A. B., & Pepin, J. (2015). Cultural competence: A constructivist definition. *Journal of Transcultural Nursing*, *26*(1), 9–15. https://doi.org/10.1177/1043659614541294

Goodman, N. (2012). Training for cultural competence. *Industrial and Commercial Training*, *44*(1), 47–50. https://doi.org/10.1108/00197851211193426

Gregory, D., Harrowing, J., Lee, B., Doolittle, L., & O'Sullivan, P. S. (2010). Pedagogy as influencing nursing students' essentialized understanding of culture. *International Journal of Nursing Education Scholarship*, *7*(1). https://doi.org/10.2202/1548-923X.2025

Guerrero, L. (2011). Women and leadership. In W. Rowe & L. Guerrero (Eds.), *Cases in leadership* (Vol. 17, pp. 380–412). SAFE Publications. https://journals.sagepub.com/doi/10.1177/097168581001700108

Hansen, N. D., Pepitone-Arreola-Rockwell, F., & Greene, A. F. (2000). Multicultural competence: Criteria and case examples. *Professional Psychology: Research and Practice*, *31*(6), 652–660. https://doi.org/10.1037/0735-7028.31.6.652

Health Research & Educational Trust. (2013). *Becoming a culturally competent Health Care Organization*. Health Research & Educational Trust. www.hpoe.org

Hof, C., & Richters, A. (1999). Exploring intersections between teenage pregnancy and gender violence: Lessons from Zimbabwe. *African Journal of Reproductive Health*, *3*(1), 51. https://doi.org/10.2307/3583229

Hofstede, G. J., Pedersen, P. B., & Hofstede, G. (2002). *Exploring culture: Exercises, stories and synthetic cultures*. Nicholas Brealey.

Holmes, D. (2014). David Napier: Cultivating the role of culture in health. *The Lancet*, *384*(9954), 1568. https://doi.org/10.1016/S0140-6736(14)61937-1

Hunt, V., Prince, S., Dixon-Fyle, S., & Yee, L. (2018). *Delivering through diversity*. McKinsey&Company. https://dln.jaipuria.ac.in:8080/jspui/bitstream/123456789/14237/1/Delivering-through-diversity_full-report.pdf

Hwang, M. C., & Jang, I. (2017). A study of predictors influencing access to health care service by immigrant wives: Focusing on different analysis using Andersen Behavioral Model. *Journal of Governance Studies*, *12*(1), 31–57.

International Labour Organization. (2021). *Empowering women at work: Government laws and policies for gender equality*. International Labour Office, United Nations.

IUCN. (2020, January 23). *Gender and the environment: What are the barriers to gender equality in sustainable ecosystem management? | IUCN*. https://www.iucn.org/news/gender/202001/gender-and-environment-what-are-barriers-gender-equality-sustainable-ecosystem-management

Jeong, G. H., Park, H.-S., Kim, K. W., Kim, Y. H., Lee, S. H., & Kim, H.-K. (2016). A concept analysis of cultural nursing competence. *Korean Journal of Women Health Nursing*, *22*(2), 86. https://doi.org/10.4069/kjwhn.2016.22.2.86

Jongen, C., McCalman, J., & Bainbridge, R. (2018). Health workforce cultural competency interventions: A systematic scoping review. *BMC Health Services Research*, *18*(1), 232. https://doi.org/10.1186/s12913-018-3001-5

Kaffes, I., Moser, F., Pham, M., Oetjen, A., & Fehling, M. (2016). Global health education in Germany: An analysis of current capacity, needs and barriers. *BMC Medical Education*, *16*(1), 304. https://doi.org/10.1186/s12909-016-0814-y

Kechero, Y. (2008). Gender responsibility in smallholder mixed crop–livestock production systems of Jimma zone, South West Ethiopia. *Livestock Research for Rural Development*, *20*.

Kim, S. H., Kim, K. W., & Bae, K. E. (2014). Experiences of Nurses Who Provide Childbirth Care for Women with Multi-cultural Background. *Journal of Korean Public Health Nursing*, *28*(1), 87–101. https://doi.org/10.5932/JKPHN.2014.28.1.87

Kimani, M. (2008, April 15). *Women struggle to secure land rights*. Africa Renewal. https://www.un.org/africarenewal/magazine/april-2008/women-struggle-secure-land-rights

Kristjanson, P., Waters-Bayer, A., Johnson, N., Tipilda, A., Njuki, J., Baltenweck, I., Grace, D., & MacMillan, S. (2014). Livestock and Women's Livelihoods. In A. R. Quisumbing, R. Meinzen-Dick, T. L. Raney, A. Croppenstedt, J. A. Behrman, & A. Peterman (Eds.), *Gender in Agriculture* (pp. 209–233). Springer Netherlands. https://doi.org/10.1007/978-94-017-8616-4_9

Lachance, J. A., & Oxendine, J. S. (2015). Redefining leadership education in graduate public health programs: Prioritization, focus, and guiding principles. *American Journal of Public Health, 105*(S1), S60–S64. https://doi.org/10.2105/AJPH.2014.302463

Lamson, M. (2018, July 3). *10 tips to develop your firm's cultural competence.* Inc.Com. https://www.inc.com/melissa-lamson/cultural-competence-your-most-valuable-business-asset.html

Langer, A., Meleis, A., Knaul, F. M., Atun, R., Aran, M., Arreola-Ornelas, H., Bhutta, Z. A., Binagwaho, A., Bonita, R., Caglia, J. M., Claeson, M., Davies, J., Donnay, F. A., Gausman, J. M., Glickman, C., Kearns, A. D., Kendall, T., Lozano, R., Seboni, N., … Frenk, J. (2015). Women and Health: The key for sustainable development. *The Lancet, 386*(9999), 1165–1210. https://doi.org/10.1016/S0140-6736(15)60497-4

Lau, F., Antonio, M., Davison, K., Queen, R., & Devor, A. (2020). A rapid review of gender, sex, and sexual orientation documentation in electronic health records. *Journal of the American Medical Informatics Association, 27*(11), 1774–1783. https://doi.org/10.1093/jamia/ocaa158

Loop, P., & DeNicola, P. (2019, February 18). You've committed to increasing gender diversity on your board. Here's how to make it happen. *Harvard Business Review.* https://hbr.org/2019/02/youve-committed-to-increasing-gender-diversity-on-your-board-heres-how-to-make-it-happen

Lynn, K. A., McKinnon, T., Madigan, E., & Fitzpatrick, J. J. (2021). Assessment of global health competence of nursing faculty in prelicensure programs. *Journal of Nursing Education, 60*(1), 20–24. https://doi.org/10.3928/01484834-20201217-05

McGregor, K., Glover, M., Gautam, J., & Jülich, S. (2010). Working sensitively with child sexual abuse survivors: What female child sexual abuse survivors want from health professionals. *Women & Health, 50*(8), 737–755. https://doi.org/10.1080/03630242.2010.530931

Mews, C., Schuster, S., Vajda, C., Lindtner-Rudolph, H., Schmidt, L. E., Bösner, S., Güzelsoy, L., Kressing, F., Hallal, H., Peters, T., Gestmann, M., Hempel, L., Grützmann, T., Sievers, E., & Knipper, M. (2018). Cultural competence and global health: Perspectives for medical education – Position paper of the GMA committee on cultural competence and global health [Text/html]. *GMS Journal for Medical Education, 35*(3), Doc28. https://doi.org/10.3205/ZMA001174

Muzychenko, O. (2008). Cross-cultural entrepreneurial competence in identifying international business opportunities. *European Management Journal, 26*(6), 366–377. https://doi.org/10.1016/j.emj.2008.09.002

National Association of Social Workers. (2001). *NASW standards for cultural competence in social work practice.* National Association of Social Workers. https://catholiccharitiesla.org/wp-content/uploads/NASW-Cultural-Competence-in-Social-Work-Practice.pdf

Odeny, M. (2013, April 8). *Improving Access to Land and strengthening Women's land rights in Africa.* Annual World Bank Conference on Land and Poverty 2013, Washington, DC. https://landwise-production.s3.amazonaws.com/2022/03/odeny_improving-access-to-land-in-africa_2013-1.pdf

Oh, L., Linden, J. A., Zeidan, A., Salhi, B., Lema, P. C., Pierce, A. E., Greene, A. L., Werner, S. L., Heron, S. L., Lall, M. D., Finnell, J. T., Franks, N., Battaglioli, N. J., Haber, J., Sampson, C., Fisher, J., Pillow, M. T., Doshi, A. A., & Lo, B. (2021). Overcoming barriers to promotion for women and underrepresented in medicine faculty in academic emergency medicine. *Journal of the American College of Emergency Physicians Open, 2*(6), e12552. https://doi.org/10.1002/emp2.12552

Ohana, S., & Mash, R. (2015). Physician and patient perceptions of cultural competency and medical compliance. *Health Education Research*, *30*(6), 923–934. https://doi.org/10.1093/her/cyv060

Orfan, S. N. (2021). High school English textbooks promote gender inequality in Afghanistan. *Pedagogy, Culture & Society*, *31*(3), 403–418. https://doi.org/10.1080/14681366.2021.1914148

Paez, K. A., Allen, J. K., Beach, M. C., Carson, K. A., & Cooper, L. A. (2009). Physician cultural competence and patient ratings of the patient-physician relationship. *Journal of General Internal Medicine*, *24*(4), 495–498. https://doi.org/10.1007/s11606-009-0919-7

Pocock, N. S., Chan, Z., Loganathan, T., Suphanchaimat, R., Kosiyaporn, H., Allotey, P., Chan, W.-K., & Tan, D. (2020). Moving towards culturally competent health systems for migrants? Applying systems thinking in a qualitative study in Malaysia and Thailand. *PLoS One*, *15*(4), e0231154. https://doi.org/10.1371/journal.pone.0231154

Purnell, L. D. (2016). The Purnell model for cultural competence. In *Intervention in mental health-substance use* (pp. 57–78). CRC Press.

Quisumbing, A. R., & Pandolfelli, L. (2010). Promising approaches to address the needs of poor female farmers: Resources, constraints, and interventions. *World Development*, *38*(4), 581–592. https://doi.org/10.1016/j.worlddev.2009.10.006

Regitz-Zagrosek, V. (2012). Sex and gender differences in health: Science & Society Series on Sex and Science. *EMBO Reports*, *13*(7), 596–603. https://doi.org/10.1038/embor.2012.87

Safran, D. G., Rogers, W. H., Tarlov, A. R., McHorney, C. A., & Ware, J. E. (1997). Gender differences in medical treatment: The case of physician-prescribed activity restrictions. *Social Science & Medicine*, *45*(5), 711–722. https://doi.org/10.1016/S0277-9536(96)00405-4

Salcedo-La Viña, C. (2020, March 3). *Beyond title: How to secure land tenure for women*. World Resources Institute. https://www.wri.org/insights/beyond-title-how-secure-land-tenure-women

Salcedo-La Viña, C. (2021, August 18). *3 benefits of women's collective land rights*. World Resources Institute. https://www.wri.org/insights/3-benefits-womens-collective-land-rights

Samar, S., Aqil, A., Vogel, J., Wentzel, L., Haqmal, S., Matsunaga, E., Vuolo, E., & Abaszadeh, N. (2014). Towards gender equality in health in Afghanistan. *Global Public Health*, *9*(sup1), S76–S92. https://doi.org/10.1080/17441692.2014.913072

Samulowitz, A., Gremyr, I., Eriksson, E., & Hensing, G. (2018). "Brave Men" and "Emotional Women": A theory-guided literature review on gender bias in Health Care and gendered norms towards patients with chronic pain. *Pain Research and Management*, 2018, 1–14. https://doi.org/10.1155/2018/6358624

Schachter, C. L., Radomsky, N. A., Stalker, C. A., & Teram, E. (2004). Women survivors of child sexual abuse. How can health professionals promote healing? *Canadian Family Physician*, *50*(3), 405.

Schein, E. H. (2010). *Organizational culture and leadership* (4th ed). Jossey-Bass.

Schwanke, D.-A. (2013). Barriers for women to positions of power: How societal and corporate structures, perceptions of leadership and discrimination restrict women's advancement to authority. *Earth Common Journal*, *3*(2). https://doi.org/10.31542/j.ecj.125

SenGupta, S., Hopson, R., & Thompson-Robinson, M. (2004). Cultural competence in evaluation: An overview. *New Directions for Evaluation*, *2004*(102), 5–19. https://doi.org/10.1002/ev.112

Shafiq, M. (2008). Analysis of the role of women in livestock of Balochistan, Pakistan. *Journal of Agriculture and Social Science*, *4*(1), 18–22.

Sharifi, N., Adib-Hajbaghery, M., & Najafi, M. (2019). Cultural competence in nursing: A concept analysis. *International Journal of Nursing Studies*, *99*, 103386. https://doi.org/10.1016/j.ijnurstu.2019.103386

Shiva, V., & Bandyopadhyay, J. (1986). The evolution, structure, and impact of the Chipko Movement. *Mountain Research and Development*, *6*(2), 133. https://doi.org/10.2307/3673267

Singh, M., & Mishra, D. K. D. (2019). Eco-feminism and folk media: A case study of the Chipko Movement. *International Journal of Scientific & Technology Research*, *8*(11).

Smedley, B. D., Stith, A. Y., & Nelson, A. R. (2003). Interventions: Cross-cultural education in the health professions. In *Unequal treatment: Confronting racial and ethnic disparities in Health Care* (p. 12875). National Academies Press. https://doi.org/10.17226/12875

Southall, H. G., DeYoung, S. E., & Harris, C. A. (2017). Lack of cultural competency in international aid responses: The Ebola outbreak in Liberia. *Frontiers in Public Health*, *5*. https://doi.org/10.3389/fpubh.2017.00005

Swanepoel, F., Stroebel, A., & Moyo, S. (Eds.). (2010). *The role of livestock in developing communities: Enhancing multifunctionality*. SunBonani Media. https://doi.org/10.18820/9781928424819

Swedish International Development Cooperation Agency (SIDA). (2009). *Quick guide to what and how: Increasing women's access to land*. SIDA. https://www.oecd.org/dac/gender-development/47566053.pdf

Tajeu, G. S., Cherrington, A. L., Andreae, L., Prince, C., Holt, C. L., & Halanych, J. H. (2015). "We'll Get to You When We Get to You": Exploring potential contributions of Health Care Staff behaviors to patient perceptions of discrimination and satisfaction. *American Journal of Public Health*, *105*(10), 2076–2082. https://doi.org/10.2105/AJPH.2015.302721

The Green Belt Movement. (2024). *Our history*. https://www.greenbeltmovement.org/who-we-are/our-history

Thom, D. H., & Tirado, M. D. (2006). Development and validation of a patient-reported measure of physician cultural competency. *Medical Care Research and Review*, *63*(5), 636–655. https://doi.org/10.1177/1077558706290946

Torres, N. (2018, August 10). Research: Having a black doctor led black men to receive more-effective care. *Harvard Business Review*. https://hbr.org/2018/08/research-having-a-black-doctor-led-black-men-to-receive-more-effective-care#:~:text=They%20found%20that%20black%20men,better%20communication%20and%20more%20trust.

Truong, M., Paradies, Y., & Priest, N. (2014). Interventions to improve cultural competency in healthcare: A systematic review of reviews. *BMC Health Services Research*, *14*(1), 99. https://doi.org/10.1186/1472-6963-14-99

Tzeng, H.-M., & Yin, C.-Y. (2006). Nurses' fears and professional obligations concerning possible human-to-human Avian Flu. *Nursing Ethics*, *13*(5), 455–470. https://doi.org/10.1191/0969733006nej893oa

UN Women. (2022). *Remove the barriers*. UN Women – Headquarters. https://www.unwomen.org/en/news/in-focus/csw61/remove-the-barriers

UN Women, IDLO, The World Bank, Task Force on Justice. (2019). *Justice for women: High-level group report*. https://www.unwomen.org/sites/default/files/Headquarters/Attachments/Sections/Library/Publications/2020/Justice-for-women-High-level-group-report-en.pdf

U.S. Agency for International Development. (2011, May). *Green Belt movement revives watershed in Kenya | Archive*. https://2012-2017.usaid.gov/global-waters/may-2011/green-belt

Vlassoff, C. (2007). Gender differences in determinants and consequences of health and illness. *Journal of Health, Population, and Nutrition*, *25*(1), 47–61.

Williams, B. (2001). Accomplishing cross cultural competence in youth development programs. *Journal of Extension*, *39*(6), 1–6.

Wilson, J., Ward, C., & Fischer, R. (2013). Beyond culture learning theory: What can personality tell us about cultural competence? *Journal of Cross-Cultural Psychology*, *44*(6), 900–927. https://doi.org/10.1177/0022022113492889

World Economic Forum. (2017, January 11). *Women own less than 20% of the world's land. It's time to give them equal property rights*. World Economic Forum. https://www.weforum.org/agenda/2017/01/women-own-less-than-20-of-the-worlds-land-its-time-to-give-them-equal-property-rights/

World Health Organization. (2017, February 3). *Determinants of health*. https://www.who.int/news-room/questions-and-answers/item/determinants-of-health

World Health Organization. (2023, October 5). *Universal Health Coverage (UHC)*. https://www.who.int/news-room/fact-sheets/detail/universal-health-coverage-(uhc)

World Health Organization. (2024). *Constitution of the World Health Organization*. https://www.who.int/about/governance/constitution

Zahoor, A., Fakher, A., Ali, S., & Sarwar, F. (2013). Participation of rural women in crop and livestock activities: A case study of Tehsil Tounsa Sharif of southern Punjab (Pakistan). *International Journal of Advanced Research in Management and Social Sciences*, *2*(12), 98–121.

12 Gender, Infectious Diseases, and One Health

Neil Vezeau, Robyn Alders, Rosa Costa, Manisha Dhakal, Jennie Gordon, and Stewart Sutherland

INTRODUCTION

Gender equity and inclusion are critical concepts in understanding and practicing One Health (OH). Gender is an important component of the human-animal-environment interface due to the different roles and expectations that exist between genders. Gender profoundly impacts and is impacted by the most pertinent issues humanity faces, including emerging infectious diseases, climate change, and other OH issues. To more effectively and equitably serve our global community, OH practitioners must recognize gender gaps and identify approaches to address those gaps. They must analyze how gender interfaces with OH threats to ultimately become transformative agents in promoting gender equality and equity in all aspects of their work.

WHY GENDER AND OH?

The Biomedical and Sociocultural Relevance of Sex and Gender

Sex is related to, but distinct from, gender (World Health Organization, 2021). Both are important in understanding determinants of health (Mauvais-Jarvis et al., 2020). Sex is based on a collection of biological traits pertaining to anatomy, physiology, genetics, and hormones. Sex is most often categorized as female and male, with intersex individuals having traits characteristic of both sexes for a given species (World Health Organization, 2021). Sex traits are exhibited across many multicellular eukaryotic forms of life including humans, non-human animals, and plants (Whitfield, 2004). Depending on the definition applied, people with intersex traits are estimated to comprise 0.018%–1.7% of the population (Sax, 2002). However, this can vary across different geographies and ethnic groups (Wiersma, 2004).

Gender is a sociocultural construct considered unique to humans, and is based on sociocultural norms, identities, and relations between different gender identities,

DOI: 10.1201/9781003232223-12

including hierarchical power dynamics (World Health Organization, 2021). Gender is commonly described along a masculinity-femininity spectrum, often correlated with biological maleness or femaleness, respectively (Gendered Innovations, 2020). Many people exist with gender identities outside of this correlation with the masculine-feminine binary, and an estimated 1% of adults worldwide do not identify with the gender correlated with their biological sex at birth (Masterson, 2021). For most diseases, the effects of sex and gender have not yet been segregated, and biomedical research has traditionally focused on men (Mauvais-Jarvis et al., 2020). Previous and ongoing research, initiatives, and methodological approaches equate sex and gender, and many treat each as only binary. Some research referenced below has taken a limited binary interpretation of sex and gender.

Sex influences epigenetic, hormonal, and cellular properties to modify the expression of biology and disease. Gender can act on patient-physician relationships and other psychological and social dynamics to alter health outcomes (Mauvais-Jarvis et al., 2020). These factors contribute to significant knowledge gaps because biomedical research has historically underrepresented females, women, and gender minorities.

Males and females, in humans as in other animal species, have different physiologic characteristics. This includes neurocognitive properties, hormonal systems, and the resultant impacts on the musculoskeletal system, among others (Mauvais-Jarvis et al., 2020). Females may go through pregnancy and childbirth, which can involve complications such as psychological prepartum and postpartum depression (Diego et al., 2004). Differences also extend to immunological properties and the propensity for various diseases. Some of the most salient examples include the higher rate of autoimmune diseases in females, while males have a higher risk of cardiovascular disease (Mauvais-Jarvis et al., 2020).

In addition to biological differences, gender roles and societal expectations can impact behaviors that affect health, including care-seeking and caregiving behaviors, access to healthcare, and societal power dynamics. Women are notably more likely to prioritize giving care than others, often to the detriment of their own health and well-being (Scarpetta et al., 2021). Additionally, they can often face barriers to healthcare access due to poverty and transportation that men may not be as likely to face (Dahab & Sakellariou, 2020). This results in gendered differences in access to healthcare services that negatively impact women's ability to receive timely diagnoses and treatments. Additionally, gender minority individuals, such as those identifying as transgender and non-binary, have higher rates of depression, anxiety, and suicidality than their cisgender counterparts (Lipson et al., 2019). Social marginalization and increased barriers to care may contribute to this (World Health Organization, 2021).

Gender-Based Occupational and Behavioral Risks

Exposures to environmental hazards are also likely to vary due to different gendered occupational settings, behaviors, and other lifestyle conditions. These can include differential exposure to toxins, environmentally reservoired pathogens, as well as physical threats from violence and industrial equipment (Eng et al., 2011; Fares-Otero et al., 2020; Oruganti et al., 2023). Differences in susceptibility to disease, both transmissible and not, extend to lifestyle habits, with men being more likely to smoke and drink (World Health Organization, 2021). Dietary patterns can also differ between genders (Masella & Malorni, 2017). For example, women are more likely to skip meals and undergo dietary restrictions. Further impacting women are various cultural norms restricting their physical activity, or limiting the time or available spaces for exercise (The Lancet Public Health, 2019).

Men engage in some riskier sexual practices such as lower contraceptive use and have, on average, more sexual partners, adversely increasing the spread of sexually transmitted diseases (Centers for Disease Control and Prevention, 2021; World Health Organization, 2021). Additionally, it is reported that women are more likely to experience depression, among other mental health conditions, which can impact behavior (Mauvais-Jarvis et al., 2020).

Intersectionality: Age, Economics, and Social Status

Gender interacts with a variety of demographic and socioeconomic factors of a given person. This phenomenon is known as "intersectionality" (World Health Organization, 2021). Multiple intersectional identities of an individual such as gender, race, sexual orientation, and immigration status can have adverse effects on their health outcomes. This makes intersectionality especially important for women from marginalized communities and areas, who may face, among other things, additional barriers to accessing healthcare and an increased risk of intimate partner violence (Harari & Lee, 2021).

Economic factors impact men and women differently. Perhaps the most important factors related to economics are poverty, wealth/income equality, and education (O'Neil et al., 2020). Women are more likely to experience poverty and have limited access to healthcare services (Boudet et al., 2018; O'Neil et al., 2020). Rates of homelessness in Organisation for Economic Co-operation and Development (OECD) member countries have been on the rise among women, youth, and the elderly, although single men are still most likely to be unhoused (OECD – Social Policy Division – Directorate of Employment, Labour and Social Affairs, 2021).

Age additionally impacts genders differently. As mentioned above, men and women have higher risks of diseases at different ages. For example, the chances of women suffering from osteoporosis or breast cancer, both positively correlated with age, are greater than for men (Alswat, 2017; Ly et al., 2013).

Gender Gaps in OH Research Policy and Implementation

Gender gaps are pervasive and reflect inequalities between genders at a structural level. They prevent OH and other interventional approaches from ensuring the most effective, equitable, inclusive, and sustainable health outcomes.

The most prominent gender gaps in OH research policy and implementation can be organized within the following categories: unequal representation, unequal compensation and recognition of work, inadequate attention to gendered health issues, limited gender-sensitive research, and unequal access to education and training (Amuguni et al., 2018; Garnier et al., 2020, 2022).

Women are chronically and broadly underrepresented throughout societal leadership and power structures in both the public and private sectors (Gurkin, 2019; UN Women, 2024). Many OH-related entities, including decision-making bodies and technical committees, are no exception (Zinsstag et al., 2023). This inherently limits women's ability to influence policy decisions, many of which will adversely affect them the most, and is a problem not unique to OH (OECD, 2021). Even when women are present in work, leadership, and decision-making, they are statistically less likely to receive equal pay, and often recognition, for equivalent work when compared to colleagues who are men (Gould et al., 2016).

Research often neglects health factors that disproportionately affect women. Examples include intimate partner and other gendered violence, reproductive health, and maternal and child health, especially, but not exclusively, in low-income regions (Fares-Otero et al., 2020; Kendall & Langer, 2015). This lesser attention dovetails with limited research methods to disaggregate the effects of gender. Research has traditionally failed to take into account sex and gender differences to determine how their pathophysiology and epidemiology differ along those lines (Mauvais-Jarvis et al., 2020). Seemingly established biomedical knowledge has been called into question due to the unnecessary exclusion of female cell lines or women from previous research (Mauvais-Jarvis et al., 2020).

In many settings, women may have less ability to empower themselves. Worldwide, women are less able than men to access education and training, limiting their ability to participate in OH initiatives and advance their careers. Empowering women with education and inclusion in the political process is positively correlated with a nation's life expectancy and environmental health (Cataldo et al., 2023).

Creating Equitable Policies and Interventions

As we move forward to create policies and conduct research in OH and global development, it is mission-critical to not perpetuate gender inequities. Participants in such endeavors should be representative of the population being studied or being affected by a policy (AFROHUN, 2020).

With regard to gender policies, programs, and projects, attitudes toward gender can be categorized as blind, aware, exploitative, accommodating, and transformative (AFROHUN, 2020). *Gender-blind* practices ignore the differing dynamics between genders. They tend to ignore the differing roles and needs of different genders and are thus liable to maintain preexisting gendered circumstances (UN Statistics Wiki, 2020). *Gender-aware* attitudes are the opposite, and actively are aware of, address, and analyze socioeconomic, political, cultural, etc., differences between genders, though they may not elevate marginalized genders

(AFROHUN, 2020). *Gender-responsive, gender-sensitive, gender-accommodating,* and *gender-transformative* practices are all types of gender-aware practices (AFROHUN, 2020; University of Basel, 2021). *Gender-exploitative* programming is a harmful practice that takes advantage of and reinforces harmful gendered stereotypes and status quos, whether knowingly or not (AFROHUN, 2020). Gender-transformative programming actively works to increase gender equity by examining and empowering the position of marginalized gender groups.

Achieving gender-aware and transformative goals requires specific approaches and a robust monitoring and evaluation (M&E) framework. Transformative methodological approaches should focus on gender sensitization of leadership and project staff, as well as ensuring equitable participation across genders (AFROHUN, 2020). M&E goals are accomplished with certain indicators, metrics, and assessment tools, especially disaggregating data such as age and socioeconomic indicators by sex and gender. The development of a gender strategy, followed by encoding these in policy, can guide the implementation of gender-sensitive activities.

Project leaders should consider potential intervention impacts with respect to age, socioeconomic status, and how this might differentially affect men and women. Throughout this process, care should be taken to foster a culture of diversity, equity, and inclusivity by identifying and mitigating biases where they exist. Being part of a collaborative approach, OH practitioners should seek partnerships with local communities and organizations, especially those led by women, to ensure that their perspectives and needs are taken into account (AFROHUN, 2020). Many lessons can be taken from inclusive approaches pioneered in global development and global health. For research-based projects, and interventions gathering M&E data, findings should be disseminated in a manner that is transparent, accessible, and respectful. This allows for a subsequent community feedback process. At all stages, any policies and interventions implemented need to be based on principles of fundamental human rights and dignity, inclusive of gender equality.

EXPERIENCES WITH GENDER AND INFECTIOUS DISEASES

Experience from Epidemics Before COVID-19

Previous disease outbreaks have given us ample experience to identify how infectious disease dynamics vary across different genders. Previous influenza pandemics have borne that women's mortality is higher than men's in such outbreaks (Klein et al., 2012). Across Ebola outbreaks from 1995 to 2014, more women were reported to have Ebola than men (Nkangu et al., 2017). This is likely due to increased exposure from their cultural caretaker roles for the sick, including predominance in nursing roles most likely to have the highest patient contact.

Important to pandemic preparedness and response efforts are gender analyses and gender-sensitive approaches. Integrating gender analysis helps target populations more vulnerable to disease, including women, children, and elderly people

(Cooke et al., 2017). Such analysis aids in ensuring that their specific needs are addressed in outbreak response efforts (Smith, 2019). Gender-sensitive approaches recognize the nonequivalent social realities and needs of different genders and seek to promote equity or analyze inequity (Hidrobo et al., 2020). They can help ensure that women and girls have access to essential services and that gendered health needs such as reproductive and sexual health issues are addressed more than by using purely biomedical approaches to health (Im & Meleis, 2001). Gender-sensitive analysis can inform decision-making around resource allocation to help ensure that response efforts are equitable and take into account differing roles and needs across genders (Hidrobo et al., 2020). Gender-sensitive approaches can encourage greater engagement and participation of women and girls in outbreak response efforts and can help to ensure that their perspectives and experiences are taken into account (Amuguni et al., 2018).

COVID-19 Pandemic Livelihoods, Food Security, and Unpaid Care Work

As the most impactful global pandemic in recent memory, we can look to our experience with coronavirus disease (COVID)-19 to see how outbreaks affect different genders differently across societal and regional contexts. Women have been more likely to lose their jobs or face reduced working hours and pay. This is particularly true in industries that have been heavily affected by the pandemic such as hospitality, care work, and food service – all of which have high concentrations of female workers (Barua, 2022). Additionally, women have taken on a disproportionate share of caregiving responsibilities during the pandemic, particularly for children and elderly relatives, leading to increased workload and stress (Scarpetta et al., 2021). Increased stressors such as lockdowns and social distancing measures can lead to an increase in domestic violence, with women bearing the brunt of the abuse (UN Women, 2023).

Women have been more likely to have their healthcare needs neglected or postponed due to the strain on healthcare systems, and may have less access to COVID-19 vaccines and treatments due to gender and socioeconomic disparities (Ghouaibi, 2021).

Antimicrobial Resistance

Antimicrobial resistance (AMR) is a gendered issue in several ways. Namely, women's more limited access to healthcare and AMR reciprocally act on one another in a vicious cycle. Women play important roles in providing and seeking healthcare for members of their families. Women may have more limited decision-making power and increased social and economic barriers in seeking healthcare for themselves or their children. This can lead to inappropriate use of antimicrobials, including self-medication, incomplete treatment courses, or sharing medications. This contributes to the development of AMR (Jones et al., 2022).

Women have a variety of occupational risks, increasing their exposure to AMR. One of the most salient is due to the number of women in the global healthcare workforce (Shannon et al., 2019). Exposure to antimicrobial-resistant pathogens is an inherent risk in healthcare. Certain occupations, such as healthcare workers and agricultural workers, may have a higher risk of occupational exposure to antimicrobials, which can contribute to the development of AMR (Graham et al., 2019; Shibabaw et al., 2013). This is made worse in limited-resource areas, where adequate personal protective equipment and other control measures could be lacking (Siow et al., 2020).

The raising of livestock, more common among women in low- and middle-income countries (LMICs), also brings a variety of occupational risks (Chemnitz & Becheva, 2021). Those in such low-resource settings may have limited access to expert veterinary care that will judiciously prescribe antimicrobials (Ayukekbong et al., 2017). This naturally leads to increased antimicrobial use in food-producing animals including growth promotion, infectious disease prevention, and treatment (ReAct et al., 2020). Any such misuse or overuse of antimicrobials contributes to AMR for zoonotic, environmental, and foodborne pathogen exposure (Ayukekbong et al., 2017).

Lack of education and information can result in inappropriate use of antimicrobials, leading to the development of AMR. This includes the importance of completing a full course of treatment and avoiding unnecessary use (Ayukekbong et al., 2017). Gender disparities in education and information can impact women's ability to engage in decision-making processes related to healthcare and contribute to the perpetuation of AMR (Friedson-Ridenour et al., 2019; Jones et al., 2022).

Additionally, women and girls suffer from a disproportionately high burden of sexually transmitted infections, urinary tract infections, and reproductive tract infections compared to men and boys (Van Gerwen et al., 2022). Non-judicious antimicrobial use to treat such conditions contributes to AMR (ReAct et al., 2020).

CURRENT LANDSCAPE

Sex and Gender Equity in Research Organizationally and Globally

Numerous preeminent organizations that fund and conduct global development and biomedical research and programming require or incentivize the integration of sex and gender considerations into both research and intervention proposals (White et al., 2021). The National Institutes of Health (NIH) requires the consideration of sex as a biological variable in all studies and encourages the integration of gender analysis in research proposals (National Institutes of Health, 2015). The European Commission (EC) Directorate-General for Research and Innovation requires the integration of gender considerations in all research proposals under its Horizon 2020 framework (European Commission, 2020). The Canadian Institutes of Health Research (CIHR) has a policy on gender and health research that requires the integration of gender and sex considerations into all research proposals (Canadian Institutes of Health Research, 2019). The Bill & Melinda Gates Foundation requires the integration of gender considerations in all proposals that address global health and development, and this is also asked for in work funded by the Wellcome Trust (Wellcome, 2023). Other examples relevant to OH include the U.S. Agency for International Development (USAID) and the International Development Research Centre (IDRC) (International Development Research Centre, 2024; USAID, 2023).

Synergizing with these organizational efforts, a global call to action has been issued for gender to be included in the research impact assessment undertaken by research funders, institutions, and evaluators in order to inform more equitable health policy and practice (Ovseiko et al., 2016). It exists to ensure gender is considered in all stages of the research process, from design, to implementation, to evaluation. The goal is to improve the relevance and impact of research, and to ensure that research outcomes and benefits are equitably distributed among all genders similar to the CIHR, EC, and NIH (Canadian Institutes of Health Research, 2019; European Commission, 2020; National Institutes of Health, 2015; Ovseiko et al., 2016). The Call to Action encourages researchers, research institutions, funders, and policymakers to incorporate a gender perspective in their work and to mainstream gender into research impact assessment frameworks. This initiative aims to increase awareness of the importance of gender in research and to promote gender-sensitive research practices that will lead to more effective and equitable outcomes (Ovseiko et al., 2016).

Organizational Gender Empowerment Policies

In addition to funding requirements, several global development and health organizations have developed gender and women's empowerment policies to ensure that their programs and initiatives are gender-sensitive and responsive to the needs and perspectives of women and girls. The World Health Organization (WHO) has a gender policy that recognizes the importance of integrating a gender perspective into all of its work and emphasizes the importance of considering the different health needs and experiences of women and men (World Health Organization, 2002). The United Nations Development Programme (UNDP) has a gender equality strategy that outlines its commitment to advancing gender equality and empowering women and girls in all of its programs and initiatives (United Nations Development Programme, 2022). The United Nations International Children's Educational Fund (UNICEF) has a gender policy that recognizes the importance of considering the different needs and experiences of girls and boys and promoting gender equality in all of its programs and initiatives (UNICEF, 2021). The World Bank Group has a gender and development policy that recognizes the importance of gender equality and women's empowerment in promoting economic growth and reducing poverty (The World Bank, 2024). Global Fund to Fight AIDS, Tuberculosis and Malaria has a gender equality strategy that outlines its commitment to ensuring that its programs are gender-sensitive and responsive to the needs and perspectives of women and girls (The Global Fund, 2023).

Organizational Gender Committees

Several organizations have formed entire committees or units to address gendered needs in research and general equity. The European Association of Scientific Editors has formed a Gender Policy Committee to improve sex and gender reporting practices across all scientific fields. The CIHR established the CIHR Institute of Gender and Health, which includes a Gender and Health Research Advisory Committee. Its purpose is to further integrate gender and sex considerations into health research (Canadian Institutes of Health Research, 2003). The NIH has an Office of Research on Women's Health, which provides guidance and recommendations for the integration of sex and gender considerations into research proposals and programs (National Institutes of Health, 2023). The WHO has a Gender, Women and Health Unit that is responsible for promoting gender equality and women's empowerment in health and developing policies and programs that address the different health needs and experiences of women and men (World Health Organization, 2021). The Bill & Melinda Gates Foundation also features a Gender Equality Division (Bill & Melinda Gates Foundation, 2024b).

Project Tools for Measuring Gender-Related Changes

Accounting for and measuring gendered features of research or programming is increasingly accomplished using gender analysis frameworks (Eba et al., 2020). These provide a structured approach to understanding and addressing gender-related issues in health research and interventions (Friedson-Ridenour et al., 2019). Such frameworks often include multiple indicators that can be used to identify and analyze gender-related differences, disparities, and needs in different health contexts. There are several tools available to measure gender-related changes in health research and intervention projects. Many academics have created toolsets, and government-grant-funded work has been conducted to create and provide such project aids.

The Sex and Gender Equity in Research Guidelines are a set of recommendations for the integration of sex and gender considerations into health research. They provide guidance on best practices for sex and gender reporting in scientific publications and can be used to assess the quality of sex and gender reporting in research projects (European Association of Science Editors, 2024).

Gender Impact Assessment tools assess the potential impact of interventions on different genders. These tools can be used to identify gender-related outcomes and disparities in health and to develop gender-sensitive interventions that are more likely to be effective and just (European Institute for Gender Equality, 2024b; Golemac Powell et al., 2021; OECD, 2024).

Gender data collection and analysis tools are used to collect data on gender-related outcomes and experiences in health research and interventions. These tools can be used to identify and analyze gender-related differences, disparities, and needs in different health contexts and to develop more effective and equitable health interventions. The Bill & Melinda Gates Foundation has created a collection of some such tools (Bill & Melinda Gates Foundation, 2024a).

GAPS AND STRATEGIES FOR INTEGRATION OF GENDER INTO OH

Mapping of Stakeholders: Gender Roles, Responsibilities, and Power Dynamics

Collecting and utilizing the voices and perspectives of different stakeholders is important in gender-sensitive OH interventions. Participatory approaches can help these to be duly taken into account. As a component of understanding relevant actors, stakeholder and resource mapping or analysis is an important part of effecting positive change to correct gender disparities (Amuguni et al., 2019). Stakeholder analyses can be used to identify and map the different individuals or groups involved in health research and capacity-building. This includes the gender roles and responsibilities of these stakeholders. There exist many methods for stakeholder mapping to achieve these goals (One Health Commission, 2022).

Engaging in stakeholder consultations serves to gather information on the gender roles and responsibilities of different stakeholders in health research and capacity-building (Amuguni et al., 2018, 2019; Garnier et al., 2020). This can also help to identify the gender-related power dynamics that exist between different stakeholders and to assess the impact of these power dynamics on the gender-related outcomes of health research and interventions (Cediel-Becerra et al., 2022).

Gender and sex-disaggregated data can help to identify gender-related disparities and needs in research capacity-building interventions (Amuguni et al., 2019; Bagnol et al., 2016; Eba et al., 2020). Collection and analysis of such data can aid the mapping of gender roles and responsibilities of different stakeholders and assess the impact of these roles and responsibilities on the gender-related outcomes of health research and interventions.

A gender audit is a process used to assess gender sensitivity of research or interventions to identify opportunities for improvement (European Institute for Gender Equality, 2024a). This can help to identify the gendered dynamics within and between different stakeholders. It can help predict the impact of such dynamics on any gender-related outcomes.

Increasing Gender Capacities of OH Practitioners

As gender becomes a more recognized determinant of health, increasing the gender capacities of OH practitioners is increasingly necessary (Amuguni et al., 2018; Friedson-Ridenour et al., 2019). Providing gender training and education to OH practitioners can increase an understanding of all the dynamics discussed previously in this chapter (Amuguni et al., 2019). It will allow these learners to better respond to the gendered needs of communities globally. This includes promoting and training on the use of the various gender-sensitive guidelines and tools mentioned in this chapter. This can help to ensure that OH practitioners are equipped to be responsive to the different needs and perspectives of different genders.

Being able to understand different perspectives is vital to advancing gender equity in OH work (Laing et al., 2023; Togami et al., 2023). Encouraging the evolving community of practice of OH practitioners to interface with gender experts and other stakeholders can help to ensure that gender considerations are integrated into OH research and interventions. This can help to ensure that this broad community of project personnel have access to the knowledge, skills, and expertise needed to be responsive to the needs of different genders. OH practitioners should encourage and promote gender-sensitive data collection and analysis to aid the integration of gender considerations into interventions and research. These techniques are likely to be among the most effective to increase gender capacities of OH practitioners.

Integrating Sex and Gender into Education

Educators should make gender a part of learning outcomes and integrate gender into the curriculum (Amuguni et al., 2019; Laing et al., 2023). Mention of gender will ideally be in the context of the overall subject matter of the course or lesson, be it AMR, infectious disease, risk analysis, or otherwise. If possible, gender should not be introduced for only fleeting sections of the instruction and not mentioned again (Amuguni et al., 2018).

Educational materials and aids, such as power points, should be created with gender sensitivity in mind. If case studies are merited for inclusion in specific lessons, they should include examples reflective of different experiences had between genders. Also important in education is audience inclusion (Amuguni et al., 2018). Instructors should encourage students to actively engage in discussions, reflections, and other interactive activities. This will help these learners better apply lessons in real-world scenarios.

If possible, bringing in guest speakers with diverse experiences, perspectives, and areas of expertise relative to gendered health can greatly help to improve audience learning and quality of education (Amuguni et al., 2019). Those with expertise in gender and health can provide students with additional insights and perspectives on the importance of gender in the real-world contexts they may hope to work in someday.

Additionally, monitoring the progress of learners is an integral part of any self-contained educational process (Berrian et al., 2018). This can come in the form of regular quizzes or tests, pre-tests or surveys, as well as student evaluations at the conclusion of a course or lesson. As with types of projects mentioned above, having a robust M&E system implemented will greatly aid the quality of an initiative over time.

TABLE 12.1
Sex Chromosome Combinations in Mammals Can Vary, with Some Combinations Being More Common Than Others

Sex Chromosome Combination	Observed in Non-human Mammals (Y/N)	Observed in Humans (Y/N)
XY	Y	
XX	Y	
XXY	Y	
XXX	Y	
46XY+46XX	Y	

Note: This table can be used to facilitate discussion by asking discussants if the combinations reported to occur in non-human mammals can also occur in humans (Skuse et al., 2018).

Integrating Sex and Gender into Education: Example

In some settings, discussions relating to gender diversity could be linked to discussions on the various sex chromosome combinations (e.g., XY, XX, XXY, XXX, 46XY+46XX) observed in non-human mammals (Szczerbal & Switonski, 2016). Where discussions about such topics in humans may not be culturally acceptable, the discussion need not go beyond asking if such combinations are also seen in humans – they are (Table 12.1) (Skuse et al., 2018). Such combinations do occur in humans and each will impact differently on an individual's genotype and phenotype. This approach may help to ensure that all participants in the discussion are aware that sex determination at the genetic level and the phenotypic level is complex; indeed, one researcher has listed up to nine different components of an individual's sexual identity: external genital appearance, internal reproductive organs, structure of the gonads, endocrinologic sex, genetic sex, nuclear sex, chromosomal sex, psychological sex, and social sex (Moore, 1968).

CASE STUDY 1: IMPLICATIONS OF NOT WORKING WITH TRANSGENDER COMMUNITIES DURING PUBLIC HEALTH CRISES

Transgender individuals face discrimination and violence because of a deeply rooted gender binary system, stereotypes, and sociocultural norms largely introduced during the colonial era (ISPI, 2020). Still to-date, in too many situations, gender means women and men and female and male. Many stakeholders continue to ignore the existence of transgender, non-binary, and intersex individuals. There are also large gaps in legal recognition of gender. Many transgender people are forced by authorities to choose a non-binary legal status rather than male and female. In Nepal, the government requires medical proof to access citizenship for transgender people and non-binary people. Because transgender people do not have a legal document recognizing their gender identity, they continue to face discrimination in all service settings including health services.

Due to the devastating situation caused by the COVID-19 pandemic, transgender people of Nepal have become more vulnerable. Transgender people are never prioritized as a vulnerable group during such pandemic situations. Thus, this community was completely forgotten by the government when designing and implementing programs to address the impact of COVID-19 infections.

The government of Nepal announced the distribution of food relief packages during the COVID-19 pandemic. At the start, many transgender people were happy and went to receive the food packages. But their happiness ended suddenly as the government required proof of citizenship to access that relief assistance. Many transgender people do not have a citizenship identification card consistent with their gender identity. Consequently, they returned empty handed. Many trans-women do not have a citizenship card and others have ID cards which do not match their gender.

During the COVID-19 pandemic, organizations such as the Blue Diamond Society managed to distribute antiretroviral (ARV) medicine through a home-based care approach (USAID, 2009). Because of the travel restrictions, it was difficult to distribute the ARV medicine. Our peer navigators started to distribute ARV using bicycles to travel up to 4 hours. Such work required not only athleticism, but also bravery as evidenced by an incident that happened in the Bara District of Nepal. In this district, a peer navigator was brutally beaten by police. That case was brought to the attention of the Blue Diamond Society, which lodged a complaint about the beating and advocated for change to government authorities. Subsequently, the National Center of AIDS and STD Control and the Home Ministry issued a letter to the Nepal Police and local government authorities to facilitate the easy supply of ARV medicine without any harassment toward peer navigators including members of the transgender community.

Similarly, the transgender, non-binary, and intersex communities have not been prioritized during the COVID-19 vaccination campaigns. This contributes to low vaccination coverage within the community. The Blue Diamond Society continues to advocate for:

1. Breaking the binary gender system
2. Legal gender recognition as per self-determination without requiring a sex change operation certificate

3. Prioritization of transgender people during humanitarian crises in relation to relief distribution and medicine distribution for infectious diseases, including human immunodeficiency virus (HIV) and acquired immunodeficiency syndrome (AIDS), sexually transmitted infections (STIs), etc.
4. National policy and laws that are inclusive of transgender people and non-binary people
5. Strategy and action plans regarding infectious diseases, including HIV and AIDS, that also address the specific needs of transgender people, especially during humanitarian crises

CASE STUDY 2: INDIGENOUS PERSPECTIVES ON GENDER AND OH IN RELATION TO INFECTIOUS DISEASE CONTROL

INDIGENOUS PHILOSOPHIES PREDATE THE OH MOVEMENT BY THOUSANDS OF YEARS

Firstly, it is important to recognize that Indigenous communities have understood and valued the interconnectedness of humans, animals, plants, and ecosystems for thousands of years. Indigenous peoples from Australia, Canada, and New Zealand "have a holistic view of health that encompasses the physical, mental, emotional, and environmental spectrum of wellbeing" (Figure 12.1) (Sutherland & Adams, 2019). In Australia for example, for Aboriginal and Torres Strait Islanders, "health" is not focused solely on the physical well-being of an individual, but refers to the social, emotional, and cultural well-being of the whole community and the health of their ecosystem (Gee et al., n.d.; Sutherland & Adams, 2019). Further, their holistic understanding of health and well-being involves the whole community throughout the entire life-course, including broad issues such as social justice, equity, and rights, as well as traditional knowledge, traditional healing, and connection to country (Commonwealth of Australia, 2013; Swensen et al., 1995). For Aboriginal and Torres Strait Islanders, the concept of health "encompass[es] mental health and physical, cultural and spiritual health. Land is central to well-being" (Swenson et al., 1995). Personal totems link an individual to the land, air, and geographical characteristics while also defining their roles, responsibilities, and their relationship with each other and creation (Turner & McKenzie, 2022). Similarly in Ecuador, according to Andean beliefs, the Indigenous perspective of health and illness focuses on a balance between the four connected bodies in each human being: the physical, spiritual, social, and mental (Bautista-Valarezo et al., 2020). For ancestral communities in Ecuador, health is a complex

FIGURE 12.1 The holistic Aboriginal and Torres Strait Islander perspective on social and emotional well-being (State of Queensland (Queensland Health), 2016).

term that incorporates the health of the environment (including community and nature) and the individual. In Canada also, "Indigenous Peoples across Canada are unified in their belief that the wellbeing of an individual is directly connected to the wellbeing of the land" (Hillier et al., 2021).

PERCEPTIONS OF GENDER AND GENDER ROLES

Cultural practices and beliefs concerning gender, gender roles, and relationships vary among Indigenous societies; therefore, the examples in this case study should not be generalized to all Indigenous communities. However, important similarities in both cultural beliefs and gender roles and health status do exist across communities from different geographies.

Colonization by European powers had a significant impact on perceptions of gender in many countries. Prior to the introduction of "rigid European rigid binaries, within the Dagaaba tribe of Ghana, Burkina Faso, and the Ivory Coast, gender identity was determined differently. Shaman Malidoma Somé of the Dagaaba says that gender to the tribe is not dependent upon sexual anatomy" (Collins, 2017). Collins also reported that "The Igbo of Nigeria, also in Western Africa, appear to assign gender around age 5" and "In Central Africa, the Mbuti do not designate a specific gender to a child until after puberty, in direct contrast to Western society."

In Canada, "women were never considered inferior in Aboriginal society until Europeans arrived" (Aboriginal Justice Inquiry of Manitoba, 1999). European economic and cultural expansion has been especially destructive for Aboriginal women with their value as equal partners in tribal society completely undermined. Colonization also complicates the health landscape in Latin America where, "identifying indigenous women's needs and interests is complicated by the intersection of sex and ethnicity. Their sex and gender roles may determine some of their health needs, but others are determined by their status as indigenous people within a larger population, and it is not always easy to separate the two variables for analysis and intervention. Indigenous women themselves may not identify gender inequities as a concern – preferring instead to address issues that affect their entire community" (Pan-American Health Organization, 2004).

In Australia also, "Traditionally, Indigenous men and women maintained distinct gendered realities. Colonization and the subsequent introduction of the patriarchal system altered these realities, negatively impacting on Indigenous men's and women's health and wellbeing in a cumulative and continuing way," especially in southern Australia (Fredericks et al., 2017). And similarly, to Indigenous communities across the world, Australian Indigenous communities have multiple ways of defining gender roles and behaviors and there are wide variations across groups and communities. Some roles are specific to men and women, and these have been described as being different but equally important (Aboriginal Health Council of South Australia, Ltd, 2019).

GENDER, HISTORY, AND INDIGENOUS KNOWLEDGE SYSTEMS

Indigenous knowledge systems (IKS) have been built over tens of thousands of years, and frequently this specialist knowledge is frequently held by different genders in line with their roles within communities (Appleton et al., 2011). According to some Indigenous lore, this knowledge cannot be shared with individuals of a different gender or those who are not members of the Indigenous community. Reluctance to share knowledge can also be associated with: past poor practices by researchers where IKS were accessed and used by non-Indigenous individuals and groups without due involvement and acknowledgment of the rightful holders of this knowledge; and mistrust of government and outside agencies created during the colonial era.

Tips for working with Indigenous communities include:

Getting to Know Where You Are Working

- Research relevant Indigenous organizations in the area where you are working.
- Familiarize yourself with the area, geographically, demographically, and historically.
- Learn the name of the local Indigenous language and ask to be taught the basics.
- Be aware and respectful of relevant extended family and kinship structures.
- Introduce yourself to Elders in the Community.

Cultural and Gender Sensitivity

- When organizing meetings with community members, discuss whether or not the topic of conversation is suitable for everyone or if the issue of Men's and Women's knowledge must be taken into account. It may require another staff member to attend and run the alternative session.
- There may be times when non-indigenous men and women may be asked to leave the room during discussions relating to Indigenous Men's or Women's knowledge and practices. It is important to not take offense to this as it indicates that sensitive or Indigenous-specific issues will be discussed.

Communication

- Respect, acknowledge, actively listen, and respond to the needs of Indigenous people and communities in a culturally appropriate manner.
- Demonstrate respect for Elders and leaders in the community and involve them in important decision-making processes.
- Establish community advisory groups with local Elders and Indigenous organizations, or access existing groups to ensure culturally relevant and sensitive development and delivery of activities.
- Respect the use of silence and don't mistake it for misunderstanding a topic or issue.

- Always wait your turn to speak; always consult with Indigenous people if unsure.
- Be aware that words might have different meanings in different communities.
- Use clear, uncomplicated language. Do not use jargon.
- Be mindful of potential language barriers; speak clearly and as loud as necessary but do not shout (New South Wales Department of Community Services, 2009).

Tips for engaging with Indigenous communities in relation to their IKS. Indigenous communities want:

- Control over who uses IKS and how it's used.
- Protection to prevent unauthorized use of their knowledge and impose sanctions against misappropriation.
- Recognition as owners of their IK, i.e., intellectual property rights remain with the community.
- Respect as owners of the IK and the cultural protocols associated with it (IP Australia, 2023).
- Data should be returned to the community for their own use.

KEY RECOMMENDATION

Employ respectful, culturally sensitive, participatory approaches to ensure that Indigenous communities themselves are managing the discussion, defining their understanding of "health," "OH," and "challenges/problems" and implementing actions.

CASE STUDY 3: GENDERED ASYMMETRY OF ACCESS TO KNOWLEDGE FOR BRUCELLOSIS CONTROL AMONG PASTORAL COMMUNITIES IN NORTH-WEST CÔTE D'IVOIRE

INTRODUCTION

In a context of national prioritization of brucellosis as a disease to control in Côte d'Ivoire (Côte d'Ivoire Ministère de la santé publique et al., 2018), assessing the knowledge of at-risk people, such as pastoral communities, is critical. Addressing the gendered knowledge asymmetry on brucellosis is key in designing pathways of intervention and response to the disease (Babo et al., 2022). This might be possible through efforts to optimize the health of humans, animals, and the environment conceptualized as the OH approach.

Thus, this case study aims to describe mechanisms of knowledge production and transfer on brucellosis according to gender and specialization, by assessing the way knowledge affects behaviors of pastoral communities for disease control.

INTERVENTION

A community-based, cross-sectional survey was conducted among a pastoral community of the Folon region in northwest Côte d'Ivoire. The study included transhumance pastoralists, sedentary livestock owners, shepherds, and their wives. By using mixed methods, 26 semi-structured interviews were conducted, and 320 questionnaires were completed. Statistical analysis with chi-square (χ^2) comparison tests was performed to compare variables between men and women. Findings were interpreted through the concept of specialization of the social exclusion theory.

OUTCOMES

Gender influences access to information on brucellosis and transfer of knowledge on brucellosis appeared gender-biased, especially from veterinarians toward men in the community. The social labor division and interventions of veterinarians through awareness reinforce the knowledge gap on brucellosis between men and women. Men and women consume raw milk, while only men in general handle animal discharges with bare hands.

CONCLUSION

To improve the control of brucellosis, knowledge on best practices should be shared with pastoral communities using the OH approach that encourages mutual learning. Innovative strategies based on gender daily tasks such as safe dairy processing by women and safe animal husbandry to expand their herd for men can be the entry point for the prevention of brucellosis.

CONCLUSION

Gender is a significant OH issue due to its influence on the health and well-being of individuals, communities, and ecosystems. Gender intersects with various factors such as biological, social, economic, and environmental determinants of health. Understanding and addressing the impact of gender on health and disease can lead to more equitable and effective health outcomes and practices. This includes even in emergency and disaster situations where women experience domestic violence, sexual assault, psychological and health problems coupled with social and financial deprivation (Chowdhury et al., 2022). Ultimately, considering the issue of gender in the context of OH is a necessary next step to further recognize and act upon the systemic complexity that drives our natural world and societies alike. Wherever possible, we must transform these systems and structural barriers – breaking down the silos that OH has aimed to deconstruct since the approach's inception.

To do this, we must set the script for future work. At a fundamental level, we must bake gender into how we converse, conceptualize, strategize, and take action. It is essential to include gender training components and gender analysis in as many sectors and activities as possible. This will require accountability frameworks with specific gender indicators and assessment tools included in activity planning from the outset. Moving forward, we need to synergize the emerging communities of practice utilizing the described methods, scale up best practices of mixed-methods approaches that incorporate more qualitative and participatory techniques, and create and track a clear funding stream as well as any other resources to support gender mainstreaming and gender-transformative action.

ACKNOWLEDGMENTS

The authors would like to acknowledge Hellen Amuguni BVSc, PhD, for developing the outline for this chapter. RA would like to recognize the knowledge shared by gender experts over the years, especially Brigitte Bagnol, PhD, and the work of the founding members of the Women for One Health network.

REFERENCES

Aboriginal Health Council of South Australia, Ltd. (2019). *The aboriginal gender study final report*. Aboriginal Health Council of South Australia and the Lowitja Institute. https://ahcsa.org.au/resources/AHC4831-Gender-Study-online.pdf

Aboriginal Justice Inquiry of Manitoba. (1999). *Aboriginal women*. Manitoba, Canada. https://www.ajic.mb.ca/volumel/chapter13.html

AFROHUN. (2020, January 11). *Gender*. https://afrohun.org/lesson/gender/

Alswat, K. A. (2017). Gender disparities in osteoporosis. *Journal of Clinical Medicine Research*, *9*(5), 382–387. https://doi.org/10.14740/jocmr2970w

Amuguni, J. H., Kyewalabye, E., Mugisha, A., Talmage, R., Bagnol, B., & Shah, N. (2019). *Gender, One Health and infectious disease training guide*. OHCEA. https://afrohun.org/wp-content/uploads/2021/01/GENDER-ONE-HEALTH-AND-INFECTIOUS-DISEASE-TRAINING-GUIDE-.pdf

Amuguni, J. H., Mugisha, A., Kyewalabye, E., Bagnol, B., Talmage, R., Bikaako, W., & Naigaga, I. (2018). EnGENDERing One Health and addressing gender gaps in Infectious disease control and response: Developing a Gender, One Health and Emerging Pandemics threat short course for the public health workforce in Africa. *Advances in Social Sciences Research Journal*, *5*(5). https://doi.org/10.14738/assrj.55.4640

Appleton, H., Fernandez, M. E., Hill, C. L. M., & Quiroz, C. (2011). Gender and indigenous knowledge. In S. Harding (Ed.), *The postcolonial science and technology studies reader* (pp. 211–224). Duke University Press. https://doi.org/10.1215/9780822393849-014

Ayukekbong, J. A., Ntemgwa, M., & Atabe, A. N. (2017). The threat of antimicrobial resistance in developing countries: Causes and control strategies. *Antimicrobial Resistance & Infection Control, 6*(1), 47. https://doi.org/10.1186/s13756-017-0208-x

Babo, S. A. Y., Fokou, G., Yapi, R. B., Mathew, C., Dayoro, A. K., Kazwala, R. R., & Bonfoh, B. (2022). Gendered asymmetry of access to knowledge for brucellosis control among pastoral communities in north-west Côte d'Ivoire. *Pastoralism, 12*(1), 28. https://doi.org/10.1186/s13570-022-00241-9

Bagnol, B., Clarke, E., Li, M., Maulaga, W., Lumbwe, H., McConchie, R., De Bruyn, J., & Alders, R. G. (2016). Transdisciplinary project communication and knowledge sharing experiences in Tanzania and Zambia through a One Health lens. *Frontiers in Public Health, 4*. https://doi.org/10.3389/fpubh.2016.00010

Barua, A. (2022, January 21). *Gender equality, dealt a blow by COVID-19, still has much ground to cover*. Deloitte Insights. https://www2.deloitte.com/xe/en/insights/economy/impact-of-covid-on-women.html

Bautista-Valarezo, E., Duque, V., Verdugo Sánchez, A. E., Dávalos-Batallas, V., Michels, N. R. M., Hendrickx, K., & Verhoeven, V. (2020). Towards an indigenous definition of health: An explorative study to understand the indigenous Ecuadorian people's health and illness concepts. *International Journal for Equity in Health, 19*(1), 101. https://doi.org/10.1186/s12939-020-1142-8

Berrian, A. M., Smith, M. H., Van Rooyen, J., Martínez-López, B., Plank, M. N., Smith, W. A., & Conrad, P. A. (2018). A community-based One Health education program for disease risk mitigation at the human-animal interface. *One Health, 5*, 9–20. https://doi.org/10.1016/j.onehlt.2017.11.002

Bill & Melinda Gates Foundation. (2024a). *Data collection methods*. Gender Equality Toolbox. https://www.gatesgenderequalitytoolbox.org/measuring-empowerment/data-collection-methods/

Bill & Melinda Gates Foundation. (2024b). *Gender equality*. Bill & Melinda Gates Foundation. https://www.gatesfoundation.org/our-work/programs/gender-equality

Boudet, A. M. M., Buitrago, P., Scott, K., & Suarez, P. (2018). *Gender differences in poverty and household composition through the life-cycle: A global perspective* (Policy Research Working Paper 8360; Poverty and Equity Global Practice and the Gender Global Theme). World Bank Group. https://documents.worldbank.org/en/publication/documents-reports/documentdetail/135731520343670750/gender-differences-in-poverty-and-household-composition-through-the-life-cycle-a-global-perspective

Canadian Institutes of Health Research. (2003, April 30). *Institute of Gender and Health—CIHR*. https://cihr-irsc.gc.ca/e/8673.html

Canadian Institutes of Health Research. (2019, August 21). *How to integrate sex and gender into research—CIHR*. https://cihr-irsc.gc.ca/e/50836.html

Cataldo, C., Masella, R., & Busani, L. (2023). Gender gap reduction and the one health benefits. *One Health, 16*, 100496. https://doi.org/10.1016/j.onehlt.2023.100496

Cediel-Becerra, N. M., Prieto-Quintero, S., Garzon, A. D. M., Villafañe-Izquierdo, M., Rúa-Bustamante, C. V., Jimenez, N., Hernández-Niño, J., & Garnier, J. (2022). Woman-sensitive One Health perspective in four tribes of indigenous people from Latin America: Arhuaco, Wayuú, Nahua, and Kamëntsá. *Frontiers in Public Health, 10*, 774713. https://doi.org/10.3389/fpubh.2022.774713

Centers for Disease Control and Prevention. (2021, November 8). *NSFG - Listing N - Key statistics from the National Survey of Family Growth*. https://www.cdc.gov/nchs/nsfg/key_statistics/n-keystat.htm

Chemnitz, C., & Becheva, S. (2021). *Meat Atlas 2021: Facts and figures about the animals we eat*. Heinrich-Böll-Stiftung; Friends of the Earth Europe.

Chowdhury, T. J., Arbon, P., Kako, M., Muller, R., Steenkamp, M., & Gebbie, K. (2022). Understanding the experiences of women in disasters: Lessons for emergency management planning. *Australian Journal of Emergency Management, 10.47389/37*(No 1), 72–77. https://doi.org/10.47389/37.1.72

Collins, S. (2017, October 10). The splendor of gender non-conformity in Africa. *Medium.* https://medium.com/@janelane_62637/the-splendor-of-gender-non-conformity-in-africa-f894ff5706e1

Commonwealth of Australia. (2013). *National Aboriginal and Torres Strait Islander health plan 2013–2023.* https://www.health.gov.au/resources/publications/national-aboriginal-and-torres-strait-islander-health-plan-2013-2023?language=en

Cooke, M., Waite, N., Cook, K., Milne, E., Chang, F., McCarthy, L., & Sproule, B. (2017). Incorporating sex, gender and vulnerable populations in a large multisite health research programme: The Ontario Pharmacy Evidence Network as a case study. *Health Research Policy and Systems, 15*(1), 20. https://doi.org/10.1186/s12961-017-0182-z

Côte d'Ivoire Ministère de la santé publique, Centers for Disease Control and Prevention (U.S.), & United States. Agency for International Development. (2018, June 21). *One Health Zoonotic Disease Prioritization for Multi-Sectoral Engagement in Côte d'Ivoire Abidjan, Côte d'Ivoire, January 25–26, 2017.* https://stacks.cdc.gov/view/cdc/56540

Dahab, R., & Sakellariou, D. (2020). Barriers to accessing maternal care in low income countries in Africa: A systematic review. *International Journal of Environmental Research and Public Health, 17*(12), 4292. https://doi.org/10.3390/ijerph17124292

Diego, M. A., Field, T., Hernandez-Reif, M., Cullen, C., Schanberg, S., & Kuhn, C. (2004). Prepartum, postpartum, and chronic depression effects on newborns. *Psychiatry: Interpersonal and Biological Processes, 67*(1), 63–80. https://doi.org/10.1521/psyc.67.1.63.31251

Eba, B., Wieland, B., Flintan, F., Njiru, N., & Baltenweck, I. (2020). *Gender and One Health context analysis for HEAL.* ILRI.

Eng, A., 'T Mannetje, A., McLean, D., Ellison-Loschmann, L., Cheng, S., & Pearce, N. (2011). Gender differences in occupational exposure patterns. *Occupational and Environmental Medicine, 68*(12), 888–894. https://doi.org/10.1136/oem.2010.064097

European Association of Science Editors. (2024). *The SAGER guidelines.* EASE. https://ease.org.uk/communities/gender-policy-committee/the-sager-guidelines/

European Commission. (2020). *Gender—H2020 online manual.* https://ec.europa.eu/research/participants/docs/h2020-funding-guide/cross-cutting-issues/gender_en.htm

European Institute for Gender Equality. (2024a, June 5). *Gender audit.* https://eige.europa.eu/gender-mainstreaming/tools-methods/gender-audit?language_content_entity=en

European Institute for Gender Equality. (2024b, June 5). *What is gender impact assessment | European Institute for Gender Equality.* https://eige.europa.eu/gender-mainstreaming/toolkits/gender-impact-assessment/what-gender-impact-assessment?language_content_entity=en

Fares-Otero, N. E., Pfaltz, M. C., Estrada-Lorenzo, J.-M., & Rodriguez-Jimenez, R. (2020). COVID-19: The need for screening for domestic violence and related neurocognitive problems. *Journal of Psychiatric Research, 130*, 433–434. https://doi.org/10.1016/j.jpsychires.2020.08.015

Fredericks, B., Daniels, C., Judd, J., Bainbridge, R., Clapham, K., Longbottom, M., Adams, M., Bessarab, D., Collard, L., Andersen, C., Duthie, D., & Ball, R. (2017). *Gendered Indigenous health and wellbeing within the Australian health system: A review of the literature.* Central Queensland University. https://doi.org/10.4226/145/5A79471112A55

Friedson-Ridenour, S., Dutcher, T. V., Calderon, C., Brown, L. D., & Olsen, C. W. (2019). Gender analysis for One Health: Theoretical perspectives and recommendations for practice. *EcoHealth, 16*(2), 306–316. https://doi.org/10.1007/s10393-019-01410-w

Garnier, J., Savic, S., Boriani, E., Bagnol, B., Häsler, B., & Kock, R. (2020). Helping to heal nature and ourselves through human-rights-based and gender-responsive One Health. *One Health Outlook, 2*(1), 22. https://doi.org/10.1186/s42522-020-00029-0

Garnier, J., Savić, S., Cediel, N., Barato, P., Boriani, E., Bagnol, B., & Kock, R. A. (2022). Mainstreaming gender-responsive One Health: Now is the time. *Frontiers in Public Health, 10*, 845866. https://doi.org/10.3389/fpubh.2022.845866

Gee, G., Dudgeon, P., Schultz, C., Hart, A., & Kelly, K. (n.d.). Aboriginal and Torres Strait Islander social and emotional wellbeing. In P. Dudgeon, H. Milroy, & R. Walker (Eds.), *Working together: Aboriginal and Torres Strait Islander mental health and wellbeing principles and practice* (2nd ed.). Commonwealth Government of Australia. https://www.telethonkids.org.au/globalassets/media/documents/aboriginal-health/working-together-second-edition/wt-part-1-chapt-4-final.pdf

Gendered Innovations. (2020). *Gender*. European Commission, Standford University, National Science Foundation. https://genderedinnovations.stanford.edu/terms/gender.html

Ghouaibi, A. (2021, September 23). *COVID-19's impact on women's health and rights*. World Economic Forum. https://www.weforum.org/agenda/2021/09/lessons-must-be-learned-from-covid-19-s-impact-on-women-s-health-and-rights/

Golemac Powell, A., Chitanava, M., Tsikvadze, N., Gaprindashvili, N., Keshelava, D., & Lobjanidze, M. (2021). *Gender impact assessment methodology*. UN Women. https://iset-pi.ge/storage/media/other/2021-09-30/0325d040-21fa-11ec-8cea-71acc4a69d98.pdf

Gould, E., Schieder, J., & Geier, K. (2016). *What is the gender pay gap and is it real?* Economic Policy Institute. https://www.epi.org/publication/what-is-the-gender-pay-gap-and-is-it-real/

Graham, D. W., Bergeron, G., Bourassa, M. W., Dickson, J., Gomes, F., Howe, A., Kahn, L. H., Morley, P. S., Scott, H. M., Simjee, S., Singer, R. S., Smith, T. C., Storrs, C., & Wittum, T. E. (2019). Complexities in understanding antimicrobial resistance across domesticated animal, human, and environmental systems. *Annals of the New York Academy of Sciences, 1441*(1), 17–30. https://doi.org/10.1111/nyas.14036

Gurkin, C. (2019). *Board diversity—Strategies to increase representation of women and minorities* (GAO-19-637T). United States Government Accountability Office. https://www.gao.gov/products/gao-19-637t

Harari, L., & Lee, C. (2021). Intersectionality in quantitative health disparities research: A systematic review of challenges and limitations in empirical studies. *Social Science & Medicine, 277*, 113876. https://doi.org/10.1016/j.socscimed.2021.113876

Hidrobo, M., Kumar, N., Palermo, T., Peterman, A., & Roy, S. (2020). *Gender-sensitive social protection: A critical component of the COVID-19 response in low- and middle-income countries*. International Food Policy Research Institute. https://doi.org/10.2499/9780896293793

Hillier, S. A., Taleb, A., Chaccour, E., & Aenishaenslin, C. (2021). Examining the concept of One Health for indigenous communities: A systematic review. *One Health, 12*, 100248. https://doi.org/10.1016/j.onehlt.2021.100248

Im, E., & Meleis, A. I. (2001). An international imperative for gender-sensitive theories in women's health. *Journal of Nursing Scholarship, 33*(4), 309–314. https://doi.org/10.1111/j.1547-5069.2001.00309.x

International Development Research Centre. (2024, March 4). *Gender equality: Making a transformative impact.* https://idrc-crdi.ca/en/research-in-action/gender-equality-making-transformative-impact

IP Australia. (2023). *Indigenous knowledge.* Understanding Intellectual Property. https://www.ipaustralia.gov.au/understanding-ip/indigenous-knowledge

ISPI. (2020, March 9). *Third gender rights in South Asia: What's new?* ISPI. https://www.ispionline.it/en/publication/third-gender-rights-south-asia-whats-new-25354

Jones, N., Mitchell, J., Cooke, P., Baral, S., Arjyal, A., Shrestha, A., & King, R. (2022). Gender and antimicrobial resistance: What can we learn from applying a gendered lens to data analysis using a participatory arts case study? *Frontiers in Global Women's Health, 3*, 745862. https://doi.org/10.3389/fgwh.2022.745862

Kendall, T., & Langer, A. (2015). Critical maternal health knowledge gaps in low- and middle-income countries for the post-2015 era. *Reproductive Health, 12*(1), 55. https://doi.org/10.1186/s12978-015-0044-5

Klein, S. L., Hodgson, A., & Robinson, D. P. (2012). Mechanisms of sex disparities in influenza pathogenesis. *Journal of Leukocyte Biology, 92*(1), 67–73. https://doi.org/10.1189/jlb.0811427

Laing, G., Duffy, E., Anderson, N., Antoine-Moussiaux, N., Aragrande, M., Luiz Beber, C., Berezowski, J., Boriani, E., Canali, M., Pedro Carmo, L., Chantziaras, I., Cousquer, G., Meneghi, D., Gloria Rodrigues Sanches Da Fonseca, A., Garnier, J., Hitziger, M., Jænisch, T., Keune, H., Lajaunie, C., … Häsler, B. (2023). Advancing One Health: Updated core competencies. *CABI One Health*, ohcs20230002. https://doi.org/10.1079/cabionehealth.2023.0002

Lipson, S. K., Raifman, J., Abelson, S., & Reisner, S. L. (2019). Gender minority mental health in the U.S.: Results of a national survey on college campuses. *American Journal of Preventive Medicine, 57*(3), 293–301. https://doi.org/10.1016/j.amepre.2019.04.025

Ly, D., Forman, D., Ferlay, J., Brinton, L. A., & Cook, M. B. (2013). An international comparison of male and female breast cancer incidence rates. *International Journal of Cancer, 132*(8), 1918–1926. https://doi.org/10.1002/ijc.27841

Masella, R., & Malorni, W. (2017). Gender-related differences in dietary habits. *Clinical Management Issues, 11*(2). https://doi.org/10.7175/cmi.v11i2.1313

Masterson, V. (2021, June 21). *6 charts that reveal global attitudes to LGBT+ and gender identities in 2021.* World Economic Forum. https://www.weforum.org/agenda/2021/06/lgbt-gender-identity-ipsos-2021-survey/

Mauvais-Jarvis, F., Bairey Merz, N., Barnes, P. J., Brinton, R. D., Carrero, J.-J., DeMeo, D. L., De Vries, G. J., Epperson, C. N., Govindan, R., Klein, S. L., Lonardo, A., Maki, P. M., McCullough, L. D., Regitz-Zagrosek, V., Regensteiner, J. G., Rubin, J. B., Sandberg, K., & Suzuki, A. (2020). Sex and gender: Modifiers of health, disease, and medicine. *The Lancet, 396*(10250), 565–582. https://doi.org/10.1016/S0140-6736(20)31561-0

Moore, K. L. (1968). The sexual identity of athletes. *JAMA: The Journal of the American Medical Association, 205*(11), 787. https://doi.org/10.1001/jama.1968.03140370089020

National Institutes of Health. (2015, June 9). *NOT-OD-15-102: Consideration of sex as a biological variable in NIH-funded research.* https://grants.nih.gov/grants/guide/notice-files/not-od-15-102.html

National Institutes of Health. (2023). *About—Office of Research on Women's Health.* https://orwh.od.nih.gov/about

New South Wales Department of Community Services. (2009). *Working with aboriginal people and communities—A practice resource*. New South Wales. https://www.community.nsw.gov.au/—data/assets/pdf_file/0017/321308/working_with_aboriginal.pdf

Nkangu, M. N., Olatunde, O. A., & Yaya, S. (2017). The perspective of gender on the Ebola virus using a risk management and population health framework: A scoping review. *Infectious Diseases of Poverty*, *6*(1), 135. https://doi.org/10.1186/s40249-017-0346-7

OECD. (2021). *Policy framework for gender-sensitive public governance* (C/MIN(2021)21; Meeting of the Council at Ministerial Level). OECD. https://www.oecd.org/mcm/Policy-Framework-for-Gender-Sensitive-Public-Governance.pdf

OECD. (2024). *Toolkit for mainstreaming and implementing gender equality*. https://www.oecd-ilibrary.org/governance/toolkit-for-mainstreaming-and-implementing-gender-equality-2023_51fee19f-en

OECD - Social Policy Division - Directorate of Employment, Labour and Social Affairs. (2021). *HC3.1. Homeless Population* (OECD Affordable Housing Database). OECD. https://www.oecd.org/els/family/HC3-1-Homeless-population.pdf

One Health Commission. (2022). *One Health tools and toolkits*. https://www.onehealthcommission.org/en/resources__services/one_health_tools__toolkits/

O'Neil, A., Russell, J. D., Thompson, K., Martinson, M. L., & Peters, S. A. E. (2020). The impact of socioeconomic position (SEP) on women's health over the lifetime. *Maturitas*, *140*, 1–7. https://doi.org/10.1016/j.maturitas.2020.06.001

Oruganti, P., Root, E., Ndlovu, V., Mbhungele, P., Van Wyk, I., & Berrian, A. M. (2023). Gender and zoonotic pathogen exposure pathways in a resource-limited community, Mpumalanga, South Africa: A qualitative analysis. *PLoS Global Public Health*, *3*(6), e0001167. https://doi.org/10.1371/journal.pgph.0001167

Ovseiko, P. V., Greenhalgh, T., Adam, P., Grant, J., Hinrichs-Krapels, S., Graham, K. E., Valentine, P. A., Sued, O., Boukhris, O. F., Al Olaqi, N. M., Al Rahbi, I. S., Dowd, A.-M., Bice, S., Heiden, T. L., Fischer, M. D., Dopson, S., Norton, R., Pollitt, A., Wooding, S., … Buchan, A. M. (2016). A global call for action to include gender in research impact assessment. *Health Research Policy and Systems*, *14*(1), 50. https://doi.org/10.1186/s12961-016-0126-z

Pan-American Health Organization. (2004). *Gender, equity, and indigenous women's health in the Americas*. Regional Office of the World Health Organization. https://www3.paho.org/hq/dmdocuments/2011/gdr-gender-equity-and-indigenous-women-health-americas.pdf

ReAct, Institute of Development Studies, & Triple Line Consulting Ltd. (2020). *Scoping the significance of gender for antibiotic resistance*. Swedish International Development Cooperation Agency. https://tripleline.com/insight/article/scoping-the-significance-of-gender-for-antibiotic-resistance

Sax, L. (2002). How common is lntersex? A response to Anne Fausto-Sterling. *The Journal of Sex Research*, *39*(3), 174–178. https://doi.org/10.1080/00224490209552139

Scarpetta, S., Pearson, M., Queisser, M., & Frey, V. (2021). *Caregiving in crisis: Gender inequality in paid and unpaid work during COVID-19* (OECD Policy Responses to Coronavirus (COVID-19)) [OECD Policy Responses to Coronavirus (COVID-19)]. https://doi.org/10.1787/3555d164-en

Shannon, G., Minckas, N., Tan, D., Haghparast-Bidgoli, H., Batura, N., & Mannell, J. (2019). Feminisation of the health workforce and wage conditions of health professions: An exploratory analysis. *Human Resources for Health*, *17*(1), 72. https://doi.org/10.1186/s12960-019-0406-0

Shibabaw, A., Abebe, T., & Mihret, A. (2013). Nasal carriage rate of methicillin resistant Staphylococcus aureus among Dessie Referral Hospital Health Care Workers; Dessie, Northeast Ethiopia. *Antimicrobial Resistance and Infection Control*, *2*(1), 25. https://doi.org/10.1186/2047-2994-2-25

Siow, W. T., Liew, M. F., Shrestha, B. R., Muchtar, F., & See, K. C. (2020). Managing COVID-19 in resource-limited settings: Critical care considerations. *Critical Care*, *24*(1), 167, s13054-020-02890–x. https://doi.org/10.1186/s13054-020-02890-x

Skuse, D., Printzlau, F., & Wolstencroft, J. (2018). Sex chromosome aneuploidies. In *Handbook of clinical neurology* (Vol. 147, pp. 355–376). Elsevier. https://doi.org/10.1016/B978-0-444-63233-3.00024-5

Smith, J. (2019). Overcoming the 'tyranny of the urgent': Integrating gender into disease outbreak preparedness and response. *Gender & Development*, *27*(2), 355–369. https://doi.org/10.1080/13552074.2019.1615288

State of Queensland (Queensland Health). (2016). *Queensland health aboriginal and Torres Strait Island Mental Health Strategy 2016–2021*. State of Queensland. https://www.health.qld.gov.au/—data/assets/pdf_file/0030/460893/qhatsi-mental-health-strategy.pdf

Sutherland, S., & Adams, M. (2019). Building on the definition of social and emotional wellbeing: An indigenous (Australian, Canadian, and New Zealand) viewpoint. *Ab-Original*, *3*(1), 48–72. https://doi.org/10.5325/aboriginal.3.1.0048

Swensen, G., Serafino, S., & Thomson, N. (1995). *Suicide in Western Australia, 1983–1992*. State Health Purchasing Authority, Health Department of Western Australia. https://healthinfonet.ecu.edu.au/key-resources/publications/8606/?title=Suicide+in+Western+Australia++1983+1992&contenttypeid=1&contentid=8606_1

Szczerbal, I., & Switonski, M. (2016). Chromosome abnormalities in domestic animals as causes of disorders of sex development or impaired fertility. In R. Payan Carreira (Ed.), *Insights from animal reproduction*. InTech. https://doi.org/10.5772/62053

The Global Fund. (2023). *Technical brief—Gender equality—Allocation period 2023–2025*. https://www.theglobalfund.org/media/5728/core_gender_infonote_en.pdf

The Lancet Public Health. (2019). Time to tackle the physical activity gender gap. *The Lancet Public Health*, *4*(8), e360. https://doi.org/10.1016/S2468-2667(19)30135-5

The World Bank. (2024). *Gender*. World Bank. https://www.worldbank.org/en/topic/gender

Togami, E., Behravesh, C. B., Dutcher, T. V., Hansen, G. R., King, L. J., Pelican, K. M., & Mazet, J. A. K. (2023). Characterizing the One Health workforce to promote interdisciplinary, multisectoral approaches in global health problem-solving. *PLOS ONE*, *18*(5), e0285705. https://doi.org/10.1371/journal.pone.0285705

Turner, A., & McKenzie, K. (2022, June 30). *The role of Totems in conservation, kinship, and spiritual connectivity with the land*. Flow-MER. https://flow-mer.org.au/the-role-of-totems-in-conservation-kinship-and-spiritual-connectivity-with-the-land/

UN Statistics Wiki. (2020, December 28). *Glossary of terms—Integrating a gender perspective into statistics*. https://unstats.un.org/wiki/display/genderstatmanual/Glossary+of+terms

UN Women. (2023). *The shadow pandemic: Violence against women during COVID-19*. UN Women – Headquarters. https://www.unwomen.org/en/news/in-focus/in-focus-gender-equality-in-covid-19-response/violence-against-women-during-covid-19

UN Women. (2024, May 31). *Facts and figures: Women's leadership and political participation*. UN Women – Headquarters. https://www.unwomen.org/en/what-we-do/leadership-and-political-participation/facts-and-figures

UNICEF. (2021). *UNICEF gender policy 2021–2030*. United Nations. https://www.unicef.org/reports/unicef-gender-policy-2021-2030

United Nations Development Programme. (2022). *Gender equality strategy 2022–2025*. UNDP. https://genderequalitystrategy.undp.org/#

University of Basel. (2021). *"Gender sensitive" or "gender responsive"?* Gender Gaps in (Re)Productive Economy. https://tales.nmc.unibas.ch/en/gender-and-labour-in-the-global-south-38/gender-gaps-in-re-productive-economy-211/gender-sensitive-or-gender-responsive-1115

USAID. (2009). *Asia regional consultation on MSM HIV care and support*. https://healtheducationresources.unesco.org/library/documents/asia-regional-consultation-msm-hiv-care-and-support

USAID. (2023). *USAID 2023 gender equality and women's empowerment policy*. U.S. Agency for International Development.

Van Gerwen, O. T., Muzny, C. A., & Marrazzo, J. M. (2022). Sexually transmitted infections and female reproductive health. *Nature Microbiology*, *7*(8), 1116–1126. https://doi.org/10.1038/s41564-022-01177-x

Wellcome. (2023). *Diversity and inclusion*. Wellcome. https://wellcome.org/who-we-are/diversity-and-inclusion

White, J., Tannenbaum, C., Klinge, I., Schiebinger, L., & Clayton, J. (2021). The integration of sex and gender considerations into biomedical research: Lessons from international funding agencies. *The Journal of Clinical Endocrinology & Metabolism*, *106*(10), 3034–3048. https://doi.org/10.1210/clinem/dgab434

Whitfield, J. (2004). Everything you always wanted to know about sexes. *PLoS Biology*, *2*(6), e183. https://doi.org/10.1371/journal.pbio.0020183

Wiersma, R. (2004). True hermaphroditism in southern Africa: The clinical picture. *Pediatric Surgery International*, *20*(5). https://doi.org/10.1007/s00383-004-1200-0

World Health Organization. (2002). *Integrating gender perspectives in the work of WHO: WHO Gender Policy*. https://iris.who.int/handle/10665/67649

World Health Organization. (2021, May 24). *Gender and health*. https://www.who.int/news-room/questions-and-answers/item/gender-and-health

Zinsstag, J., Muhummed, A., Nuvey, F., Keita, Z., Pelikan, K., Kaiser-Grolimund, A., Crump, L., & Heitz-Tokpa, K. (2023). Impressions from the 7th World One Health Congress in Singapore. *CABI One Health*, ohcs20230004. https://doi.org/10.1079/cabionehealth.2023.0004

13 One Health Education, Training, and Capacity Building

Jane Blake, Leah Goodman, Elsie Kiguli-Malwadde, Cheryl Stroud, and Deborah Thomson

INTRODUCTION TO ONE HEALTH EDUCATION, TRAINING, AND CAPACITY BUILDING

In 2010, a group of students from ecology, medicine, veterinary medicine, and global public health concluded that education would play a particularly important role in realizing the One Health concept but that a shortage of collaborative student programs, insufficient environmental training for health professionals, and a lack of institutional support were impeding progress (Barrett et al., 2011).

Today, as noted in previous chapters, One Health awareness has grown exponentially over the past few decades and is being embraced all over the world (One Health Commission, 2023). This trend applies to One Health-related research and to One Health framed education, training, and capacity building (ETC). ETC programs are expanding to provide One Health degree-granting undergraduate and graduate curricula, certificate-based training programs, and capacity building initiatives (One Health Commission, 2024h; Togami et al., 2018; Von Borries et al., 2020).

The interdisciplinary nature of the One Health approach is applicable to all forms and levels of ETC. One Health experts advocate for involvement of One Health concepts early in education (primary school and beyond) to ensure the general population has a core knowledge base regarding health (Von Borries et al., 2020). This multidisciplinary concept of "oneness," the interconnectedness of all life, must emphasize the value of biodiversity and potential consequence of zoonotic diseases when biodiversity is lost. Reaching beyond academia, the One Health concept must be incorporated in public messaging and become the default approach to government, industry, policy, and research. One Health education, not only in academia but for the general public, is needed from cradle to grave. This chapter will focus on the incorporation of One Health education in "academic" settings.

The veterinary sector first recognized the One Health concept and has been the most active participant in One Health ETC although traditional silos do still exist. However, additional professional sectors are increasingly beginning

DOI: 10.1201/9781003232223-13

to incorporate One Health into educational curricula. As early as 2008, Kahn, Kaplan, Monath, and Steele suggested that schools of medicine and schools of veterinary medicine should provide their students the opportunities to learn about how animal and human health can impact each other. The expansion of One Health ETC programs has since escalated as revealed by the seminal work of the One Health Commission and colleagues beginning in 2015 (McKenzie et al., 2016; One Health Commission, 2016b, 2023, 2024h, 2024e; Reid et al., 2016; Rwego et al., 2016; Sikkema & Koopmans, 2016; Stroud et al., 2016; Wu et al., 2016), lending to a notable upward trend in One Health education since the early 2000s.

In this chapter, we discuss various international ETC programs that expand global One Health Workforce capacity, explore their design, identify their approach to monitoring and evaluating student outcomes, and identify potential gaps. We conclude this chapter with a discussion of future directions in One Health ETC. The programs we discuss represent a sample and may not be all inclusive; our intent is to provide an overview rather than a full literature review and to potentially spur revelation of additional programs not yet compiled. While One Health research plays a significant role in capacity building and education and ensures that human, animal, and environmental health questions are evaluated in an integrated and holistic manner (Lebov et al., 2017), the scope of this chapter is limited to ETC programs that may or may not include a research component. Further, this chapter focuses on the formal One Health ETC programs for which information was available and may not include grassroots or less formal efforts conducted by One Health advocates across the world.

One Health Education

In this discussion, the term "One Health education" will refer to primary-secondary school education (Von Borries et al., 2020) as well as undergraduate or graduate programs in academic, degree-granting institutions and related One Health education associated with non-profit organizations. Examples of formal university programs include a PhD, a master's, or bachelor's degree in One Health or Global Health, a Master of Public Health (MPH) with One Health concentrations, and One Health components of a medical or veterinary program. While additional academic programs exist covering planetary health and environmental health related to the intersection of human impact on the environment and vice versa, in this chapter we focus on One Health education that addresses the full spectrum of issues included under the One Health purview, such as comparative medicine, translational research, antimicrobial resistance, and the human-animal bond, that would not likely be included in planetary health curricula (Devinsky et al., 2018). Examples of specific education programs include the One Health undergraduate minor at Berry College (2015), the undergraduate class "One Health: Human, Animal, and Environment" at the Federal University of Minas Gerais in Brazil (Pettan-Brewer et al., 2021), the One Health undergraduate major offered at Fontbonne University (2024), various degrees offered at the University of Edinburgh School of Veterinary Studies (The Royal (Dick) School of Veterinary Studies, 2024), and the Master's in Global Health Delivery (MGHD) provided at the University of Global Health Equity (2024). Additional programs and details are described in Section "Program Design for One Health Education, Training and Capacity Building." We also include educational programs overseen by non-governmental organizations (NGOs) and certificates, both free-standing or earned, as a part of a formal degree program.

One Health Training

We define training as programs that provide discrete exercise on specific One Health skills. Existing training programs may include free lectures, self-paced modules, such as the online One Health Training Course for working professionals provided by the One Health Workforce Academies (OHWA) (One Health Workforce Academies, 2024a), and free lectures and non-degree certificates, such as the One Health Course Series provided by the World Health Organization (WHO) (OpenWHO, 2024) and the World Small Animal Veterinary Association's One Health Certificate program (dvm360, 2020). Additional training programs and details are described in Section "Program Design for One Health Education, Training and Capacity Building."

One Health Capacity Building

We utilize the United Nations (UN) definition of capacity building: "the process of developing and strengthening the skills, instincts, abilities, processes and resources that organizations and communities need to survive, adapt, and thrive in a fast-changing world" (United Nations, n.d.). The UN links this definition to Sustainable Development Goal 17: Revitalizing the Global Partnership for Sustainable Development and notes that capacity building should be transformational and sustained over time. Capacity building aims to empower people and/or institutions to maintain and sustain key skills.

For One Health, capacity building efforts take various forms such as train-the-trainer (ToT) programs, curriculum twinning projects between universities, and network collaborations. Specific examples include Ethiopia's One Health – Zoonotic Disease: A Training of Trainers Course (Ethiopian Agriculture Training Portal, 2021), the ToT program called the Certified Lesson Leaders Program provided by One Health Lessons for adults to educate the public (children and adults) about One Health (One Health Lessons, 2023), and One Health curriculum twinning projects such as the collaboration between Ethiopia's University of Gondar veterinary school and the Ohio State University's Global One Health Initiative since 2016 (Office of International Affairs, The Ohio State University, 2024). Additionally, we consider continuing medical or veterinary education to be capacity building, as they are ingrained in system-wide accreditation requirements, thus strengthening overall sustainable development. Capacity building endeavors can be performed by academic institutions, government programs, NGOs, or by

TABLE 13.1
Definitions of One Health Education, Training, and Capacity Building

Type	Definition	Examples
Education	Formal learning in primary, secondary, and higher academic institutions	• Primary and secondary school education • Undergraduate or graduate degrees • Certificates in degree-granting programs
Training	Part of capacity building, discrete education on specific One Health skills	• Lectures • Self-paced online modules • Non-degree certificates • Ad hoc One Health trainings and workshops
Capacity Building	Process of developing and strengthening the skills, instincts, abilities, processes, and resources that organizations and communities need to survive, adapt, and thrive in a fast-changing world	• Train the Trainer (ToT) • Twinning • Continuing Medical Education (CME) • Continuing Education (CE) • Continuing Professional Development (CPD) • Nursing Continued Professional Development (NCPD) • Certified Health Education Specialists (CHES) • Master Certified Health Education Specialists (MCHES)

any grassroots organization focused in One Health. Additional details on these, and more capacity building programs, are described in Section "Program Design for One Health Education, Training and Capacity Building."

Table 13.1 summarizes our definitions of One Health education, capacity building, and training.

EXAMPLES OF EXISTING PROGRAMS AND/OR CURRICULA

One Health ETC programs are growing in number but are still somewhat nascent. That said, the education sector is arguably seeing the most growth in number of One Health learning opportunities, especially within degree-granting programs.

Existing One Health Education Programs

As of this writing, there are hundreds of institutions around the world offering either primary and secondary school curricula, undergraduate minors and majors, masters degrees, certificates, or PhDs focused on the One Health approach (Berg & Olsen, 2016). Many universities were highlighted in a 2016 special issue of *Infection Ecology and Epidemiology* highlighting One Health Training, Research, and Outreach around the world (One Health Commission, 2016a). More recently, a 2022 review of organizations embracing and working to further One Health identified 289 organizations, including 126 civil society organizations, 133 academic institutions, and 30 governmental organizations (Laaser et al., 2022).

For this chapter, we reviewed universities, colleges, non-profit organizations, and associations, offering in-person, online, and summer field work. The reviewed programs are located in Australia, Africa, Canada, the Caribbean, Central and South America, Europe, Asia, and the United States. In the United States alone, over 50 One Health educational programs were identified at more than 35 institutions; this represents an approximate 3-fold increase since 2016 (Stroud et al., 2016). Many One Health "activities" at universities have expanded into full degree-granting programs as detailed in an overview of One Health education and activities in Brazil, Chile, and Colombia (Pettan-Brewer et al., 2021). For example, in 2020, a One Health course for non-biology undergraduate students (law, psychology, journalism, and engineering) was given at the Adolfo Ibáñez University in Chile (Cianfagna et al., 2021). On the primary-secondary school education front, notable resources include One Health Lessons, a non-profit organization (Barrett et al., 2011) that provides high-quality age-appropriate lessons about One Health to primary and secondary school students as well as to adults in the community and at universities and (One Health Commission, 2023) trains adults of various disciplines and backgrounds to teach One Health in classrooms around the world.

Additional notable programs include the One Health Commission's One Health Education – United States working group (One Health Commission, 2024f) that explores ways to take One Health to primary and secondary school educators. This latter group compiles One Health educational resources for teachers (One Health Commission, 2024l) and students and has a One Health survey for primary/secondary teachers currently in the field that not only assesses teacher familiarity with One Health but also informs teachers about the concept and why it is important for their students. Other existing programs include the Kansas State Olathe Campus K-12 One Health educational program (Kansas State University, 2024) and the St. Louis Zoo Institute for Conservation Medicine's (ICM) One Health education program built around local box turtles that teaches children about One Health while getting them outdoors to look for and work with turtles (Saint Louis Zoo, 2021). One Health Colombia at the University of Cordoba takes learning material about One Health into primary/secondary classrooms (Pettan-Brewer et al., 2021) while One Health Cares, Inc. does similar work in Fukuoka, Japan (FAVA OneHealth Fukuoka Office, 2023).

HIGHLIGHT: SAINT LOUIS ZOO ICM'S CHILDREN'S ONE HEALTH TRAINING (SAINT LOUIS ZOO, 2021)

The Saint Louis Zoo ICM takes a holistic approach to wildlife conservation, public health, and sustainable ecosystems to ensure healthy animals and healthy people. In 2012, the ICM team and partners developed the St.

Louis Box Turtle Project. The project received National Science Foundation funding to develop teacher workshops and field trip opportunities to take students into nature to track box turtles in rural and urban research sites. This program is now a flagship STEM (Science, Technology, Engineering and Mathematics) program for the Saint Louis Zoo.

The Saint Louis Box Turtle Project includes a classroom visit to introduce conservation issues that box turtles face and to prepare students for their field experience. To date, the team has taken over 1,500 students into natural areas to track box turtles using radio telemetry. Many of the students are from underserved, urban schools near Forest Park. For many students, this is their first-time exploring nature and seeing a turtle in the wild. Students not only gain experience using scientific equipment alongside a field biologist, but they also gain a better understanding of their connection to nature, a fundamental part of One Health and the foundation for developing conservation mindfulness.

In 2016, the Ferguson-Florissant School District asked ICM to expand the project to include their site, Little Creek Nature Area. This 98-acre natural area is owned and managed by the school district. This expansion has allowed the team to collaborate with teachers and administrators to use the Saint Louis Box Turtle Project as a part of the district's STEM curriculum and experiential learning opportunity.

Based on our review of One Health education, veterinary schools are more likely to incorporate One Health into their curriculum while medical schools continue to lag behind. As surveyed by Georgetown University in 2021, only 56% of 133 U.S. medical schools include One Health-related subject matter in the curriculum (Docherty & Foley, 2021). Of note, the University of Hawaii at Manoa's medical school offers a Certificate of Distinction in One Health (University of Hawai'i, 2024) and its undergraduate sector also has a One Health Interdisciplinary Undergraduate Certificate (University of Hawai'i, 2024). A list of education programs reviewed for this chapter is included in Table 13.2.

HIGHLIGHT: MEDICAL SCHOOLS AND SCHOOLS OF PUBLIC HEALTH – MAKING ONE HEALTH A REQUIREMENT

While the One Health concept has been highlighted at many fora and is endorsed by a number of major medical and public health organizations such as the Association of American Medical Colleges (AAMC) and the American Public Health Association (APHA) (American Public Health Association, 2017), many medical and public health educators may not yet be familiar (Rabinowitz et al., 2017). One Health awareness has increased in human health circles in the past few years in light of recent emerging zoonotic disease outbreaks including West Nile virus (1999), zoonotic avian influenza A (2003), Severe Acute Respiratory Syndrome coronavirus 2 (SARS-CoV-2, 2003), pandemic H1N1 influenza (2009), Middle East Respiratory Syndrome (MERS, 2012), Ebola virus (2014), Zika virus (2015), and SARS-CoV-2-causing COVID-19 (2019) (Allen-Scott et al., 2015; Boyle, 2019; Holmes, 2022; Institute of Medicine, 2009; Taylor et al., 2001). The World Medical Association along with the World Veterinary Association held their first joint conference on One Health in 2015 in Spain (World Medical Association, 2015). Yet, as zoonotic disease outbreaks become more frequent, threatening global health security and economic stability, early detection and mitigation are imperative. Veterinarians are critical for early recognition of zoonotic diseases that might initially show up in animals. But human health and public health practitioners are the first line of defense to detect and alert when they present in humans. Therefore, physicians and public health workers must be able to recognize and diagnose emerging zoonotic diseases that may appear in their exam rooms or communities. It is important that medical educators, in both schools of medicine and public health (One Health Commission, 2024b), include the full scope of One Health in their curricula to prepare human health professionals for future outbreaks, ensuring they know what questions to ask their patients and which signs to recognize. Medical students, along with other health-focused students, have publicly expressed interest in learning more about One Health (Elmahi et al., 2022).

Generally, when One Health is presented in medical schools, it is taught in a siloed manner scattered through microbiology, internal medicine, public health, etc., and is usually only included by faculty who are familiar with the concept, which can be limiting (Hilliard, 2015). The ideal would be to expose medical students to a dedicated One Health course in their first year and for medical and public health school faculty to be keenly aware of incorporating the One Health way of thinking into all subjects so it is deliberately applied across the full training experience. Consistently incorporating One Health into all subjects will instill in future clinicians this holistic way of approaching patients.

Overall, One Health education in medical schools is in its infancy and lags behind veterinary schools; many veterinary schools have already incorporated One Health as a central part of their curricula (Linder et al., 2020; Rabinowitz et al., 2017). Adoption of the One Health concept and its application to clinical care requires educating and training of health professionals; therefore, core competencies for One Health training have been proposed (One Health Commission, 2024m; Rabinowitz et al., 2017). We acknowledge the challenge of revising medical curriculum due to the magnitude of information already required; however, medical school faculty can incorporate One Health examples within the material already presented without a huge effort. Similarly, requiring students to complete an online Intro to One Health for Human Health Clinicians lecture or course in their first year would not add much burden to the curriculum and can be achieved through online and asynchronous approaches. The emphasis must shift from solely curative care to incorporate prevention with a public health approach (One Health Commission, 2024b).

Progress is ongoing in Europe, the Americas, Africa, and Asia along with recommendations on the next steps for improving One Health thinking in medicine (Iatridou et al., 2021; One Health Commission, 2016a; Rwego et al., 2016). Both academic and non-academic institutions are creating networks to do research, conduct trainings, and hold conferences on the One Health concept. Innovative approaches such as interprofessional electives, school clubs, working groups, and research networks are being adopted to increase awareness. Though the work is in its early stages globally, awareness is expanding and slowly translating into concrete One Health education and research activities.

TABLE 13.2
MPH with Specialization in Global Environmental Sustainability and Health Competencies at Johns Hopkins University

Competency	Course Number(s) and Name(s)
GESH1. Define climate change and describe multiple ways climate change will have an impact on public health	180.611 The Global Environment, Climate Change and Public Health
GESH2. Define behaviors related to climate change and environmental sustainability, identify factors that affect them, and design a behavior change intervention to address either climate change or environmental sustainability	224.689 Health Behavior Change at the Individual, Community and Household Levels
GESH3. Effectively communicate a position/opinion on a climate and/or sustainability topic	188.688 Global Environmental Sustainability and Health
GESH4. Characterize the role of food systems in sustainability and the environment	180.620 Introduction to Food Systems and Public Health OR 180.606 Case Studies in Food Production and Public Health OR 180.655 Baltimore Food Systems: A Case Study of Urban Food Environments
GESH5. Apply principles of systems thinking to understand the various causes and effects of climate change on the environment and public health	180.611 The Global Environment, Climate Change and Public Health

Source: Johns Hopkins Bloomberg School of Public Health, 2024

One Health Training

While training is a subset of capacity building, the authors chose to discuss discrete training programs separately, as many are provided for free, often for no educational credit, and are not always a component of an overall capacity building approach (One Health Commission, 2024h). However, these training programs provide a large impact in building awareness and inspiring actions by One Health advocates. Additionally, while most of the programs listed in Table 13.2 do not provide official credit, they often provide certificates and can be used to study for certification exams and other such efforts.

Like overall capacity building, One Health trainings are provided by governments, NGOs, One Health networks, and academic institutions. For example, the Africa One Health University Network (AFROHUN) provides One Health Training Modules to assist in workforce development; Kerala Veterinary and Animal Sciences University, Centre for One Health Education Advocacy Research & Training (COHEART) provides adult workshops in India (Kerala Veterinary and Animal Sciences University, 2023); and OHWA provides an online One Health Training Course for working professionals (One Health Workforce Academies, 2024b).

HIGHLIGHT: UNIVERSITY OF GENEVA GLOBAL HEALTH AT THE HUMAN-ANIMAL-ECOSYSTEM INTERFACE COURSERA COURSE (UNIVERSITY OF GENEVA, 2024)

The University of Geneva, Institute Pasteur, University of Montreal, and Centre Virchow-Villermé/University Paris Descartes have an 8-week course where students explore and learn about past and current Global Health Challenges at the Human-Animal-Ecosystem Interface: zoonotic emerging infections (e.g., Ebola, Nipah, MERS, Avian Influenza), antimicrobial resistance, neglected tropical diseases (e.g., rabies, leishmaniasis, zoonotic TB), snakebites, and other human-animal conflicts, etc. The course covers epidemiology, social anthropology, disease ecology, veterinary sciences, global health policy, etc., and approaches such as One Health, EcoHealth, and Planetary Health. It also covers innovative tools and frameworks used to study and address Global Health challenges of the UN Sustainable Development Goals.

The course involves more than 30 experts from over 20 academic and research institutions and international organizations including the WHO, clinicians from the University Hospitals of Geneva, and epidemiologists from Institute Pasteur. The format utilizes video lectures filmed in different parts of the world and settings (from the field to the lab and office) combined with the latest open readings and interactive activities in the discussion forum and video conferences.

The development of this course was led by Dr. Rafael Ruiz de Castañeda, Dr. Isabelle Bolon, and Prof. Antoine Flahault from the Institute of Global Health of the University of Geneva.

One Health Capacity Building

As noted above, capacity building is a broad field that includes developing skills, instincts, abilities, processes, and resources to implement and sustain One Health approaches. As such, capacity building can be implemented by multiple public, private, and academic institutions. Capacity building programs focus on providing a wide range of training resources, from e-learning online modules, continuing education credits, internships/fellowships, curriculum twinning, and other learning resources for existing or future One Health practitioners to gain knowledge and maintain skills (One Health Commission, 2024h).

Although this chapter cannot cover all capacity building efforts, there are at least 25 large One Health capacity building programs and organizations across the world, many of which are run by governments, non-governmental or non-profit organizations such as the U.S. Centers for Disease Control and Prevention, the WHO, the Food and Agriculture Organization (FAO), the World Organisation for Animal Health (WOAH), One Health Lessons, the One Health Commission, and others (One Health Commission, 2024h).

One Health networks and academic institutions play a large role in providing capacity building resources across the world, often on a regional basis (Streichert et al., 2022). Example networks include AFROHUN, the Southeast Asia One Health University Network (SEAOHUN), the One Health Regional Network for the Horn of Africa, the New England Public Health Training Center, the Task Force for Global Health, the Network for EcoHealth and One Health (NEOH), the One Health Commission's Global One Health Community (One Health Commission, 2024a), and the Capacity Development Support Facility (CDSF) in Ethiopia and others (Khan et al., 2018). Universities involved in capacity building may also provide degrees in One Health or simply have research, workshops, or committees related to One Health, such as the International Veterinary Student Association's (IVSA) Standing Committee on One Health (SCOH), a student-led One Health committee that hosts events to connect disciplines. A list of example programs is included in Table 13.2.

HIGHLIGHT: ONE HEALTH CENTRAL AND EASTERN AFRICA (OHCEA) (AFROHUN, 2023)

OHCEA is a network of 14 Public Health and Veterinary Higher Education Institutions that are located in six countries in the Eastern and Central Africa region (Democratic Republic of Congo (DRC), Ethiopia, Kenya, Rwanda, Tanzania, and Uganda), a region that includes the Congo Basin that is considered to be a hot spot for emerging and re-emerging infectious diseases. OHCEA is funded by the U.S. Agency for International Development (USAID) through RESPOND component of the Emerging Pandemic Threats Program. The collaborating institutions include University of Minnesota and Tufts University. This initiative was spearheaded by Makerere University School of Public Health (MakSPH) in Uganda, in collaboration with Muhimbili University of Health and Allied Sciences (MUHAS), School of Public Health in Tanzania since 2005. This has been expanded and has become a continental initiative; responding to this continental shift, the network leadership agreed to change the name of the network from OHCEA to AFROHUN in 2019. The network is working to transform the training environment and approaches in universities, in a bid to develop a workforce without disciplinary barriers.

PROGRAM DESIGN FOR ONE HEALTH ETC

One Health ETC programs range widely in their design and implementation. Between 2008 and 2011, One Health core competencies were developed by separate groups and were synthesized in Rome in 2012 to a key set of domains: management, communication and informatics, values and ethics, leadership, team and collaboration, roles and responsibilities, and systems thinking (AFROHUN, 2020; Allen-Scott et al., 2015; Binot et al., 2015; Fenwick, 2016; Frankson et al., 2016, 2016; Larsen, 2021; Lee & Global OHCC Working Group, 2013; Lucey et al., 2017; Mor et al., 2018; Rabinowitz et al., 2017; Togami et al., 2018). Despite the existence of such core competencies, a literature review of One Health degree-granting programs in the United States in 2018 found that while One Health education programs were on the rise, competency-based education models were not widely practiced or used by educational institutions and several key areas were underrepresented, as noted in Figure 13.1 (Togami et al., 2018).

However, there are several efforts underway to institutionalize One Health core competencies which will benefit students and improve monitoring and evaluation (M&E). First, in October 2016, the Council on Education for Public Health, the accrediting body for U.S. public health schools, added One Health to its accreditation criteria (CEPH, 2024). This core addition was to be incorporated by the end of 2018. In 2021, an informal survey of schools of public health was conducted to assess how well this criterion was being met and if a collection of One Health resources for public health educators would be useful. In response to the informal

UNDERREPRESENTED | WELL REPRESENTED

Key areas identified in *less than 25%* of total programs	**Identified in *25% to <50%***	**Identified in *50% to <75%***	**Identified in *75% or more***
• Plant biology	• Zoonoses	• Food safety / food security	• Epidemiology
• Antimicrobial resistance	• Geography / GIS	• Agriculture / livestock	• Environmental health / ecology
• Law	• Emerging infectious diseases	• Policy	
	• Economics	• Vector-borne diseases / entomology	
	• Toxicology	• Social and behavioral sciences	
	• Conservation / wildlife		

FIGURE 13.1 Representation of One Health-related competencies across educational institutions. (Adapted from Togami et al., 2018.)

recommendations, a compilation of One Health education resources for public health (and other) educators was created and made freely available (One Health Commission, 2024b). In addition, physicians are increasingly advocating for the inclusion of One Health in medical school education, especially post-COVID-19 (Lucey et al., 2017; Machalaba et al., 2021).

In 2019, Steel et al. explored opinions and insights into desired knowledge, attitudes, and practices of effective One Health clinical practitioners and concluded that educational interventions that foster interprofessional communication and collaboration will be necessary to successfully bring about the cultural change required to achieve effective One Health practice [in Australia], and thus expedite improved human, animal, and environmental health outcomes (Steele et al., 2019).

Laing et al. (2023), affiliated with the NEOH, has updated core competencies for all institutions that are building or expanding their One Health programs (Laing et al., 2023). Competencies include nontraditional topics such as gender inclusion and public speaking to allow the next generation of One Health professionals to better communicate with various audiences and to achieve community buy-in for One Health activities.

Finally, as noted in Figure 13.2, in 2018 Togami et al. (2018). recommended several improvements: (i) making clearly stated core competencies, (ii) including proficiency in at least one health science, (iii) educating future professionals in the One Health arena in both represented and underrepresented disciplines, and (iv) continuing to focus on practical and applied training. These recommendations constitute important foundational elements for updated future One Health core competencies being brought forward by Laing et al. (2023).

Program Design: Education

Broadly speaking, undergraduate majors and minors, certificates, and courses in One Health cover human, animal, and ecosystem health and include curricula in One Health concepts, biological sciences, infectious diseases, environmental sciences, global health, ecology, chemistry, epidemiology, food and agriculture, data collection and analysis, communications, ethics, policy, culture, social justice, and more. It should be noted that most U.S. bachelor's programs require four

RECOMMENDED CORE COMPETENCIES FOR ONE HEALTH EDUCATION

Health Knowledge	Global & Local Issues in Humans, Animals, Plants, & the Environment	Professional Characteristics
Objective To demonstrate knowledge of established and evolving transdisciplinary One Health sciences, including those relevant to public health, animal health, environmental sciences, and modern agriculture	**Objective** To demonstrate an understanding of historical, cultural, political, economic, and scientific aspects of complex and emerging health problems that are amenable to the One Health approach	**Objective** To demonstrate the ability to understand and apply principles of research and evaluation methods to policy and health program implementation, as well as apply scientific findings to real-life situations
• Characterized the etiology, evolution, and ecology of infectious disease agents of people, animals, and plants that are of importance to health. • Describe the main transmission routes for toxins, pathogens, and resistance genes, including human-animal-plant-environmental exposures, as well as vector-borne, waterborne, and airborne cycles. • Explain epidemiological principles used to characterize problems that involve human, animal, plant, and environmental components. • Understand scientific principles such as biological complexity, genetic diversity, and interactions of systems from individuals to ecosystems that influence modern complex challenges in human, animal, plant, and environmental health. • Identify common cultural and socioeconomical determinants and effects of illness, including poverty, residential geography, cultural practices, education, nutrition, and resource security. • Explain how biosurveillance diagnostics, and therapeutic countermeasures are deployed. • Describe interventions used to prevent disease and improve human, animal, and environmental health at the individual, community, and population levels.	• Describe the biological principles, scope, and complexity of disease in people, animals, plants, and the environment. • Understand the effects of global change on health and how both local and global factors affect disease transmission within and between countries. • Identify and understand the origins and determinants of health (human, animal, plant, and the environment) as related to disease. • Compare and contrast health and non-health consequences of diseases and exposures, including social and behavioral, economic, and political effects across global regions. • Recognize major challenges and opportunities to improve health in a global and local context through practical and applied training. • Demonstrate a basic understanding of pre- and post-production food safety. • Understand the structure and responsibilities of the public health system, including the local, state, and national levels of government. • Describe the relationship among various key One Health stakeholders locally and globally.	• Describe the benefits and challenges of a multidisciplinary, integrative approach when implementing studies regarding health concerns at the human-animal-plant-environment interface. • Effectively communicate, both orally and in writing, scientific findings to the scientific community, non-health-related academics, public audiences, media, and policy makers. • Demonstrate scientific quantitative skills, such as the ability to evaluate experimental design, interpret scientific findings, and develop discussions, as well as provide implementable recommendations. • Demonstrate the ability to build and manage a transdisciplinary team and apply principles to conduct ethical, scientifically sound research that will inform policy. • Develop a plan to translate research findings and new discoveries into health policies, community programs, interventions, and public education in a manner that is sustainable, culturally relevant, and economically feasible.

FIGURE 13.2 Recommended core competencies for One Health education. (Adapted from Togami et al., 2018.)

years of study whereas in South America, Europe, Africa, the United Kingdom, and Asia, veterinary and medical school is considered a bachelor's degree and requires five to six years of study.

Master's programs focused on One Health are most often in the form of an MPH and Master of Science (MS) and vary from one to five years to complete (One Health Commission, 2024g, 2024h). For example, the Masters in One Health at the University of Alaska at Fairbanks provides students an option of two possible concentrations: Community Advocacy Concentration (Social Sciences focus) or Biomedical Concentration (Veterinary Medicine and Biology focus) (University of Florida, 2024a). Existing One Health-specific MPH degrees require core MPH courses, such as Epidemiology, Public Health Policy and Management, Environmental Health, Biostatistics, Sociocultural Aspects of Public Health, but focus the bulk of studies on One Health courses. MPH degrees that have a One Health specialization or track vary in requirements as noted in Table 13.2, but usually require completion of the MPH Core Curriculum and at least 18 One Health credits for the specialization. Specific examples include the full-time MPH International Program at Hanoi Medical University specialization in One Health and Public Health and the Johns Hopkins University MPH concentration in Global Environmental Sustainability and Health (Johns Hopkins Bloomberg School of Public Health, 2024).

Doctoral degrees with One Health concentrations, courses, or requirements include PhDs, DVMs, and MDs (University of Florida, 2024b). PhD programs range in the number of post-baccalaureate credit hours required and focus on research or applied learning (One Health Commission, 2024j). The first PhD with a concentration in One Health was created at the University of Florida in its School of Public Health (University of Florida, 2024b); this is a research-oriented degree that emphasizes working across public health, veterinary health, and environmental health disciplines. The program requires 90 total credit hours, 24 of which are One Health concentration courses, with the rest being core public health courses such as quantitative methods and statistics courses; professional issues and teaching courses; supervised research; and dissertation research plus student-chosen and advisor-approved electives (University of Florida, 2024c).

As noted previously, when One Health is included in medical education, it tends to be as an elective, sometimes created by the students themselves (Hilliard, 2015) including those at Georgetown University (Docherty & Foley, 2021), the University of Hawaii, and Western University (Western University, 2024). Conversely, the World Veterinary Association released a position statement in 2023 (World Veterinary Association, 2023) that conveys the overall wish for One Health education in veterinary medicine. The position statement is divided into several sections pertaining to the stages of one's veterinary career: before veterinary school, during veterinary school, at continuing education events used to maintain a veterinary license, and a message to accreditation bodies. The ultimate aims of the position statement are to (i) accept One Health-minded people into the profession through acceptance at veterinary school and (ii) encourage more veterinarians to practice medicine with a One Health mindset and to participate in transdisciplinary, multisectoral work and learning events (e.g., conferences) (World Veterinary Association, 2023).

HIGHLIGHT: UNIVERSITY OF ALASKA AT FAIRBANKS (UAF) (UNIVERSITY OF ALASKA – FAIRBANKS, 2024)

The One Health Master's Degree (OHM) at the UAF emphasizes community-centered problem solving through training and practice. This interdisciplinary degree cultivates skills in communication, epidemiology, conflict resolution, cultural awareness, and data collection. The OHM targets those seeking to work in the community to address One Health and those seeking to pursue a medical or veterinary degree through two concentrations: a Community Advocacy Concentration and a Biomedical Concentration. The degree is comprised of 19 core credits with an additional 12 credits per concentration.

This hands-on approach allows students to engage with key stakeholders and community members to develop and implement realistic management plans for communities across the Circumpolar North. Through the array of courses offered, this program acknowledges that a successful One Health professional requires knowledge of the social, cultural, governmental, historical, and scientific realities that influence One Health issues and their solutions. In addition, UAF requires students to take environmental economics regardless of their concentration. The OHM does not rely on student's in-depth scientific knowledge but instead their ability to communicate effectively and utilize a diverse, multidisciplinary perspective.

Program Design: Training

Most One Health training opportunities are not affiliated with larger capacity building programs but are run out of universities or One Health networks (One Health Commission, 2024h). Many are free or minimal cost lectures designed to introduce One Health concepts; however, content and topic areas tend to be more general, not based on core competencies. Formats vary, ranging from basic lectures, to introductory courses/seminars provided online, to field work for working professionals during university vacation periods (One Health Commission, 2024k).

Several online and in-person One Health courses have been developed around the globe (One Health Commission, 2024d, 2024c); for example, Hanoi Medical School's Global Health seminar series provides a general One Health overview lecture (Hanoi University, 2022). Notable online coursera-type series include Kahn's Bats, Ducks, and Pandemics course, University of Geneva's Global Health at the Human-Animal-Ecosystem Interface course (University of Geneva, 2024), and University of Basel's One Health: Connecting Humans, Animals, and the Environment course (FutureLearn, 2023). While similar in topic areas, these offerings range in course design and learning objectives, as noted in Table 13.3.

In addition, there are several notable fields or hybrid short courses offered during university vacation periods (One Health Commission, 2024k). For example, the Center for One Health Education, Advocacy, Research, and Training (COHEART), Kerala Veterinary and Animal Sciences University, hosts a winter training entitled "Application of One Health Concepts for Control of Emerging Zoonoses and Health Threats," a 21-day program for professionals focused on assimilating the concept of One Health and how to approach health threats especially during outbreak of emerging zoonoses. The course includes theory sessions, hands-on training on disease diagnosis and field visits, and interactive sessions on the One Health approach and economic impact (Kerala Veterinary and Animal Sciences University, 2023).

TABLE 13.3
Online Training Courses

Week	Bats, Ducks, and Pandemics (Kahn)	Global Health at the Human-Animal-Ecosystem Interface (Geneva)	One Health: Connecting Humans, Animals, and the Environment (Basel)
1	Getting Started. One Health, Public Health, Basic Epidemiology	Global Health at the Human-Animal-Ecosystem Interface: The Need for Intersectoral Approaches	Theoretical foundations of One Health
2	Public Policy, Environmental and Ecosystem Health, National Governments, International Organizations	Emerging Infectious Diseases	One Health quantitative methods
3	Human Nutrition, Basic Microbiology, Food Safety, Food Security	Antimicrobial Resistance & Zoonotic Foodborne Infectious Diseases	One Health case studies I: One Health in practice
4	Examining Leadership, Corruption, Communication, and Healthcare Access	Zoonotic Neglected Infectious Diseases	One Health qualitative and mixed methods
5	Antimicrobials: Antibiotics, Antimicrobial Resistance, Bacteriophages, Vaccines and Antivirals	Conflicts & Injuries Innovation & Opportunities	One Health case studies II
6	Containing Disease Outbreaks, Prion Diseases, Bacterial Diseases, and Viral Diseases	Health Benefits at the Human-Animal-Ecosystem Interface	Beyond One Health: ecosystem approaches to health and summary of course
7		Management of Ecosystems under Global Changes: Implications for Human Health	

HIGHLIGHT: RX ONE HEALTH PROGRAM DESIGN (UNIVERSITY OF CALIFORNIA – DAVIS, 2024B)

The Rx One Health Field Institute from the University of California – Davis, is a summer, field-based experiential learning course focused on One Health core competencies for graduate students and early career professionals from all disciplines. Rx One Health participants develop skills in laboratory and research methods, ecosystem dynamics, biodiversity conservation, epidemiology, infectious disease surveillance, biosafety, biosecurity, food safety, agriculture, food security, hydrology, marine ecology, communications, community and stakeholder engagement, ethics, teamwork, and leadership. Through hands-on experiences, case studies, group discussions, and field exercises, Rx One Health develops participants' skills for addressing complex challenges using the One Health approach, which recognizes that the health of people, animals, and their environments are interconnected, and that problem solving to address complex challenges is best achieved through transdisciplinary collaboration. Thematic modules include: One Health Foundations & Leadership; Zoonotic Disease & Biosecurity; Human Health & Health Equity; Food Security, Agriculture, & Nutrition; Wildlife Conservation; Environmental Health & Conservation; Research Methods & Implementation.

Program Design: Capacity Building

One Health capacity building resources and program designs range from international initiatives with specific targets and desired competencies, such as the Global Health Security Agenda (GHSA) Workforce Development and Zoonotic Disease Action Packages, to networks that utilize existing or design future One Health competencies, such as SEAOHUN and AFROHUN One Health academies. Additionally, many government-sponsored capacity building programs are designed with specific agency or department goals in mind, such as global health security capacity building efforts by the United States, UK, Australia, and others. Other capacity building approaches include workshops designed around specific topic areas; ToT programs that teach individuals how to educate others in One Health concepts; internships (One Health Lessons, 2023); and twinning programs where people are mentored on One Health concepts by established institutions. Though many capacity building programs are designed around the 2012 core competencies synthesized in Rome (Frankson et al., 2016), they can vary widely in design and subject matter.

To partially address this wide variability in design, the One Health Quadripartite One Health Joint Plan of Action (OH JPA) aims to strengthen capacity building through an integrated framework. This effort was launched by the Quadripartite (the United Nations Environment Programme (UNEP), FAO, WHO, and WOAH) in October 2022. Action track 1 of the OH JPA, "Enhancing

One Health capacities to strengthen health systems," requires establishing the foundations of One Health capacities and competencies, an analysis of existing gaps and the ability to design, plan, and implement leadership, decision-making, strategies, and governance; sustainable frameworks, infrastructures, and competencies; affordable economic models and financial mechanisms; and M&E processes (FAO et al., 2022). Activity 1.1.2 is focused on defining the One Health institutional and workforce capacities, developing methodologies and tools to assess national One Health performances, and identifying needs. The plan has several capacity-related deliverables expected over the next three years, including:

- Definition of One Health competencies and capacities at institutional and individual levels;
- Mapping and integration of existing methodologies and tools, and new methodologies and tools and pilot tests for:
  - National capacities for One Health and the performance of systems at the human-animal-plant-environment interface
  - One Health competencies
  - Workforce learning needs assessment
- Support for the application of tools and assessments provided
- Identified learning needs
- Identified opportunities to strengthen One Health coordination

A salient example of international cooperation in capacity building is the GHSA, an effort by nations, international organizations, and civil society to accelerate progress toward a world safe and secure from infectious disease threats; to promote global health security as an international priority; and to spur progress toward full implementation of the WHO International Health Regulations (IHR) 2005, WOAH Performance of Veterinary Services (PVS) pathway, and other relevant global health security frameworks (CDC, 2024). An innovative idea borne of the U.S. CDC, GHSA was launched in 2014 by a group of 44 countries and organizations and utilized action packages to target specific outcomes for preventing, detecting, and responding to infectious diseases (Global Health Security Agenda, 2016). While all action packages within the GHSA aim to take a One Health approach, two packages are of relevance to One Health capacity building:

- Zoonotic Disease Action Package: Requires implementation of joint IHR and PVS training programs for human and animal health services. Monitors activities through implementation of WHO's IHR Monitoring Framework and the OIE's PVS Pathway.
- Workforce Development: The Workforce Development Action Package is targeted for one trained field epidemiologist per 200,000 population, and one trained veterinarian per 400,000 animal units (or per 500,000 population), who can systematically cooperate to meet relevant IHR and PVS core competencies in the countries. This action package prioritizes building field epidemiology capacity using a One Health approach.

Another excellent example of an international capacity building effort designed around existing core competencies is the activities undertaken by SEAOHUN and AFROHUN, both associated with the One Health Workforce – Next Generation (OHW-NG) project currently led by the University of California at Berkeley. Both SEAOHUN and AFROHUN have been working over the past decade to build teams and training content that incorporate a One Health approach to problem solving and capacity building. They provide training modules in the seven existing core competencies: (i) Collaboration and Partnership; (ii) Communication and Informatics; (iii) Culture, Beliefs, Values, and Ethics; (iv) Leadership; (v) Management; (vi) Policy, Advocacy, and Regulation; and (vii) Systems Thinking. In addition, their modules include seven One Health technical areas: (i) One Health Concepts and Knowledge; (ii) Fundamentals of Infectious Disease; (iii) Infectious Disease Management; (iv) Epidemiology and Risk Analysis; (v) Fundamentals of Public Health; (vi) Ecosystem Health; and (vii) Behavior Change (One Health Workforce Academies, 2024a).

Additionally, part of USAID-sponsored OHW-NG project (2019–2024) (University of California – Davis, 2024a), SEAOHUN and AFROHUN are part of a Global Consortium of world-renowned partners based across North America, Africa, and Southeast Asia. Regional training is implemented using a hub and spokes model in which AFROHUN Secretariat links out to Country Chapters and the SEAOHUN Secretariat links out to Country One Health University Networks (OHUNs). According to their Year 1 report, during its first year, the OHW-NG project achieved the following:

- Curricula survey instruments were developed with 13 categories of One Health training activities and 26 variables (338 item matrix). A glossary of 33 One Health training-related terms and phrases was produced.
- A table of training activities that capture 18 One Health competency domains was produced and distributed to country managers and national coordinators to record specific exemplar training activities in specific academic units.
- A survey questionnaire was developed for self-assessment of progress toward institutionalization of One Health training activities at the national and regional levels, according to the CLASS (Calibrated, Lifelong, Adaptable, Scalable, and Sustainable) schematic scores.

CALL OUT: ONE HEALTH LESSONS PROGRAM DESIGN

One Health Lessons, established in 2020, is a successful One Health educational grassroots non-profit organization designed to educate children and adults about One Health and has created a global One Health educational movement. Within 2 years, it has educated over 25,000 students, including children as young as 3 years, about One Health. Multiple age-appropriate lessons are taught in-person and online and are available in over 30 languages. Lessons have been taught in-person and online at primary, secondary, and tertiary schools, community events, religious institutions, youth gatherings, and more in Europe, Africa, Middle East, Asia, Oceania, North America, South America, and the Caribbean.

In addition, over 150 adults from over 80 countries have completed the organization's systematic ToT program (One Health Lessons, 2023) (known as the Certified Lesson Leader Program) to (i) improve their One Health communication skills, (ii) serve as One Health influencers in their communities, and (iii) educate vulnerable populations about vaccines and public health risks in relation to the environment, plants, and animals. The trainees in this program develop their communication skills by teaching children and other community members about One Health, thereby sensitizing communities about One Health challenges that may arise in the future. The organization has also trained over 60 interns from various disciplines including law, international relations, veterinary medicine, global health, medicine, microbiology, dentistry, environmental health sciences, public health, American Sign Language, and more. The One Health Lessons' Internship Program was designed to improve the leadership, communication, and networking skills of promising future leaders in One Health.

M&E OF ONE HEALTH ETC

The NEOH (COST Action, 2015), a 2014–2018 European cooperation in science and technology (COST) action, developed an evaluation framework to assess the "One Health-ness" of a health initiative based on the dimensions that can affect the outcomes (Rüegg et al., 2018). Elements include:

1. Defining and describing the One Health initiative and its context (i.e., the system, its boundaries, and the One Health initiative as a subsystem), providing information for the further Elements.
2. Assessing expected outcomes based on the theory of change (TOC) of the initiative, and collecting unexpected outcomes emerging in the context of the initiative.
3. Assessing the "One Health-ness," i.e., the implementation of operations and infrastructure contributing to the One Health initiative.
4. Comparing the degree of "One Health-ness" and the outcomes produced.

While the NEOH framework is useful for measuring One Health overarching programs and initiatives, it is not designed to monitor and evaluate student learning outcomes from ETC. Similarly, the GHSA measures capacity building progress by each country according to each action package, using the WHO's Joint External Evaluation (JEE) tool indicators.

According to the European University Association's 2018 trends study, the implementation of learning outcomes has progressed steadily in the past decade. According to the report, 76% of responding higher education institutions had developed learning outcomes for all courses and respondents suggested more positive attitudes toward curricula based on learning outcomes than in earlier years (One Health Commission, 2024f). That said, learning outcomes for One Health across ETC education and academic sectors rely on traditional M&E approaches such as student retention rates, university rankings, exams, surveys, and pre/post-testing of One Health knowledge.

Here, we highlight examples of program-specific approaches to M&E across the ETC domains. First, the One Health Commission aligns its "Guide to Developing One Health Lessons for K-12" (Gaebel & Zhang, 2018), which notes that the One Health organizing framework supports many of the U.S. Next Generation Science Standards (NGSS), and can easily be included in developing curriculum that meets state and Common Core requirements (One Health Commission, 2022). The One Health Commission also provides a free conference survey developed by its One Health Education Task Force (One Health Commission, 2024m) on its website. A One Health Veterinarians without Borders project in Southeast Asia, entitled "Building EcoHealth Capacity in Asia" (Hall & Le, 2015), utilized progress markers (pre- and post-workshop), online questionnaires, interviews, small group discussions, gap analysis, performance indicators, and SWOT (Strength-Weakness-Opportunity-Threat) analyses (The Community Tool Box, 2024).

In 2021, Cianfagna et al. used One Health education programs as a starting point to collect a global list of institutions potentially carrying out education in the links between biodiversity and health and then analyzed the offerings from these institutions to determine the degree of integration of biodiversity and health interlinkages (Cianfagna et al., 2021).

Finally, the Vietnam One Health University Network (VOHUN, part of SEAOHUN as previously mentioned) conducted a cross-sectional survey in early 2019 to assess the effectiveness of their trainings. Out of 188 health and veterinary professionals from 55 provinces, more than 80% of participants reported they did not know what One Health was before attending the training, but 89.9% of them have since shared One Health knowledge with their colleagues. The same survey noted that the three most valuable One Health core competencies for health and veterinary professionals were "Communication and Informatics" (80%), "Collaboration and Partnership" (75%), and "System Thinking" (44%) (Tb et al., 2022).

CALL OUT: ONE HEALTH LESSONS M&E (ONE HEALTH LESSONS, 2023)

One Health Lessons has a M&E system that is based on surveys from full-time teachers, interns, and trainees. For each of its components as described in the previous call out box, there are surveys to evaluate the impact of the training at the individual and group levels.

All interns are individually mentored to improve their communication, networking, and leadership skills. In order to do this accurately, all interns complete pre-internship surveys to help their mentor tailor the style of mentorship to meet the needs of the individual intern. The survey evaluates their comfort levels pertaining to different aspects of their specific role as an intern. For instance, all interns must speak in public and explain One Health in less than eight seconds during their internship (i.e., improve communication skills) because they are responsible for increasing One Health awareness to new audiences. If an intern notes that they are very uncomfortable with public speaking, then their mentor will focus on this area when individually mentoring the intern. Once the internship is completed, all interns complete a post-intern survey with similar questions to help the mentor gauge the overall improvement and confidence of the intern.

Another survey system is utilized with the ToT program, known as the Certified Lesson Leaders Program, as mentioned in a previous call out box. For the first two years of the Program, volunteers who want to teach lessons about One Health and improve their own communication skills undergo a 4-hour training program. The first hour is spent at an orientation session that teaches the participant how to communicate One Health to different ages. The second hour is to watch a recording of a lesson and pass a quiz that evaluates teaching and communication techniques of the observed teacher. The third hour is to observe a live lesson and the fourth hour is to have the volunteer teach the lesson with another volunteer observing and assisting when needed. Each individual volunteer is invited by a full-time supervisor (e.g., teacher, scout leader) to speak as a guest scientist to their students. The full-time supervisor completes a survey evaluating the impact of the volunteer and the learning material covered at the classroom level. Of special note, the full-time supervisors are asked about the level of interest and engagement by the students and if the supervisors want to continue teaching about One Health in their own time.

Through such surveys, the impact on individual interns and entire classrooms is evaluated and future mentorship and lessons are adjusted to meet the needs of those various audiences.

CONCLUSION

As noted throughout this chapter, One Health ETC programs and awareness have grown exponentially over the past two decades (One Health Commission, 2023). Indeed, the COVID-19 pandemic shed even further light onto the connection between human, animal, and planetary health. We expect that more One Health ETC programs will emerge to protect the health and well-being of people, animals, plants, and their shared environment.

To capitalize on the momentum surrounding COVID-19 and its One Health implications, we have several recommendations for consideration for the future of ETC programs. First, medical and osteopathic and chiropractic schools, physician's assistant, nursing, pharmacy, microbiology, and occupational therapy programs should prioritize incorporating One Health into their educational curricula. Additionally, public health academic institutions should increase their involvement in One Health training and capacity building efforts. Other academic disciplines, such as the social sciences, engineering, ecology, pharmacology, economics, and chemistry, should include One Health in their curriculum. Finally, we recommend mainstreaming One Health by promoting the teaching of One Health in all primary and secondary schools; this will prepare youth to become "global citizens," teaching them that every action they take, every moment of every day, has an impact on the Earth and its ecosystems. This reframing of attitudes about the place of humans in relation to the planet starts at a young age and can have long-lasting positive effects on the Earth and its people.

The future directions of One Health ETC will be dictated by the needs of communities as determined by community leaders and policymakers, but require a multidisciplinary approach that incorporates human, animal, and environmental factors. This will require the One Health community with increase education and awareness not just in the academic sector, but also to the public and policymakers. Only by utilizing such a multi-pronged, top-down (from policy) and bottom-up (from grassroots) approach will we be able to fully implement One Health and realize all of its benefits to the health and well-being of all living things on the planet. Through institutionalization of One Health in academic and political systems (One Health Commission, 2024i), more One Health actions are expected to become solidified to protect the health and well-being of people, animals, plants, and their shared environment.

REPRESENTATIVE TABLE OF ONE HEALTH PROGRAMS

To prepare this chapter, the authors reviewed a sampling of One Health ETC programs and literature reviews. While this table does not cover all existing programs nor does it represent a comprehensive literature review, we share it here to provide a starting point for students and One Health practitioners to find additional information. Indeed, a follow-on literature review to Cianfagna et al.'s (2021) examination of programs concentrated on biodiversity and human health interlinkages in higher education offerings (Cianfagna et al., 2021) will focus on One Health programs in Latin America and will uncover many mission programs that this chapter simply could not detail. Readers are urged to visit the online One Health Opportunities webpage to submit any programs that are not yet represented (One Health Commission, 2024h). Please excuse any inconsistencies or missing programs as this is an evolving landscape and we will continue to update our live database (Table 13.4).

TABLE 13.4
Sampling of One Health Education, Training, and Capacity-Building Programs

Organization	Title	Region	Links	Type	Column 1
Africa One Health Network (AfOHNET)	One Health Network Workshop	Africa	https://afohnet.org/conference2022	Training	
Africa One Health University Network (AFROHUN)	One Health Training Modules	Africa	https://afrohun.org/course/onehealthmodules/	Capacity Building	
Africa One Health University Network (AFROHUN)	One Health Competency-Based Education Workshops	International	https://icap.columbia.edu/news-events/in-africa-and-southeast-asia-icap-trains-one-health-educators-in-competency-based-education-methods/	Capacity Building	
American Society of Tropical Medicine and Hygiene	Clinical Tropical Medicine and Travelers' Health	United States	https://www.astmh.org/education-resources/update-course	Capacity Building	
American Veterinary Medicine Association (AVMA)	One Health Training Modules	United States	https://axon.avma.org/page/one-health-courses	Capacity Building	
Animal and Plant Health Agency (APHA)	One Health Capacity Building	United Kingdom	https://improve-ohcb.com/home/	Training	
Association of American Veterinary Medical Colleges (AAVMC)/ Association for Prevention Teaching and Research (APTR)	One Health Interprofessional Education	United States	https://www.aptrweb.org/page/OneHealth	Capacity Building	https://www.aavmc.org/programs/one-health/

(*Continued*)

TABLE 13.4 (*Continued*)
Sampling of One Health Education, Training, and Capacity-Building Programs

Organization	Title	Region	Links	Type	Column 1
Auburn University College of Forestry, Wildlife and Environment	Online One Health certificate	United States	https://cfwe.auburn.edu/online-professional-graduate-certificate-programs/one-health/	Education	
Baltic University	Ecosystem Health and Sustainable Agriculture	Sweden	http://www2.balticuniv.uu.se/index.php/ecosystem-health-a-sustainable-development	Education	
Berry College	One Health undergraduate minor	United States	https://catalog.berry.edu/preview_program.php?catoid=4&poid=459	Education	
Boise State	Vertically Integrated One Health Projects	United States	https://www.boisestate.edu/vip/one-health/	Education	
Capacity Development Support Facility (CDSF)	One Health – Zoonotic Disease ToT Course	Ethiopia	https://agri-training-et.org/zoonotic-e-learning-course/	Capacity Building	
CDC TRAIN	An Introduction to One Health	United States	https://www.train.org/cdctrain/course/1095553/?activeTab=certificates	Capacity Building	
Centre for One Health Education, Advocacy, Research and Training (COHEART)	PG Diploma in One Health, PG Certificate courses in One Health Surveillance, and PG Certificate in Community-Based Disaster Management	Kerala, India	https://coheart.ac.in/	Capacity Building	

(*Continued*)

TABLE 13.4 (*Continued*)
Sampling of One Health Education, Training, and Capacity-Building Programs

Organization	Title	Region	Links	Type	Column 1
City University of Hong Kong	One Health Seminars	Asia	https://www.cityu.edu.hk/ohrp/one-health-seminars	Capacity Building	
City University of New York	PhD in Environmental and Planetary Health Sciences	United States	https://sph.cuny.edu/academics/degrees-and-programs/doctoral-programs/phd-in-environmental-and-planetary-health-sciences/#1548194101710-5203d426-acc7	Education	
Colorado State University	World Small Animal Veterinary Association One Health Certificate Course	United States	https://vetmedbiosci.colostate.edu/vth/	Capacity Building	
Colorado State University	Summer program credit	United States	https://onehealth.colostate.edu/summer-program/	Education	Summer program credit
Colorado State University	Capacity Building	United States	https://onehealth.colostate.edu/seminars/	Capacity Building	Capacity Building
Cornell University	Conservation Medicine: A Veterinary Perspective Summer Course	United States	https://courses.cornell.edu/preview_course_nopop.php?catoid=31&coid=495375	Education	
Delaware Valley University	One Health Seminar Series	United States	https://delval.edu/search-results?search_api_fulltext=one+health	Capacity Building	One Health Seminar Series

(*Continued*)

TABLE 13.4 (*Continued*)
Sampling of One Health Education, Training, and Capacity-Building Programs

Organization	Title	Region	Links	Type	Column 1
Duke University	One Health Graduate Training Program	United States	https://www.onehealthcommission.org/index.cfm/83391/72449/duke_university_one_health_graduate_training_program	Education	
Duke University	One Health Course: From Philosophy to Practice	United States	https://nicholas.duke.edu/academics/courses/one-health-philosophy-practice	Education	One Health Course: From Philosophy to Practice
Dundalk Institute of Technology School of Health and Science	One Health Seminar Series	Ireland	https://www.dkit.ie/about-dkit/academic-schools/school-of-health-and-science/one-health-at-dkit.html	Capacity Building	One Health Seminar Series
Ecohealth Alliance	EcoHealthNet Workshop	United States	https://www.ecohealthalliance.org/program/ecohealthnet	Capacity Building	
Erasmus Mundus	Joint Master's Degree in Infectious Diseases and One Health	Europe	https://erasmus-plus.ec.europa.eu/projects/search/details/610556-EPP-1-2019-1-FR-EPPKA1-JMD-MOB	Education	
Ferrum College	Undergraduate Minor in One Health	United States	http://catalog.ferrum.edu/preview_program.php?catoid=7&poid=1026&returnto=414	Education	
Fontbonne University	One Health undergraduate major	United States	https://www.fontbonne.edu/academics/college-arts-sciences/biological-behavioral-sciences-department/bachelor-of-science-in-one-health/	Education	***Use for highlight?

(*Continued*)

TABLE 13.4 (*Continued*)
Sampling of One Health Education, Training, and Capacity-Building Programs

Organization	Title	Region	Links	Type	Column 1
Fontbonne University	One Health undergraduate minor	United States	https://www.fontbonne.edu/academics/college-arts-sciences/biological-behavioral-sciences-department/one-health-minor/	Education	
George Mason University	Summer Workshop – Pandemics, Bioterrorism, and Global Health Security	United States	https://pandorareport.org/summer-workshop/	Training	
Georgetown University School of Medicine	2-week One Health elective course for third years	United States	https://www.sciencedirect.com/science/article/pii/S2352771421000215#:~:text=One%20health%20is%20the%20study,animal%20health%20and%20the%20environment.&text=The%20intersection%20between%20human%20health,good%20of%20people%20and%20environment	Education	

(Continued)

TABLE 13.4 (*Continued*)
Sampling of One Health Education, Training, and Capacity-Building Programs

Organization	Title	Region	Links	Type	Column 1
Hanoi Medical University	International Master of Public Health – One Health	Vietnam	https://spmph.edu.vn/dao-tao-sau-dai-hoc/thong-bao-tuyen-sinh-lop-thac-si-y-te-cong-cong-chuong-trinh-quoc-te-nam-2020-94/ngon-ngu-tieng-anh.html?fbclid=IwAR07RFytxAWmUTlxONpZpdna63wQuInvTrEqv6gq5R3Zd1h-Ar5CWMOJCXY	Education	
Hanoi Medical University	Global Health Seminar Series	Vietnam	https://spmph.edu.vn/en-GB/article/news/global-health-seminar-series-applying-a-one-health-approach-to-tackling-the-next-pandemic%201	Training	
Hanoi University of Public Health	Master of Public Health majoring in Environmental Health	Vietnam	https://huph.edu.vn/public/english	Education	
Helmholtz Institute for One Health (HIOH)	Capacity Building	Germany	https://www.helmholtz-hzi.de/en/the-hzi/sites/helmholtz-institute-for-one-health-hioh/about-hioh/	Capacity Building	Capacity Building

(*Continued*)

TABLE 13.4 (*Continued*)
Sampling of One Health Education, Training, and Capacity-Building Programs

Organization	Title	Region	Links	Type	Column 1
Hokkaido University	PhD in infectious diseases or veterinary medicine	Japan	https://onehealth.vetmed.hokudai.ac.jp/en/programs/courses/	Education	
Hokkaido University	One Health Ally Course	Japan	https://onehealth.vetmed.hokudai.ac.jp/en/programs/allycourse/	Education	
Ibero-American Science and Technology for Development Program (CYTED)	Finances IberoAmerican Development, including One Health Projects	South America	https://www.cyted.org/		
Indian Institute of Public Health	Capacity Building at Center for One Health Education, Research & Development (COHERD)	India	https://www.onehealthcommission.org/documents/filelibrary/resources/whos_who/COHERD_Whos_Who_Resources_21102022_0CCC9948BEA59.pdf	Capacity Building	Capacity Building at Center for One Health Education, Research & Development (COHERD)
Indonesia One Health University Network (INDOHUN)	One Health University Network	Indonesia	https://www.seaohun.org/indohun	Education	
Institute of Tropical Medicine Antwerp	Master of Science in Global One Health:	Europe	https://www.itg.be/en/study/courses/msc-in-global-one-health-diseases-at-the-human-animal-interface	Education	

(*Continued*)

TABLE 13.4 (*Continued*)
Sampling of One Health Education, Training, and Capacity-Building Programs

Organization	Title	Region	Links	Type	Column 1
International Student One Health Alliance (ISOHA)	Serves as an umbrella for international and national student organizations, One Health Clubs, BSc, MSc, and PhD students and provides education and opportunities related to One Health, through partnerships and collaborative projects	International	https://isohaonehealth.wordpress.com/		
Iowa State University College of Veterinary Medicine	One Health Lecture Series	United States	https://vetmed.iastate.edu/research-grad-studies/centers-institutes-and-initiatives/one-health/one-health-related-programs	Training	
Iowa State University College of Veterinary Medicine	MPH for Veterinarians	United States	https://www.public-health.uiowa.edu/how-to-apply-mph/	Education	
Iowa State University College of Veterinary Medicine, Center for Food Security and Public Health	Continuing education credits in One Health	United States	https://www.cfsph.iastate.edu/courses/	Capacity Building	

(*Continued*)

TABLE 13.4 (*Continued*)
Sampling of One Health Education, Training, and Capacity-Building Programs

Organization	Title	Region	Links	Type	Column 1
James Cook University	Intro to One Health Course	Australia	https://apps.jcu.edu.au/subjectsearch/#/redirecting/??subject=TM5600&year=2021&transform=subjectwebview.xslt	Education	Intro to One Health Course
James Lind Institute	Online MPH in One Health and Planetary Health	Switzerland	https://jliedu.ch/courses/mph-in-one-health-and-planetary-health/	Education	
Johns Hopkins University	One Health Lab	United States	https://publichealth.jhu.edu/departments/environmental-health-and-engineering/research-and-practice/faculty-research-interests/one-health-laboratory-at-johns-hopkins-university	Networks	
Johns Hopkins University	One Health Course	United States	https://www.jhsph.edu/courses/course/29303/2019/552.612.81/essentials-of-one-health	Education	
Johns Hopkins University	Concentration in Global Environmental Sustainability & Health	United States	https://publichealth.jhu.edu/academics/mph	Education	
Kansas State University	MPH – Infectious Diseases at Kansas State University	United States	https://www.k-state.edu/mphealth/academics/areas/disease.html	Education	

(*Continued*)

TABLE 13.4 (*Continued*)
Sampling of One Health Education, Training, and Capacity-Building Programs

Organization	Title	Region	Links	Type	Column 1
Kansas State University	K-12 Program	United States	https://olathe.k-state.edu/academics/programs-k12/programs/one-health/	Education	
Kansas State University	One Health High School Summer Online Course	United States	https://olathe.k-state.edu/academics/programs-k12/programs/one-health-summer/index.html	Education	
Kerala Veterinary and Animal Sciences University, Centre for One Health Education Advocacy Research & Training (COHEART)	Mission is to generate scientific knowledge	Hong Kong	https://coheart.ac.in/	Networks	
Kerala Veterinary and Animal Sciences University, Centre for One Health Education Advocacy Research & Training (COHEART)	Post Graduate Diploma in One Health	India	https://coheart.ac.in/education	Education	
Kerala Veterinary and Animal Sciences University, Centre for One Health Education Advocacy Research & Training (COHEART)	Multiple Certificate Programs	India	https://coheart.ac.in/training	Training	

(*Continued*)

TABLE 13.4 (*Continued*)
Sampling of One Health Education, Training, and Capacity-Building Programs

Organization	Title	Region	Links	Type	Column 1
Kerala Veterinary and Animal Sciences University, Centre for One Health Education Advocacy Research & Training (COHEART)	Post Graduate Certificate in OH Surveillance	India	https://coheart.ac.in/education	Education	
Laos One Health University Network (LAOHUN)	One Health University Network	Laos	https://www.seaohun.org/laohun	Education	
Lincoln Memorial University College of Veterinary Medicine	Curriculum is taught with the One Health philosophy	United States	https://www.lmunet.edu/college-of-veterinary-medicine/academics	Education	Entire curriculum is taught with the One Health philosophy
London School of Hygiene & Tropical Medicine (LSHTM) and the Royal Veterinary College (RVC)	Post Graduate Diploma in One Health	England	https://www.lshtm.ac.uk/study/courses/masters-degrees/one-health	Education	
London School of Hygiene & Tropical Medicine (LSHTM) and the Royal Veterinary College (RVC)	MSc One Health: ecosystems, humans, and animals	England	https://www.lshtm.ac.uk/study/courses/masters-degrees/one-health	Education	

(*Continued*)

TABLE 13.4 (*Continued*)
Sampling of One Health Education, Training, and Capacity-Building Programs

Organization	Title	Region	Links	Type	Column 1
Makarere University	One Health Institute Field Attachment	Uganda	https://chs.mak.ac.ug/news/one-health-institute-attachment	Other	Africa One Health University Network (formerly One Health Central and Eastern Africa (OHCEA)) runs online One Health modules
Makarere University	Masters in One Health	Uganda	https://sph.mak.ac.ug/academics/masters-public-health-mph	Other	
Malaysia One Health University Network (MOHUN)	One Health University Network	Malaysia	https://www.seaohun.org/myohun	Education	
Massey University of New Zealand	Global One Health programs	New Zealand	https://www.massey.ac.nz/about/colleges-schools-and-institutes/college-of-sciences/our-research/themes-and-research-strengths/one-health/	Capacity Building	
McGill University	Summer Institute in Infectious Diseases & Global Health	Canada	https://www.mcgill.ca/summerinstitute-globalhealth/	Training	

(*Continued*)

TABLE 13.4 (*Continued*)
Sampling of One Health Education, Training, and Capacity-Building Programs

Organization	Title	Region	Links	Type	Column 1
Michigan State University	Capacity Building at Institute for Global Health	United States	https://ighealth.msu.edu/one-health/	Capacity Building	Capacity Building at Institute for Global Health
Michigan State University	Capacity Building at Canadian Studies Center One Health Initiative	United States	https://canadianstudies.isp.msu.edu/initiatives/one-health/	Capacity Building	Capacity Building at Canadian Studies Center One Health Initiative
Michigan State University	K-12 4-H Lessons	United States	https://www.canr.msu.edu/uploads/236/65684/4H1689_AnimalScienceAnywhere-OneHealth_NEW.pdf	Education	K-12 4-H Lessons
Midwestern University	Master of Public Health (MPH) – Global One Health	United States	https://www.midwestern.edu/academics/degrees-and-programs/master-of-public-health.xml	Education	
Minnesota Department of Health	Minnesota One Health Antibiotic Stewardship Collaborative Resources	United States	https://www.onehealthcommission.org/documents/filelibrary/resources/opportunities/MN_DoH_One_Health_Resources_for_pdf_6D259ABAC5B50.pdf	Other	

(Continued)

TABLE 13.4 (*Continued*)
Sampling of One Health Education, Training, and Capacity-Building Programs

Organization	Title	Region	Links	Type	Column 1
Murdoch University Center for Biosecurity and OH	Graduate Certificate in OH	Australia	https://www.murdoch.edu.au/course/Postgraduate/C1145	Education	
Myanmar One Health University Network (MMOHUN)	One Health University Network	Myanmar	https://www.seaohun.org/mmohun	Education	
Nantes-Atlantic National College of Veterinary Medicine, Food Science and Engineering (ONIRIS)	Man-Imal Master's Degree : From Animal to Man	Europe	https://www.oniris-nantes.fr/formations/les-masters/man-imal	Education	
Netherlands Centre for One Health	PhD One Health Research Program	Netherlands	https://ncoh.nl/research/phd-research-programme/		
New England Public Health Training Center	Online Intro to One Health for public health professionals, nurses, vets, clinicians, environmental scientists	United States	https://www.nephtc.org/login/index.php	Capacity Building	
Nong Lam University in Ho Chi Minh city	Master of Veterinary Medicine, specialized in Public Health	Vietnam	https://vet.nlu.edu.vn/mvph_en	Education	

(*Continued*)

TABLE 13.4 (*Continued*)
Sampling of One Health Education, Training, and Capacity-Building Programs

Organization	Title	Region	Links	Type	Column 1
North Carolina State	Global One Health Academy	United States	https://provost.ncsu.edu/news/2022/09/nc-state-launches-universitywide-global-one-health-academy/	Networks	
North Carolina State University	Evolutionary Medicine Summer Institute (EMSI)	United States	https://sites.duke.edu/emsi/	Training	
Ohio State University	Curriculum Twinning	Africa	https://oia.osu.edu/units/global-one-health-initiative/training/completed-projects/curriculum-twinning/	Capacity Building	
Ohio State University	Ohio State Summer One Health Institute	United States	https://oia.osu.edu/global-one-health-initiative	Capacity Building	
Ohio State University	In-person and online Global One Health certificate	United States	https://cph.osu.edu/students/graduate-certificate-global-one-health	Education	
Ohio State University	MPH – Veterinary Public Health	United States	https://vet.osu.edu/education/programs/vph	Education	
One Health & Development Initiative (Nigeria)	Resources on One Health	Africa	https://onehealthdev.org/	Networks	
One Health Aotearoa	Postgraduate Study Opportunities	New Zealand	https://onehealth.org.nz/research/postgraduates/		
One Health Brasil	One Health Network	Brazil	https://onehealthbrasil.com/		

(*Continued*)

TABLE 13.4 (*Continued*)
Sampling of One Health Education, Training, and Capacity-Building Programs

Organization	Title	Region	Links	Type	Column 1
One Health Brasil: Ecossistemas Aquáticos: Saúde animal, humana e Ambiental (ECOHA)	Sub-network of One Health Brasil, applies One Health to aquatic ecosystems	Brazil	https://onehealthbrasil.com/		Ask Cheryl
One Health Commission	One Health Resources for Public Health Educators	Worldwide	https://www.onehealthcommission.org/en/resources__services/oh_resources_for_public_health_educators/	Networks	
One Health Commission	One Health Library	Worldwide	https://www.onehealthcommission.org/en/resources__services/one_health_library/	Networks	
One Health Commission	One Health Opportunities Bulletin Board	Worldwide	https://www.onehealthcommission.org/en/resources__services/oh_opportunities_bulletin_board/	Networks	
One Health Commission	One Health Education Program Database	Worldwide	https://www.google.com/maps/d/u/0/viewer?ll=1.2722218725854067e-14%2C44.4944252080503&z=2&mid=1ip7y8lz8icEQ0VhqHcAeXFBsiCjZXUP9	Networks	

(Continued)

TABLE 13.4 (*Continued*)
Sampling of One Health Education, Training, and Capacity-Building Programs

Organization	Title	Region	Links	Type	Column 1
One Health Commission	One Health Education Resources for Primary, Secondary, K-12	Worldwide	https://www.onehealthcommission.org/en/resources__services/one_health_education_resources/primary_secondary_k12_oh_education_resources/	Capacity Building	https://www.onehealthcommission.org/documents/filelibrary/resources/Guide_to_Developing_K12_One_Health__AE95AD314CD45.pdf
One Health European Joint Programme (OHEJP)	Continuing Education/ Continuing Professional Development Curriculum – modules for Early Career Researchers (<5 years post-PhD)	Europe	https://onehealthejp.eu/continuing-professional-development-module/		
One Health European Joint Programme (OHEJP)	Doctoral Program	Europe	https://onehealthejp.eu/the-doctoral-programme		
One Health European Joint Programme (OHEJP)	Summer Schools (undergrad, grad, early career)	Europe	https://onehealthejp.eu/the-ohejp-summer-school		

(*Continued*)

TABLE 13.4 (*Continued*)
Sampling of One Health Education, Training, and Capacity-Building Programs

Organization	Title	Region	Links	Type	Column 1
One Health European Joint Programme (OHEJP)	Partnership between food, veterinary, and medical laboratories and institutes across Europe	Europe	https://onehealthejp.eu/		
One Health Lessons	One Health Education Resources for Primary, Secondary, K-12	Worldwide	https://onehealthlessons.org/	Capacity Building	Confirm CAP
One Health Network South Asia and Massey University NZ	One Health Epidemiology Fellowship Program	Afghanistan, Bangladesh, Bhutan, and Nepal	https://ww25.onehealthnetwork.asia/node/583?subid1=20241005-0633-57b0-9961-37daa6c1f000	Capacity Building	
One Health Regional Network for the Horn of Africa	One Health e-Learning Materials	Kenya, Ethiopia, Eritrea, Somalia/Somaliland	https://onehealthhorn.net/uncategorized/new-look-horn-elearning-modules/	Capacity Building	
One Health Sweden	One Health Seminars	Sweden	https://www.slu.se/en/Collaborative-Centres-and-Projects/one-health-sweden/	Networks	

(*Continued*)

TABLE 13.4 (*Continued*)
Sampling of One Health Education, Training, and Capacity-Building Programs

Organization	Title	Region	Links	Type	Column 1
One Health Sweden	Teaching OH to K-12 teachers	Sweden	https://www.slu.se/en/Collaborative-Centres-and-Projects/one-health-sweden/one-health-education/	Education	Teaching OH to K-12 teachers
One Health Workforce Academies (OHWA)	Online One Health Training Course for working professionals	Online	https://onehealthworkforceacademies.org/training/fundamentals-of-one-health-practice/	Capacity Building	
Oregon State University	Epidemiology & Public Health Required	United States	https://vetmed.oregonstate.edu/	Education	
Oregon State University	One Health Laboratory	United States	https://vetmed.oregonstate.edu/one-health-osu-vet-med	Other	
Penn State	One Health Minor	United States	https://agsci.psu.edu/academics/undergraduate/minors/one-health	Education	
Philippine One Health University Network (PHILOHUN)	One Health University Network	Philippines	https://www.seaohun.org/philohun	Education	
Princeton University	Bats, Ducks and Pandemics: An Introduction to One Health Policy	United States	https://www.coursera.org/learn/onehealth?action=enroll	Training	

(*Continued*)

TABLE 13.4 (*Continued*)
Sampling of One Health Education, Training, and Capacity-Building Programs

Organization	Title	Region	Links	Type	Column 1
Purdue University College of Veterinary Medicine	One Health in the DVM curriculum	United States	https://vet.purdue.edu/onehealth/Curriculum.php	Education	One Health in the DVM curriculum
Ross University	MSc in One Health	Caribbean	https://veterinary.rossu.edu/postgraduate/msc-one-health	Education	
Ross University School of Veterinary Medicine	One Health Electives, DVM	Caribbean	https://veterinary.rossu.edu/research/one-health	Education	
Ross University School of Veterinary Medicine	Online Graduate certificate in One Health	Caribbean	https://veterinary.rossu.edu/postgraduate/certificate-in-one-health	Education	
Ross University School of Veterinary Medicine	Online Graduate degree in One Health	Caribbean	https://veterinary.rossu.edu/research/one-health	Education	
Schulich Medicine and Dentistry	One Health Specialization	Canada	https://www.schulich.uwo.ca/oleapopelkaonehealth/one_health/index.html	Education	The Honors Specialization in One Health, leading to a Bachelor of Medical Sciences (BMSc) degree, was created in collaboration with Western University's Departments of Sociology and Geography in the Faculty of Social Science to enrich the learning experience of students.

(*Continued*)

TABLE 13.4 (*Continued*)
Sampling of One Health Education, Training, and Capacity-Building Programs

Organization	Title	Region	Links	Type	Column 1
Shanghai Jiao Tong University School of Medicine, Chinese Center for Tropical Disease Research	Continuing Education – Global Health and One Health Education Teaching Seminar and Teacher Training Course	China	https://www.shsmu.edu.cn/sghen/Education/Continuing_Education.htm		
Shanghai Jiao Tong University School of Medicine, Chinese Center for Tropical Disease Research	Global Health and One Health compulsory course for undergraduates majoring in clinical medicine	China	https://www.shsmu.edu.cn/sghen/Education/Undergraduate_Education.htm	Education	Global Health and One Health compulsory course for all medical students
Soulsby Foundation	One Health Fellowship	United Kingdom	https://soulsbyfoundation.org/apply/	Capacity Building	
Southeast Asia One Health University Network (SEAOHUN)	One Health Course Modules	South East Asia	https://seaohunonehealth.wordpress.com/	Capacity Building	
Southern African Centre for Infectious Disease (SACIDS) Foundation for One Health	One Health Training	Africa	https://www.sacids.org/	Capacity Building	
Southern African Centre for Infectious Disease Surveillance	Master of Science in One Health Molecular Biology	Africa	https://www.sacids.org/news/call-application-master-science-one-health-molecular-biology/	Education	
St. George's University	One Health Medicine course	Caribbean	https://online.sgu.edu/courses/SGUx/OHOM1/2014/about	Education	

(*Continued*)

TABLE 13.4 (*Continued*)
Sampling of One Health Education, Training, and Capacity-Building Programs

Organization	Title	Region	Links	Type	Column 1
Tephinet	One Health Training workshops with CDC	International	https://www.tephinet.org/news/tephinet-and-cdc-host-training-adva`nce-one-health-central-america	Capacity Building	
Texas A&M	Oceans and One Health undergraduate concentration	United States	https://catalog.tamu.edu/undergraduate/galveston/liberal-studies/oceans-one-health-university-studies-bs/	Education	CANT FIND
Texas A&M University	One Health Educational Programs	United States	https://onehealth.tamu.edu/education/	Education	One Health Educational Programs
Thailand One Health University Network (THOHUN)	One Health University Network	Thailand	https://www.seaohun.org/thohun	Education	
Food and Agriculture Organization (FAO)	Workshops and Resources	Italy	https://www.fao.org/one-health/en	Capacity Building	Do they have modules?
The Quadripartite (UNEP, WOAH, FAO, WHO)	The Quadripartite	International	https://www.fao.org/one-health/en	Networks	
The Task Force for Global Health	Training Program in Epidemiology and Public Health Interventions Network (TEPHINET)	International	https://www.tephinet.org/our-network-fetps/about-fetp	Networks	
The University of Saskatchewan	One Health Certificate	Canada	https://programs.usask.ca/veterinary-medicine/certificate-one-health/index.php#CertificateinOneHealth12creditunits	Education	

(*Continued*)

TABLE 13.4 (*Continued*)
Sampling of One Health Education, Training, and Capacity-Building Programs

Organization	Title	Region	Links	Type	Column 1
Tufts University	Clinical-Translational Research One Health Fellowship	United States	https://www.tuftsctsi.org/education/tl1-fellowship-programs/	Capacity Building	Clinical-Translational Research One Health Fellowship
Tufts University School of Veterinary Medicine	Human-Animal Relationships	United States	https://vet.tufts.edu/legacy-curriculum	Education	
Tufts University School of Veterinary Medicine	Masters in Conservation Medicine (MCM) One Health approach	United States	https://grad.vet.tufts.edu/ms-conservation-medicine/	Education	
U.S. Centers for Disease Control and Prevention (CDC)	ZOHU Call Continuing Education	United States	https://www.cdc.gov/onehealth/zohu/continuingeducation.html	Capacity Building	
UK Animal and Plant Health Agency (APHA)	Capacity Building Modules	International	https://improve-ohcb.com/home/	Capacity Building	
UK Health Security Agency	health education program	United Kingdom	https://e-bug.eu/	Capacity Building	
Una Europa University Network	One Health Summer School	Europe	https://www.una-europa.eu/opportunities/one-health-summer-school	Education	One Health Summer School
Universidad Peruana Cayetano Heredia	One Health Unit – School of Public Health	Peru	https://www.onehealthcommission.org/documents/filelibrary/resources/whos_who/OHC_Whos_Who_Resources_Peruvian_Uni_6DC66C5A75362.pdf	Education	One Health Unit – School of Public Health

(*Continued*)

TABLE 13.4 (*Continued*)
Sampling of One Health Education, Training, and Capacity-Building Programs

Organization	Title	Region	Links	Type	Column 1
Universitas Gadjah Mada OH Collaborating Center	Summer Course on One Health Approach for Microbiome Identification on Komodo and Wildlife	Indonesia	https://onehealthwg.web.ugm.ac.id/komodo-summer-course/	Training	
Universitat Autonoma de Barcelona, Veterinary School	Masters in Zoonoses and One Health	Spain	https://www.uab.cat/web/estudiar/official-master-s-degrees/general-information/zoonosis-and-one-health-1096480962610.html?param1=1345694246010	Education	
University of Alaska Fairbanks	One Health master's program	United States	https://www.uaf.edu/onehealth/education/master.php	Education	
University of Alaska Fairbanks	One Health Certificate (open to public) – One Health Course: A Ten-Thousand-Year-Old View into the Future	United States	https://www.uaf.edu/onehealth/education/edx.php	Education	Center for One Health Research One Health Certificate (open to public)
University of Alaska Fairbanks	Online One Health EdX courses – One Health: A Ten-Thousand-Year-Old View into the Future	United States	https://www.edx.org/learn/climate-change/university-of-alaska-fairbanks-one-health-life-interconnected	Capacity Building	Online One Health EdX courses – One Health: A Ten-Thousand-Year-Old View into the Future
University of Arizona	One Health Research Initiative	United States	https://healthsciences.arizona.edu/connect/stories/investigating-human-animal-and-environmental-connections	Capacity Building	One Health Research Initiative

(*Continued*)

TABLE 13.4 (*Continued*)
Sampling of One Health Education, Training, and Capacity-Building Programs

Organization	Title	Region	Links	Type	Column 1
University of Arizona – Dept of Epidemiology and Biostatistics	MPH in One Health	United States	https://publichealth.arizona.edu/academics/masters/mph/one-health	Education	
University of Basel	One Health: Connecting Humans, Animals and the Environment	Switzerland	https://www.futurelearn.com/courses/one-health	Training	
University of Basel	One Health: Connecting Humans, Animals and the Environment	Switzerland	https://www.futurelearn.com/courses/one-health	Training	
University of Bern	Hidden players in the food chain Int. Bachelor and Master Summer School	Switzerland	https://edit.cms.unibe.ch/unibe/portal/microsites/micro_ircoh/content/summer_schools/bachelor_and_master_summer_schools/bachelor_summer_school_2020/scope	Training	
University of Bonn	One Health and Urban Transformation	Germany	https://www.zef.de/onehealth.html	Education	
University of Brighton	One Health Water Project Research and Education of the public	United Kingdom	https://onehealthwater.org/	Capacity Building	One Health Water Project Research and Education of the public
University of Bristol, Veterinary School	Student-led committee	United Kingdom	https://www.bristol.ac.uk/vet-school/research/inspire/onehealthbristol/	Capacity Building	
University of Calgary	One Health Training Programs	Canada	https://research.ucalgary.ca/one-health/training/one-health-trainee-chapter	Training	One Health Training Programs

(*Continued*)

TABLE 13.4 (*Continued*)
Sampling of One Health Education, Training, and Capacity-Building Programs

Organization	Title	Region	Links	Type	Column 1
University of Calgary	One Health Summer Institute	Canada	https://research.ucalgary.ca/one-health/training/one-health-summer-institute	Capacity Building	Summer Institute
University of California, Davis	Rx One Health	United States	https://rxonehealth.vetmed.ucdavis.edu/	Training	
University of California, Davis	Global Disease Biology Major	United States	https://gdb.ucdavis.edu/	Education	
University of California, Davis	Global Disease Biology Major	United States	https://gdb.ucdavis.edu/	Education	Integrated, One Health-based approach to advance student understanding of the concept(s) of disease, the societal and personal impacts of past, present, and future diseases, and the science behind disease discoveries, causes, evolution, diagnosis, treatment, and prevention
University of California, Davis One Health Institute	Strengthening the One Health Workforce	United States	https://ohi.vetmed.ucdavis.edu/	Capacity Building	One Health Workforce/ Capacity Building
University of California, Davis, USA and FAO	Emerging Pandemic Threat Program (PREDICT)	International	https://p2.predict.global/	Capacity Building	

(*Continued*)

TABLE 13.4 (*Continued*)
Sampling of One Health Education, Training, and Capacity-Building Programs

Organization	Title	Region	Links	Type	Column 1
University of California, San Francisco	Master of Science in Global Health	United States	https://globalhealthsciences.ucsf.edu/masters-program-sparks-ideas-and-inspires-action/	Education	
University of Copenhagen (UCPH) and Technical University of Denmark (DTU)	One Health International Summer Course	Denmark	https://healthsciences.ku.dk/education/summercourses/one-health/	Education	
University of Edinburgh	MSc One Health Online	Scotland	https://www.ed.ac.uk/studying/postgraduate/degrees/index.php?r=site/view&id=814	Education	
University of Edinburgh School of Veterinary Studies	Masters in One Health	United Kingdom	https://www.ed.ac.uk/vet/studying/postgraduate/taught-programmes/one-health	Education	
University of Edinburgh School of Veterinary Studies	Certificate in One Health	United Kingdom	https://www.ed.ac.uk/vet/studying/postgraduate/taught-programmes/one-health	Education	
University of Edinburgh School of Veterinary Studies	Diploma in One Health	United Kingdom	https://www.ed.ac.uk/vet/studying/postgraduate/taught-programmes/one-health	Education	
University of Edinburgh and Glasgow	Joint PhD Projects in One Health	Scotland	https://www.gla.ac.uk/research/ourresearchenvironment/prs/uofguofedinphdstudentships/onehealth/	Education	

(Continued)

TABLE 13.4 (*Continued*)
Sampling of One Health Education, Training, and Capacity-Building Programs

Organization	Title	Region	Links	Type	Column 1
University of Edinburgh and Leiden	PhD in Integrated One Health Solutions	Scotland	https://edinburgh-infectious-diseases.ed.ac.uk/teaching-and-training/phd-programmes/joint-phd-programme-integrated-one-health-solutions	Education	
University of Florida	One Health Certificate	United States	https://egh.phhp.ufl.edu/education/degree-programs/one-health-certificate/	Education	
University of Florida	Master of Health Science in Environmental and Global Health – One Health concentration	United States	https://egh.phhp.ufl.edu/education/degree-programs/mhs-one-health/	Education	
University of Florida	PhD in Public Health, One Health concentration	United States	https://edinburgh-infectious-diseases.ed.ac.uk/news-and-events/latest-news/new-phd-programme-integrated-one-health-solutions	Education	
University of Florida	One Health Certificate	United States	https://egh.phhp.ufl.edu/education/degree-programs/one-health-certificate/	Education	
University of Geneva	Online Courses	Switzerland	https://www.coursera.org/learn/global-health-human-animal-ecosystem?utm_source=recommendations&utm_medium=email&utm_campaign=15333&sfmc_id=59064775&sfmc_key=0031U00001k0kDLQAY	Training	

(*Continued*)

TABLE 13.4 (*Continued*)
Sampling of One Health Education, Training, and Capacity-Building Programs

Organization	Title	Region	Links	Type	Column 1
University of Georgia College of Veterinary Medicine	In-person Undergraduate certificate in One Health	United States	https://vet.uga.edu/education/undergraduate-programs/undergraduate-certificate-in-one-health/	Education	
University of Glasgow	MSc in One Health	Scotland	https://www.gla.ac.uk/postgraduate/taught/one-health-infectious-disease/	Education	
University of Glasgow	Post Graduate Certificate in OH	Scotland	https://www.gla.ac.uk/postgraduate/taught/one-health-infectious-disease/#programmestructure	Education	
University of Global Health Equity	One Health Option: Master's in Global Health Delivery (MGHD)	Africa	https://ughe.org/mghd-one-health-track	Education	
University of Guelph	Bachelors in One Health	Canada	https://onehealth.uoguelph.ca/bachelor-of-one-health/	Education	
University of Guelph	One Health Program Brochure	Canada	https://graduatestudies.uoguelph.ca/programs/onehealth	Education	
University of Guelph	Collaborative Specialization in One Health	Canada	https://onehealth.uoguelph.ca/collaborative-specialization-in-one-health/	Education	
University of Helsinki One Health	Helsinki One Health	Finland	https://www.helsinki.fi/en/networks/helsinki-one-health	Capacity Building	
University of Idaho	Biology of Vector-borne Diseases Course	United States	https://www.uidaho.edu/research/entities/ihhe/education/vector-borne-diseases	Training	

(*Continued*)

TABLE 13.4 (*Continued*)
Sampling of One Health Education, Training, and Capacity-Building Programs

Organization	Title	Region	Links	Type	Column 1
University of Illinois at Urbana-Champaign	DVM/MPH Joint Degree Program One Health Course	United States	https://vetmed.illinois.edu/education/dvm-mph-program/	Education	
University of Illinois College of Veterinary Medicine	Center for One Health – Illinois	United States	https://www.onehealthcommission.org/documents/filelibrary/resources/whos_who/Center_OH_Illinois_Whos_Who_templat_35D107D9A4518.pdf	Capacity Building	Center for One Health – Illinois
University of Maine	Graduate training program in One Health and the Environment for masters and doctoral students (NSF Traineeship)	United States	https://elh.umaine.edu/one-health/one-health-nrt/	Education	
University of Maine	National Science Foundation Research Traineeship (NRT) – MS Students	United States	https://elh.umaine.edu/one-health/one-health-nrt/	Education	
University of Maine	Research Experience for Undergrads	United States	https://elh.umaine.edu/one-health/reu/	Education	

(Continued)

TABLE 13.4 (*Continued*)
Sampling of One Health Education, Training, and Capacity-Building Programs

Organization	Title	Region	Links	Type	Column 1
University of Melbourne	One Health PhD Program	Australia	https://research.unimelb.edu.au/study/options/phd-programs/one-health#:~:text=The%20One%20Health%20PhD%20Program%20offers%20graduate%20researchers%3A,peers%20and%20early%20career%20researchers	Education	
University of Michigan College of Veterinary Medicine	3-week study abroad program implementing One Health in Nepal	Nepal	https://cvm.msu.edu/about/international-programs/our-programs/nepal-one-health	Education	
University of Minnesota	One Health Workforce Project (USAID)	United States	https://vetmed.umn.edu/centers-programs/global-one-health-initiative/one-health-workforce	Capacity Building	One Health Workforce Project (USAID) One Health Capacity Building
University of Missouri	Master of Public Health degree in the emphasis area of Veterinary Public Health	United States	https://healthsciences.missouri.edu/mph/	Education	
University of Nebraska	Institute of Ag and Natural Resources – Nebraska One Health Bat Activities for K-12	United States	https://nebraskaonehealth.unl.edu/bat-activities	Education	Institute of Ag and Natural Resources – Nebraska One Health Bat Activities for K-12
University of Nebraska	Nebraska One Health Science and Art Series	United States	https://nebraskaonehealth.unl.edu/art-series	Capacity Building	Nebraska One Health Science and Art Series

(*Continued*)

TABLE 13.4 (*Continued*)
Sampling of One Health Education, Training, and Capacity-Building Programs

Organization	Title	Region	Links	Type	Column 1
University of North Carolina	Graduate level One Health Course: Philosophy to Practical Application of Human, Animal and Environmental	United States	https://sph.unc.edu/courses/one-health-philosophy-to-practical-application-of-human-animal-and-environmental-health-10/	Education	Gilling School of Global Public Health Graduate level One Health Course: Philosophy to Practical Application of Human, Animal and Environmental
University of Pennsylvania	In-person One Health track	United States	https://www.publichealth.med.upenn.edu/OneHealthTrack/	Education	
University of Pretoria	Masters of Science in Global One Health	South Africa	https://www.up.ac.za/msc-global-one-health	Education	
University of Pretoria	Who's Who in One Health	South Africa	https://www.onehealthcommission.org/documents/filelibrary/resources/whos_who/U_Pretoria_CVZ_Whos_Who_temp_to_pos_886E6C1850744.pdf		

(*Continued*)

TABLE 13.4 (*Continued*)
Sampling of One Health Education, Training, and Capacity-Building Programs

Organization	Title	Region	Links	Type	Column 1
University of Queensland Veterinary School	Queensland Alliance of One Health Sciences	Australia	https://science.uq.edu.au/article/2021/11/queensland-alliance-one-health-sciences-launch-global-one-health-day-2021	Capacity Building	Based at UQ's School of Veterinary Science and jointly funded by The University of Queensland, industry, and government partners, QAOHS will aim to improve zoonotic disease public health policy through integrated health research that addresses local, national, and global health challenges.
University of Queensland Veterinary School	One Health: Animals, the Environment & Human Disease	Australia	https://programs-courses.uq.edu.au/course.html?course_code=vets1030	Education	
University of Sarajevo	MS in One Health	Bosnia/Herzegovina	https://cis.unsa.ba/bs/programi/magistarski/jedinstveno-zdravlje/	Education	
University of Surrey	One Health One Medicine; Research, Projects, Partnerships	United Kingdom	https://www.surrey.ac.uk/one-health-one-medicine	Capacity Building	University of Surrey One Health One Medicine
University of Tennessee	One Health Initiative	United States	https://onehealth.tennessee.edu/	Capacity Building	One Health Initiative

(*Continued*)

TABLE 13.4 (*Continued*)
Sampling of One Health Education, Training, and Capacity-Building Programs

Organization	Title	Region	Links	Type	Column 1
University of Tennessee	K-12 Educators Resources	United States	https://onehealth.tennessee.edu/k12-resources/	Education	K-12 Educators Resources
University of Tennessee	Undergrad and Graduate One Health Minor	United States	https://onehealth.tennessee.edu/one-health-minor/	Education	Undergrad and Graduate One Health Minor
University of Washington, Center for One Health Research	In-person One Health Graduate certificate	United States	https://deohs.washington.edu/cohr/graduate-certificate-one-health	Education	
University of Washington, Center for One Health Research	Clinical Elective in OH, OH tract in Occupational and Environmental Health	United States	https://deohs.washington.edu/cohr/medical-training	Education	Center for One Health Research Medical Training, Clinical Elective in OH, OH tract in Occupational and Environmental Health
University of Washington, Environmental & Occupational Health Sciences	One Health Graduate Degree Programs	United States	https://deohs.washington.edu/one-health	Education	Environmental & Occupational Health Sciences One Health Graduate Degree Programs
University of Washington, School of Public Health	MPH in One Health	United States	https://sph.washington.edu/students/graduate-programs/mph-program	Education	

(Continued)

TABLE 13.4 (*Continued*)
Sampling of One Health Education, Training, and Capacity-Building Programs

Organization	Title	Region	Links	Type	Column 1
University of Wisconsin-Madison	One Health Centers – Opportunities for faculty, staff, and students to address health and social issues related to One Health	United States	https://ghi.wisc.edu/one-health-centers/	Education	One Health Centers – Opportunities for faculty, staff, and students to address health and social issues related to One Health
University of Zambia	Master of Science in One Health Analytical Epidemiology	Zambia	https://www.unza.zm/academics/postgraduate-programmes/master-of-science-one-health-analytical-epidemiology	Education	
Utrecht University	In-person One Health program with focus on research	Western Europe	https://www.uu.nl/en/masters/one-health	Education	
US Agency for International Development (USAID)	One Health Assessment for Planning & Performance (OH APP) Resources	United States	https://ww38.onehealthapp.org/resources	Other	
US Department of Agriculture (USDA)	One Health Certified	United States	https://www.ams.usda.gov/services/auditing/one-health	Other	
Utrecht University	One Health Masters in Biomedical Science	Europe	https://www.uu.nl/en/masters/one-health	Education	

(*Continued*)

TABLE 13.4 (*Continued*)
Sampling of One Health Education, Training, and Capacity-Building Programs

Organization	Title	Region	Links	Type	Column 1
Utrecht University	One Health Utrecht: from molecule to population	The Netherlands	https://www.onehealthcommission.org/documents/filelibrary/resources/whos_who/UTRECHT_UNIVERSITY_One_Health_leafl_BD133B01F7433.pdf		
Vietnam One Health University Network (VOHUN)	One Health University Network	Vietnam	https://www.seaohun.org/vohun	Education	
Vietnam One Health University Network (VOHUN)	One Health Training Programs	Vietnam	https://www.seaohun.org/vohun	Capacity Building	
Virginia-Maryland College of Veterinary Medicine	MPH grounded in One Health Approach	United States	https://publichealth.vt.edu/about/signature-areas/one-health.html	Education	
Wake Forest School of Medicine	Master's Program in Comparative Medicine (COMD)	United States	https://school.wakehealth.edu/education-and-training/graduate-programs/comparative-medicine-masters	Education	
Washington University, St Louis	Course – One Health: Linking the Health of Humans, Animals, and the Environment	United States	https://www.coursicle.com/wustl/courses/EnSt/250/	Education	

(*Continued*)

TABLE 13.4 (*Continued*)
Sampling of One Health Education, Training, and Capacity-Building Programs

Organization	Title	Region	Links	Type	Column 1
Western University School of Medicine and Dentistry	One Health PhD Program	Canada	https://www.schulich.uwo.ca/pathol/gps/research_programs/programs/msc_phd/one_health.html	Education	
Western University School of Medicine and Dentistry	One Health Master Program	Canada	https://www.schulich.uwo.ca/pathol/gps/research_programs/programs/msc_phd/one_health.html	Education	
Western University School of Medicine and Dentistry	Bachelor of Medical Sciences – Honors Specialization in One Health	Canada	https://www.schulich.uwo.ca/pathol/undergraduate/bmsc/onehealth/index.html	Education	
Westminster College	One Health undergraduate major	United States	https://www.wcmo.edu/academics/programs/one-health.html	Education	https://www.wcmo.edu/academics/files/majors%20-%20program%20plan/One%20Health%20Major.pdf
Wildlife Institute	Wildlife Medicine & Conservation course	Belize	https://www.wildlife-institute.com/wmc/	Training	
World Health Organization (WHO)	Building One Health Preparedness Capacities: Implementation of the National Bridging Workshop Roadmap in Kazakhstan	International	https://www.who.int/news-room/feature-stories/detail/building-one-health-preparedness-capacities	Capacity Building	

(*Continued*)

TABLE 13.4 (*Continued*)
Sampling of One Health Education, Training, and Capacity-Building Programs

Organization	Title	Region	Links	Type	Column 1
World Health Organization (WHO)	Provides learning resources for technical staff, decision-makers, general public	International	https://openwho.org/channels/onehealth	Capacity Building	
World Organisation for Animal Health and Children Radio Foundation	Zoonosis awareness project	Africa	https://www.woah.org/en/woah-raises-community-awareness-on-zoonosis-through-rural-radio-in-western-and-central-africa/	Capacity Building	
World Organisation for Animal Health and VSF International	Project on community animal health workers	International	https://vsf-international.org/woah-vsf-project-cahws/	Capacity Building	
World Small Animal Veterinary Association	One Health Certificate Course	United States	https://vetmedbiosci.colostate.edu/vth/	Education	
Zoo New England One Health Program	Hosts OH Electives for Med Students at the Zoo	United States	https://www.zoonewengland.org/protect/inside-our-zoos/one-health-program/	Education	Hosts OH Electives for Med Students at the Zoo
Zoonoses and Emerging Livestock Systems (ZELS)	Training Program	United Kingdom	https://www.onehealthcommission.org/documents/filelibrary/resources/whos_who/62519_ZELS_for_Whos_Who_mapwebpage_CCC18FF1C72E7.pdf	Training	Training Program

REFERENCES

AFROHUN. (2020, January 15). *One Health modules* | AFROHUN. https://afrohun.org/course/onehealthmodules/

AFROHUN. (2023). *AFROHUN | Advancing One Health*. https://afrohun.org/

Allen-Scott, L. K., Buntain, B., Hatfield, J. M., Meisser, A., & Thomas, C. J. (2015). Academic institutions and One Health: Building capacity for transdisciplinary research approaches to address complex health issues at the animal–human–ecosystem interface. *Academic Medicine*, *90*(7), 866–871. https://doi.org/10.1097/ACM.0000000000000639

American Public Health Association. (2017). *Advancing a "One Health" approach to promote health at the human-animal-environment interface*. APHA. https://apha.org/Policies-and-Advocacy/Public-Health-Policy-Statements/Policy-Database/2018/01/18/Advancing-a-One-Health-Approach

Barrett, M. A., Bouley, T. A., Stoertz, A. H., & Stoertz, R. W. (2011). Integrating a One Health approach in education to address global health and sustainability challenges. *Frontiers in Ecology and the Environment*, *9*(4), 239–245. https://doi.org/10.1890/090159

Berg, L., & Olsen, B. (2016). Foreword. *Infection Ecology & Epidemiology*, *6*(1), 34094. https://doi.org/10.3402/iee.v6.34094

Berry College. (2015). *Academic catalog. Program: One Health Minor*. https://catalog.berry.edu/index.php?catoid=4

Binot, A., Duboz, R., Promburom, P., Phimpraphai, W., Cappelle, J., Lajaunie, C., Goutard, F. L., Pinyopummintr, T., Figuié, M., & Roger, F. L. (2015). A framework to promote collective action within the One Health community of practice: Using participatory modelling to enable interdisciplinary, cross-sectoral and multi-level integration. *One Health*, *1*, 44–48. https://doi.org/10.1016/j.onehlt.2015.09.001

Boyle, P. (2019, December 10). *What medical students can learn about health from animals*. AAMC. https://www.aamc.org/news/what-medical-students-can-learn-about-health-animals

CDC. (2024, May 9). *Global Health Security*. Global Health. https://www.cdc.gov/global-health/topics-programs/global-health-security.html

CEPH. (2024). *Home—Council on Education for Public Health*. https://ceph.org/

Cianfagna, M., Bolon, I., Babo Martins, S., Mumford, E., Romanelli, C., Deem, S. L., Pettan-Brewer, C., Figueroa, D., Velásquez, J. C. C., Stroud, C., Lueddeke, G., Stoll, B., & Ruiz De Castañeda, R. (2021). Biodiversity and human health interlinkages in higher education offerings: A first global overview. *Frontiers in Public Health, 9*, 637901. https://doi.org/10.3389/fpubh.2021.637901

COST Action. (2015). *Network for Evaluation of One Health | NEOH*. https://neoh.onehealthglobal.net/

Devinsky, O., Boesch, J. M., Cerda-Gonzalez, S., Coffey, B., Davis, K., Friedman, D., Hainline, B., Houpt, K., Lieberman, D., Perry, P., Prüss, H., Samuels, M. A., Small, G. W., Volk, H., Summerfield, A., Vite, C., Wisniewski, T., & Natterson-Horowitz, B. (2018). A cross-species approach to disorders affecting brain and behaviour. *Nature Reviews Neurology*, *14*(11), 677–686. https://doi.org/10.1038/s41582-018-0074-z

Docherty, L., & Foley, P. L. (2021). Survey of One Health programs in U.S. medical schools and development of a novel one health elective for medical students. *One Health*, *12*, 100231. https://doi.org/10.1016/j.onehlt.2021.100231

dvm360. (2020, March 10). *New One Health certificate program focuses on companion animals*. DVM 360. https://www.dvm360.com/view/new-one-health-certificate-program-focuses-on-companion-animals

Elmahi, O. K. O., Uakkas, S., Olalekan, B. Y., Damilola, I. A., Adedeji, O. J., Hasan, M. M., Dos Santos Costa, A. C., Ahmad, S., Essar, M. Y., & Thomson, D. J. (2022). Antimicrobial resistance and one health in the post COVID-19 era: What should health students learn? *Antimicrobial Resistance & Infection Control, 11*(1), 58. https://doi.org/10.1186/s13756-022-01099-7

Ethiopian Agriculture Training Portal. (2021). *One Health – Zoonotic disease: A Training of Trainers (ToT) course*. Ethiopian Agriculture Training Portal. https://agri-training-et.org/zoonotic-e-learning-course/

FAO, UNEP, WHO, & WOAH. (2022). *One Health Joint Plan of Action (2022–2026). Working together for the health of humans, animals, plants and the environment.* FAO; UNEP; WHO; World Organisation for Animal Health (WOAH) (founded as OIE); https://doi.org/10.4060/cc2289en

FAVA OneHealth Fukuoka Office. (2023, July 24). *One Health—FAVA*. https://fof-office.com/en/one-health/

Fenwick, S. (2016, April 25). *CORE competencies and One health—From theory to action* [Powerpoint]. USAID - One Health Workforce, Hanoi, Vietnam. https://bit.ly/2WwHKzZ

Fontbonne University. (2024). *Bachelor of Science in One Health*. https://www.fontbonne.edu/academics/college-arts-sciences/biological-behavioral-sciences-department/bachelor-of-science-in-one-health/

Frankson, R., Hueston, W., Christian, K., Olson, D., Lee, M., Valeri, L., Hyatt, R., Annelli, J., & Rubin, C. (2016). One Health core competency domains. *Frontiers in Public Health, 4*. https://doi.org/10.3389/fpubh.2016.00192

FutureLearn. (2023). *Understanding One Health—Online Course—University of Basel*. FutureLearn. https://www.futurelearn.com/courses/one-health

Gaebel, M., & Zhang, T. (2018). *Learning and teaching in the European higher education area*. European University Association. https://eua.eu/downloads/publications/trends-2018-learning-and-teaching-in-the-european-higher-education-area.pdf

Global Health Security Agenda. (2016). *Advancing the global health security agenda: Progress and early impact from U.S. Investment*. https://www.state.gov/wp-content/uploads/2019/02/1-ghsa-annual-report-2016.pdf

Hall, D. C., & Le, Q. B. (2015). Monitoring and evaluation of One Health projects; lessons from Southeast Asia. *Procedia - Social and Behavioral Sciences, 186*, 681–683. https://doi.org/10.1016/j.sbspro.2015.04.070

Hanoi University. (2022, April 29). *Global Health Seminar Series "Applying a One Health approach to tackling the next pandemic."* Global Health Seminar Series "Applying a One Health Approach to Tackling the next Pandemic." https://spmph.edu.vn/en-GB/article/news/global-health-seminar-series-applying-a-one-health-approach-to-tackling-the-next-pandemic%201

Hilliard, C. A. (2015). One Health: An introduction and initial assessment. *One Health Commission*. https://www.onehealthcommission.org/documents/news/OneHealthAnIntroductionandInitialAs_52225ADCB9E25.pdf.

Holmes, E. C. (2022). COVID-19—Lessons for zoonotic disease. *Science, 375*(6585), 1114–1115. https://doi.org/10.1126/science.abn2222

Iatridou, D., Bravo, A., & Saunders, J. (2021). One Health interdisciplinary collaboration in veterinary education establishments in Europe: Mapping implementation and reflecting on promotion. *Journal of Veterinary Medical Education, 48*(4), 427–440. https://doi.org/10.3138/jvme-2020-0019

Institute of Medicine. (2009). *Sustaining global surveillance and response to emerging zoonotic diseases* (G. T. Keusch, M. Pappaioanou, & M. C. Gonzalez, Eds.). National Academies Press.

Johns Hopkins Bloomberg School of Public Health. (2024). *Master of Public Health (MPH) Curriculum*. https://publichealth.jhu.edu/academics/mph/curriculum

Kansas State University. (2024, February 21). *One Health Kansas*. https://olathe.k-state.edu/research/one-health-newsletter/resources/

Kerala Veterinary and Animal Sciences University. (2023). *COHEART | Centre for One Health education advocacy research and training*. https://coheart.ac.in/training/10

Khan, M. S., Rothman-Ostrow, P., Spencer, J., Hasan, N., Sabirovic, M., Rahman-Shepherd, A., Shaikh, N., Heymann, D. L., & Dar, O. (2018). The growth and strategic functioning of One Health networks: A systematic analysis. *The Lancet Planetary Health*, *2*(6), e264–e273. https://doi.org/10.1016/S2542-5196(18)30084-6

Laaser, U., Stroud, C., Bjegovic-Mikanovic, V., Wenzel, H., Seifman, R., Craig, C., Kaplan, B., Kahn, L., & Roopnarine, R. (2022). Exchange and coordination: Challenges of the global One Health movement. *South Eastern European Journal of Public Health (SEEJPH), Volume XIX*, 2022. https://doi.org/10.11576/SEEJPH-6076

Laing, G., Duffy, E., Anderson, N., Antoine-Moussiaux, N., Aragrande, M., Luiz Beber, C., Berezowski, J., Boriani, E., Canali, M., Pedro Carmo, L., Chantziaras, I., Cousquer, G., Meneghi, D., Gloria Rodrigues Sanches Da Fonseca, A., Garnier, J., Hitziger, M., Jaenisch, T., Keune, H., Lajaunie, C., … Häsler, B. (2023). Advancing One Health: Updated core competencies. *CABI One Health*, ohcs20230002. https://doi.org/10.1079/cabionehealth.2023.0002

Larsen, R. J. (2021). Shared curricula and competencies in One Health and health professions education. *Medical Science Educator*, *31*(1), 249–252. https://doi.org/10.1007/s40670-020-01140-7

Lebov, J., Grieger, K., Womack, D., Zaccaro, D., Whitehead, N., Kowalcyk, B., & MacDonald, P. D. M. (2017). A framework for One Health research. *One Health*, *3*, 44–50. https://doi.org/10.1016/j.onehlt.2017.03.004

Lee, M. Y., & Global OHCC Working Group. (2013). *One Health core competency domains, subdomains, and competency examples* (Institution). USAID Respond Initiative. https://hdl.handle.net/10427/000364

Linder, D., Cardamone, C., Cash, S. B., Castellot, J., Kochevar, D., Dhadwal, S., & Patterson, E. (2020). Development, implementation, and evaluation of a novel multidisciplinary one health course for university undergraduates. *One Health*, *9*, 100121. https://doi.org/10.1016/j.onehlt.2019.100121

Lucey, D. R., Sholts, S., Donaldson, H., White, J., & Mitchell, S. R. (2017). One health education for future physicians in the pan-epidemic "Age of Humans." *International Journal of Infectious Diseases*, *64*, 1–3. https://doi.org/10.1016/j.ijid.2017.08.007

Machalaba, C., Raufman, J., Anyamba, A., Berrian, A. M., Berthe, F. C. J., Gray, G. C., Jonas, O., Karesh, W. B., Larsen, M. H., Laxminarayan, R., Madoff, L. C., Martin, K., Mazet, J. A. K., Mumford, E., Parker, T., Pintea, L., Rostal, M. K., de Castañeda, R. R., Vora, N. M., … Weiss, L. M. (2021). Applying a One Health approach in global health and medicine: Enhancing involvement of medical schools and global health centers. *Annals of Global Health*, *87*(1), 30. https://doi.org/10.5334/aogh.2647

McKenzie, J. S., Dahal, R., Kakkar, M., Debnath, N., Rahman, M., Dorjee, S., Naeem, K., Wijayathilaka, T., Sharma, B. K., Maidanwal, N., Halimi, A., Kim, E., Chatterjee, P., & Devleesschauwer, B. (2016). One Health research and training and government support for One Health in South Asia. *Infection Ecology & Epidemiology*, *6*(1), 33842. https://doi.org/10.3402/iee.v6.33842

Mor, S. M., Norris, J. M., Bosward, K. L., Toribio, J.-A. L. M. L., Ward, M. P., Gongora, J., Vost, M., Higgins, P. C., McGreevy, P. D., White, P. J., & Zaki, S. (2018). One health in our backyard: Design and evaluation of an experiential learning experience for veterinary medical students. *One Health*, *5*, 57–64. https://doi.org/10.1016/j.onehlt.2018.05.001

Office of International Affairs, The Ohio State University. (2024). *Curriculum twinning—Global One Health initiative*. https://oia.osu.edu/global-one-health-initiative/training/completed-projects/curriculum-twinning

One Health Commission. (2016a). *One Health movement news / One Health Topics "in" the news*. https://www.onehealthcommission.org/index.cfm?nodeID=38050&audienceID=1&action=search&tag=infection%20ecology%20epidemiology

One Health Commission. (2016b, November 18). *One Health education online conference*. https://www.onehealthcommission.org/en/events_since_2001/one_health_education_online_conference/

One Health Commission. (2022). *One Health tools and toolkits*. https://www.onehealthcommission.org/en/resources__services/one_health_tools__toolkits/

One Health Commission. (2023, April 25). *Who's who in One Health*. https://batchgeo.com/map/bcaa0402f3da642ed28392052ecd2152

One Health Commission. (2024a). *Join the OHC's international One Health Network*. https://www.onehealthcommission.org/en/resources__services/join_the_global_oh_community_listserv/

One Health Commission. (2024b). *OH Resources for public health educators*. https://www.onehealthcommission.org/en/resources__services/oh_resources_for_public_health_educators/

One Health Commission. (2024c). *One Health courses (Online)*. Google Docs. https://docs.google.com/spreadsheets/d/1MRZNuOh897GmzIUCOX_WjYrqiENNOAd4oXUlTjolEtM/edit?usp=embed_facebook

One Health Commission. (2024d). *One Health courses (Traditional-In Person)*. Google Docs. https://docs.google.com/spreadsheets/d/1wfmH-UqsLvmapsMzolMSaWwwBWfEeeSHPVYIvcTJMAk/edit?usp=embed_facebook

One Health Commission. (2024e). *One Health education initiatives—Overview*. https://www.onehealthcommission.org/en/programs/one_health_education_initiatives__overview/

One Health Commission. (2024f). *One Health education—US initiative*. https://www.onehealthcommission.org/en/programs/one_health_education__us_initiative/

One Health Commission. (2024g). One Health masters programs. Google Docs. https://docs.google.com/spreadsheets/d/1yKhwwUNtCSWHA8BHmAgZ3WwZzc8wAcTlaMNiuG7Vz6s/edit?usp=embed_facebook

One Health Commission. (2024h). *One Health opportunities bulletin board*. https://www.onehealthcommission.org/en/resources__services/oh_opportunities_bulletin_board/

One Health Commission. (2024i). *One Health strategic action plans*. https://www.onehealthcommission.org/en/resources__services/one_health_strategic_action_plans/

One Health Commission. (2024j). One Health-related PhD programs. Google Docs. https://docs.google.com/spreadsheets/d/1KYNOUI3zgES8aXSXACwPPm4jwx4Itw9UecCf1BwL2Oo/edit?usp=embed_facebook

One Health Commission. (2024k). *One Health-related summer programs*. Google Docs. https://docs.google.com/spreadsheets/d/1F_VDr0hdm1cxvzIPkV4bl_IJqvbsLV-XNJQnsw6bH5s/edit?usp=embed_facebook

One Health Commission. (2024l). *Primary, secondary, K-12 OH education resources*. https://www.onehealthcommission.org/en/resources__services/one_health_education_resources/primary_secondary_k12_oh_education_resources/

One Health Commission. (2024m). *Why One Health?* https://www.onehealthcommission.org/en/why_one_health/

One Health Lessons. (2023). *One Health lessons*. https://onehealthlessons.org/

One Health Workforce Academies. (2024a). *OHWA – One Health Workforce Academies*. OHWA. https://onehealthworkforceacademies.org/

One Health Workforce Academies. (2024b). *Training*. OHWA. https://onehealthworkforceacademies.org/training-material/

Open WHO. (2024). *One Health for Global Health Security*. https://openwho.org/channels/onehealth

Pettan-Brewer, C., Martins, A. F., Abreu, D. P. B. D., Brandão, A. P. D., Barbosa, D. S., Figueroa, D. P., Cediel, N., Kahn, L. H., Brandespim, D. F., Velásquez, J. C. C., Carvalho, A. A. B., Takayanagui, A. M. M., Galhardo, J. A., Maia-Filho, L. F. A., Pimpão, C. T., Vicente, C. R., & Biondo, A. W. (2021). From the approach to the concept: One Health in Latin America-experiences and perspectives in Brazil, Chile, and Colombia. *Frontiers in Public Health*, *9*, 687110. https://doi.org/10.3389/fpubh.2021.687110

Rabinowitz, P. M., Natterson-Horowitz, B. J., Kahn, L. H., Kock, R., & Pappaioanou, M. (2017). Incorporating one health into medical education. *BMC Medical Education*, *17*(1), 45, s12909-017-0883–0886. https://doi.org/10.1186/s12909-017-0883-6

Reid, S. A., McKenzie, J., & Woldeyohannes, S. M. (2016). One Health research and training in Australia and New Zealand. *Infection Ecology & Epidemiology*, *6*(1), 33799. https://doi.org/10.3402/iee.v6.33799

Rüegg, S. R., Häsler, B., & Zinsstag, J. (Eds.). (2018). *Integrated approaches to health: A handbook for the evaluation of One Health*. Brill | Wageningen Academic. https://doi.org/10.3920/978-90-8686-875-9

Rwego, I. B., Babalobi, O. O., Musotsi, P., Nzietchueng, S., Tiambo, C. K., Kabasa, J. D., Naigaga, I., Kalema-Zikusoka, G., & Pelican, K. (2016). One Health capacity building in sub-Saharan Africa. *Infection Ecology & Epidemiology*, *6*(1), 34032. https://doi.org/10.3402/iee.v6.34032

Saint Louis Zoo. (2021). *Saint Louis Zoo Institute for conservation medicine—10 year report*. Saint Louis Zoo Institute for Conservation Medicine. https://cdn2.assets-servd.host/maniacal-finch/production/documents/ICM_10_Year_Report.pdf?dm=1668783052

Sikkema, R., & Koopmans, M. (2016). One Health training and research activities in Western Europe. *Infection Ecology & Epidemiology*, *6*(1), 33703. https://doi.org/10.3402/iee.v6.33703

Steele, S. G., Toribio, J.-A., Booy, R., & Mor, S. M. (2019). What makes an effective One Health clinical practitioner? Opinions of Australian One Health experts. *One Health*, *8*, 100108. https://doi.org/10.1016/j.onehlt.2019.100108

Streichert, L. C., Sepe, L. P., Jokelainen, P., Stroud, C. M., Berezowski, J., & Del Rio Vilas, V. J. (2022). Participation in One Health networks and involvement in the COVID-19 pandemic response: A global study. *Frontiers in Public Health*, *10*, 830893. https://doi.org/10.3389/fpubh.2022.830893

Stroud, C., Kaplan, B., Logan, J. E., & Gray, G. C. (2016). One Health training, research, and outreach in North America. *Infection Ecology & Epidemiology*, *6*(1), 33680. https://doi.org/10.3402/iee.v6.33680

Taylor, L. H., Latham, S. M., & Woolhouse, M. E. J. (2001). Risk factors for human disease emergence. *Philosophical Transactions of the Royal Society of London. Series B: Biological Sciences*, *356*(1411), 983–989. https://doi.org/10.1098/rstb.2001.0888

Tb, T. N., Nguyen, H. T., Tk, T. T., Pham, P. M., & Pham, P. D. (2022). Effectiveness of one health training program for health and veterinary workers: Perspective from trainees and employers. *International Journal of Infectious Diseases*, *116*, S68. https://doi.org/10.1016/j.ijid.2021.12.160

The Community Tool Box. (2024). *Chapter 3. Assessing community needs and resources | Section 14. SWOT analysis: Strengths, weaknesses, opportunities, and threats*. University of Kansas. https://ctb.ku.edu/en/table-of-contents/assessment/assessing-community-needs-and-resources/swot-analysis/main

The Royal (Dick) School of Veterinary Studies. (2024, March 26). *One Health—MSc One Health*. The University of Edinburgh. https://www.ed.ac.uk/vet/studying/postgraduate/taught-programmes/one-health

Togami, E., Gardy, J. L., Hansen, G. R., Poste, G. H., Rizzo, D. M., Wilson, M. E., & Mazet, J. A. K. (2018). Core competencies in One Health education: What are we missing? *NAM Perspectives*, *8*(6). https://doi.org/10.31478/201806a

United Nations. (n.d.). *Capacity-building*. United Nations. Retrieved June 6, 2024, from https://www.un.org/en/academic-impact/capacity-building

University of Alaska - Fairbanks. (2024). *One Health Master's Degree (OHM) | Center for One Health Research*. https://www.uaf.edu/onehealth/education/master.php

University of California - Davis. (2024a). *One Health Institute*. https://ohi.vetmed.ucdavis.edu/

University of California - Davis. (2024b). *Rx One Health*. https://rxonehealth.vetmed.ucdavis.edu/

University of Florida. (2024a). *MHS in Environmental and Global Health with a Concentration in One Health*. https://egh.phhp.ufl.edu/education/degree-programs/mhs-one-health/

University of Florida. (2024b). *One Health Center of Excellence*. https://onehealth.phhp.ufl.edu/homepage/academics/

University of Florida. (2024c). *PhD in Public Health, One Health*. https://egh.phhp.ufl.edu/education/degree-programs/phd-in-one-health/

University of Geneva. (2024). *Global Health at the human-animal-ecosystem interface. Coursera*. https://www.coursera.org/learn/global-health-human-animal-ecosystem

University of Global Health Equity. (2024). One Health Option I MGHD. *UGHE*. https://ughe.org/mghd-one-health-track

University of Hawai'i. (2024, April 23). *One Health interdisciplinary undergraduate certificate*. https://manoa.hawaii.edu/onehealth/

University of Hawai'i. (2024, May 28). *Department of tropical medicine, medical microbiology & pharmacology*. https://manoa.hawaii.edu/tropicalmedicine/

Von Borries, R., Guinto, R., Thomson, D. J., Abia, W. A., & Lowe, R. (2020). Planting sustainable seeds in young minds: The need to teach planetary health to children. *The Lancet Planetary Health*, *4*(11), e501–e502. https://doi.org/10.1016/S2542-5196(20)30241-2

Western University. (2024). *One Health*. https://www.schulich.uwo.ca/oleapopelkaresearchgroup//one_health/index.html

World Medical Association. (2015, May). *The World Medical Association Global Conference on One Health*. The World Medical Association Global Conference on One Health, Madrid, Spain. https://www.wma.net/events-post/global-conference-on-one-health/015-gcoh-report-may-2015/

World Veterinary Association. (2023, March 29). *New WVA position statement on One Health*. https://worldvet.org/news/new-wva-position-statement-on-one-health/

Wu, J., Liu, L., Wang, G., & Lu, J. (2016). One Health in China. *Infection Ecology & Epidemiology*, *6*(1), 33843. https://doi.org/10.3402/iee.v6.33843

14 Harnessing Data, Collaborations, and Connectedness to Foster One Health

Kristen T. Honey, Leilani V. Francisco, and Tonya Nichols

In this chapter we discuss the need for a combination of data, collaborations, and connectedness to foster One Health successes in public health research. We define collaborations as bringing together the tools, disciplines, and study leaders needed to solve health challenges. We define connectedness as the networked linkages between researchers and those we aim to serve wherein ground-truthing plays a central role. Each section connects to another, highlighting the importance of tying together data needs, collaborations, and connectedness. Each includes a case study that illustrates why all three are needed in One Health efforts to be successful.

Across sections in this chapter, and for both quantitative and qualitative research, ethical conduct and ethical considerations are paramount. There is no "right" approach to ethics, as philosophers show and entire disciplines explore. Multiple ethical frameworks exist with extensive resources in data ethics, research ethics, medical ethics, communications ethics, and the emerging concept of "ethical modernization." As a One Health practitioner, select and embrace an ethical framework that works for you and communities collaborating with you. No matter which ethical frameworks you choose to use, be intentional and design with ethics in mind. Embrace insights from the philosophers and social scientists who deeply research humanity, so that even the most data-driven work has ethical design for One Health. Intentionally embody ethics at all levels from the individual – who leads by example through actions and everyday leadership – to programmatic and global decisions with decision frameworks to evaluate unknowns, biases, upstream effects, downstream effects, transparency, and other ethical considerations while recognizing that law and legality are a lower bound for ethics. We should not assume what is legal is ethical. An ethical society rests on individual responsibility, as well as our collective responsibility, so ethics must be intentionally designed into One Health data, collaborations, and connectedness across all scales: individual, local, regional, state, national, and global.

DOI: 10.1201/9781003232223-14

HARNESSING DATA

The U.S. Centers for Disease Control and Protection (CDC) defines evidence-based decision-making as "a process for making decisions about a program, practice, or policy that is grounded in the best available research evidence and informed by experiential evidence from the field and relevant contextual evidence" (Centers for Disease Control and Prevention, 2010). The importance of data in the field of health became formalized with the rise of evidence-based medicine (EBM) in the 1990s (Sackett et al., 1996) and its principal dimensions include: contextual evidence, best available evidence, and experiential evidence. However, critics have argued that there are inherent weaknesses in the model including reducing years of medical experience down to an overly algorithmic exercise and an overreliance on randomized controlled trials (Sur & Dahm, 2011). Herein lies the essence of this section, that data alone are not sufficient. In this chapter, we emphasize the need to couple data with experience through strong collaborations and connectedness to foster One Health successes in infectious disease research.

Specifically, we discuss what constitutes evidence with a focus on the benefits of mixed methods research (those that combine quantitative and qualitative approaches) as well as rigor in quantitative and qualitative research to demonstrate why coupling data with collaborations and connectedness is critical to this rigor. This is followed by a case study that provides a practical illustration of this point.

In terms of quantitative research, four of the main concepts that define rigor include: internal validity, external validity, reliability, and objectivity. Internal validity asks whether the investigator found the truth, and external validity speaks to generalizability and whether the findings would hold outside of the study sample. Reliability asks whether the study could be repeated and have the same result. Objectivity explores whether "distance" was created to eliminate bias. Similar concepts exist in qualitative research, with main concepts being: credibility, transferability, dependability, and confirmability (Lincoln & Guba, 1986). As qualitative data can be as valuable as quantitative data in One Health emerging infectious disease research, it is important to clarify the parallel concepts of rigor therein.

Credibility acts as a parallel to internal validity in quantitative research. This concept explores whether the investigator has succeeded in capturing the situation or idea. It asks whether the investigator's interpretations and descriptions ring true to the participants. Transferability parallels external validity by aiming to ensure that results are generalizable, which is established by providing thick (detailed) descriptions, including those that describe time, place, context, and culture. Reliability of findings in qualitative research is covered by the concept of dependability wherein an external investigator could "audit" the original investigator's field notes and confirm that the research process was executed properly. This can be accomplished by providing detailed description of what was done, and the rationale for why it was appropriate given what is reflected in the data. Finally, confirmability in qualitative research addresses the parallel concept of objectivity in quantitative research. Similar to dependability, this concept asks whether an external investigator could confirm the original investigator's findings by tracing back through the data collected. It ultimately asks if the research findings are supported by the data that was collected.

This discussion of what constitutes evidence in both quantitative and qualitative research is important for this chapter as it illustrates how rigorous data cannot be secured without collaboration and connectedness. For example, internal validity in quantitative research and credibility in qualitative research cannot be achieved without confirming that the investigator's interpretations ring true for participants. Collaborations provide the tools, disciplines, and study leaders needed to ensure that research is conducted by trained professionals who know the tools, considering both the strengths or pros and shortcomings or cons of tools for implementation. Similarly, connectedness is defined herein as the linkages between researchers and those we aim to serve in local communities. It would not be possible to "find the truth" without the presence of long-standing trusted relationships with the communities within which we are conducting research.

Similarly, we would not be able to ensure the transferability of qualitative research findings without collaborations that allow us to conduct and synthesize research across various settings or without connectedness between researchers and community member respondents to gain trusted access to local truths. Methods exist to assess how generalizable the research findings are from qualitative data. This involves assessing the qualitative findings outside of the study setting, including time, place, context, and culture, such that it can be reproduced by external investigators.

In the case of reliability in quantitative research and dependability in qualitative research, collaborations are needed to ensure that the right mix of scientists are dedicated to implementing the most appropriate methods for the research question at hand such that they could be replicated by an external investigator. Connectedness is needed to ensure that the methods leveraged are appropriate for the community being explored. For example, through connectedness with local scientific partners, we ensure that health interventions are tailored appropriately to the appetite and reading level of the community we aim to serve.

Finally, objectivity in quantitative research and confirmability in qualitative research rely on collaborations to ensure that scientists trained on the concepts of bias across disciplines are deployed for the data collection. Connectedness ensures that researchers see "both sides of the story" by relying on both quantitative and qualitative data to arrive at "the truth," thereby enhancing their ability to be replicated by an external investigator.

CASE STUDY: USAID PREDICT

A poignant example of the need to harness data, collaborations, and connectedness simultaneously to fight emerging infectious diseases within a One Health framework is the USAID PREDICT project. The goal of PREDICT was to "strengthen global capacity for detection and discovery of zoonotic viruses with pandemic potential" by leveraging a One Health approach to investigate the "behaviors, practices, and ecological and biological factors driving disease emergence, transmission, and spread" (US Agency for International Development, 2023). The results of this research are being harnessed to develop intervention and policy recommendations to reduce pandemic risk globally.

Components of the study included human and animal surveillance, pathogen discovery and diagnostics, modeling and analytics, behavioral risk surveillance, information management, capacity strengthening, and One Health partnerships. The interplay between data, collaborations, and connectedness was integral to the success of the project in numerous ways. In the case of data, the project included biological data collection from both humans and animals in "zoonotic hotspots" (communities within which zoonotic risk was previously identified) within predetermined radii. This was accompanied by additional data collection with community participants, including a quantitative questionnaire, semi-structured open-ended interviews, and focus group discussions to better understand potential risk behaviors and exposures. These complementary quantitative and qualitative data collection efforts demonstrate a clear acknowledgment of the value of multi-disciplinary collaborative research that explores both biomedical and behavioral risk factors for emerging infectious disease. In addition, the randomization plan for the study was developed through extensive cross-disciplinary discussions.

The complementary data collection efforts also demonstrate the need for connectedness with the communities we serve. For example, access to villages was not possible without strong trusted relationships and communications starting with village leaders and other key community stakeholders. The long-standing relationships that the PREDICT Consortium has in local communities in close to 30 countries was critical to ensuring the successful implementation and global reach of the project.

Finally, one of the most valuable outputs of the PREDICT project was the "Bat Book," a picture book that is moderated by trusted community leaders entitled, "Living Safely with Bats." The Bat Book was developed as part of a public health communication strategy related to the identification of a novel filovirus in bats in West Africa, with distant correlations to the virus responsible for the 2013–2016 Ebola outbreak. The Bat Book was developed by a consortium of public health, animal health, conservation, and disease ecology experts from 29 countries. The tool has now been adapted, translated, and used in more than 20 countries in Africa and Asia

(Martinez et al., 2022) and its deployment was timely given the impending coronavirus disease (COVID)-19 outbreak that would follow. The Bat Book elucidates the critical importance of robust, rigorous data, coupled with interdisciplinary One Health collaborations, and solid relationships with the communities we aim to serve.

HARNESSING COLLABORATIONS

Operationalizing One Health requires the interpersonal skills of collaborating for the common good – healthy humans, healthy animals, healthy plants, and healthy environment. Collaborations evolve as individuals or organizations make choices to work together and share resources (knowledge, facilities, networks, and funding). In practice, One Health collaborations are scientific and social alliances between health professionals, scientists, engineers, educators, and stakeholders that are concerned about pathogens, chemicals, or hazardous conditions that impact the health of humans, animals, plants, or the environment. Effective alliances begin with initially listening to stakeholder issues followed by transparent knowledge sharing between subject matter experts, internal and external to stakeholder entities. All collaborations work best when backed by leadership at every level and aligns with organizational missions and resources.

There are five basic questions to help guide One Health collaborations:

1. What's the problem?
2. What can we do?
3. Who can do what?
4. How can we work together?
5. Are we making a difference?

What's the Problem?

Health-related issues under the One Health umbrella are countless and massive in scope. The COVID-19 pandemic reinforced that zoonoses are One Health problems that may impact every sector of society. Adverse impacts from zoonotic diseases range from morbidity, mortality, critical workforce, and supply issues to economic stability. The COVID-19 pandemic dramatically changed global views on almost every sector – healthcare, food supply, critical infrastructure, and communications. Severe acute respiratory syndrome coronavirus 2 (SARS-CoV-2) is just one of many viruses that have pandemic potential. More than 75% of emerging infectious diseases come from animals. Scientists estimate that there are over one million viruses not yet identified (Woolhouse et al., 2012; Wu, 2020; Racaniello, 2013). How many of these yet to be identified viruses infect humans or animals is unknown and the risk of spillover is also unknown. The scope of One Health grows exponentially when looking beyond zoonoses at the health effects from anthropogenic contaminants (López-Pacheco et al., 2019), such as pharmaceuticals, plastics, and pesticides.

One Health issues are essentially "wicked problems" that cannot be solved by a single solution or single entity. Such issues require coordination, collaboration, and connectedness on many fronts. The complexity of wicked problems arises from incomplete or confounding data, the number of people or sectors involved, and the large economic burden. These problems are dynamic in nature. In more cases than not, additional problems are revealed or created when addressing the initial concern and the consequences are difficult to predict. Therefore, scoping the One Health issue is not straightforward and cannot be solved in traditional linear scientific and engineering processes.

Using the best available knowledge, One Health problems must be scoped into manageable actions. Generally, the first step is identifying the greatest risks to impacted populations of humans and animals in their shared environment. It is very important to hold listening sessions and build connections with stakeholders in the impacted community. By doing so, scientists and engineers may recommend immediate simple steps to mitigate further damage and provide some relief while sustainable solutions or interventions are being evaluated and developed. When community stakeholders are sincerely engaged and sense that scientists and engineers are working to solve *their* problem with *their* input, trust is earned and problem formulation becomes a collaborative effort.

There are many online tools to assist in problem formulation in ecohealth, agriculture, and healthcare assessments. One Health practitioners may benefit from decades of work by ecological risk assessors, who have constructed processes and frameworks using a One Health approach for problem formulation when scoping wicked environmental problems. One example is the EPA EcoBox (US EPA, 2015) developed by the U.S. Environmental Protection Agency. This toolbox is designed to guide the development and evaluation of preliminary hypotheses about why ecological effects have occurred, or may occur, from human activities (Risk Assessment Forum, 1998).

STEPS IN PROBLEM FORMULATION

1. Identify areas of concerns
2. Identify available information and potential biases
3. Select variables within quality data sets to evaluate
4. Identify interdependencies between variables
5. Construct a concise problem statement

What Can We Do?

After problem formulation, the One Health project team collaborates to define a set of objectives and specific tasks to assess and evaluate the variables at the intersection of humans, animals, and the environment. The dynamic nature of One Health aligns well with Agile (Project Management Institute, 2017) project

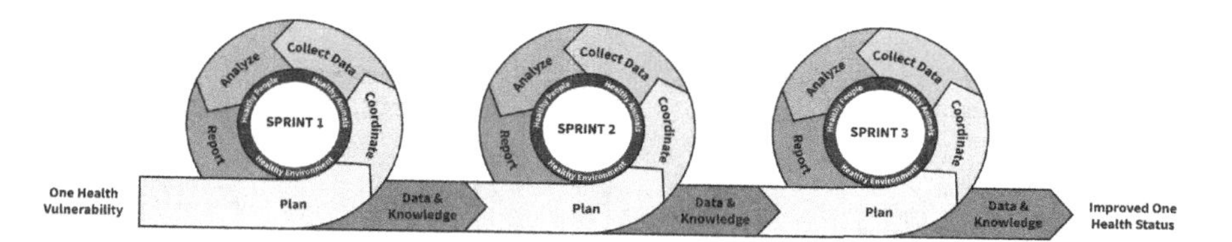

FIGURE 14.1 Agile project management cycles in a One Health approach.

management which allows new data to be reviewed and incorporated in sprints (i.e., defined time periods of a research development cycle). As more data are gathered and knowledge is gained in an iterative series of sprints, the product team is better positioned to address unforeseen scientific and technical variables as well as benefit from serendipitous findings (Figure 14.1). In "Wholesome Design for Wicked Problems," Knapp suggests that there should be a "shift in the goal from solution to intervention. Instead of seeking the answer that totally eliminates a problem, one should recognize that actions occur in an ongoing process, and further actions will always be needed" (Knapp, 2022).

Who Can Do What?

One Health collaborations benefit from recognizing that the expertise needed to address large-scale, complex problems is not concentrated in any single person or organization. Likewise, the solutions or interventions cannot be implemented by any one entity or sustained by any one stakeholder. Furthermore, there is no algorithm for collaborating between people or organizations or predetermined composition of a One Health collaborative team. One Health collaborations evolve when people share the common vision of optimal health of humans, animals, plants, and the environment. Collaborations grow with communication ease. Collaborations may begin with a simple conversation between two people on a human, animal, or ecosystem health issue. These two people may or may not be considered subject matter experts. The key is that they are comfortable communicating about a common concern and are willing to work together. As others join the conversation, it is crucial to have communication networks and tools available. As knowledge, data, and resources are shared and communicated, a non-competitive trusting relationship builds between collaborators, and an open collaborative environment drives innovation.

One Health practitioners recognize that collaborations are essential because no one person is an expert in the biology of all living species, the toxicity of all chemicals, the impacts of all environmental hazards, or the policy implications at each level of governance. A major advantage of One Health projects is the recognition and evaluation of the health connections and the interdependencies of all biological systems. The question of who needs to be involved is a moving target as the project evolves. As tasks are defined, the skillsets needed become more evident.

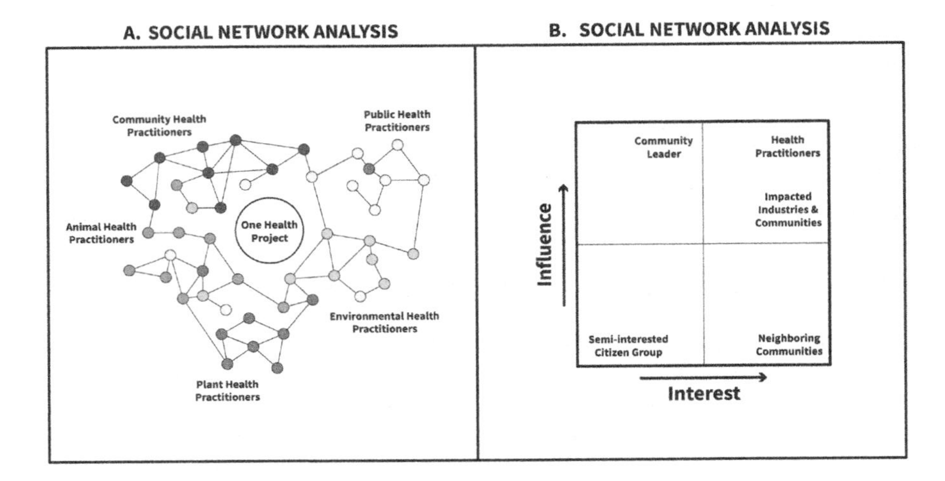

FIGURE 14.2 Social network analysis examples of connections and influence.

Early in the project, roles and responsibilities need to be assigned to persons with adequate expertise and sufficient resources to accomplish the task assignments. A social network analysis of all players helps to identify relationships and connections between individuals, groups, and assets. This approach helps to understand who is working with whom, how data are collected, groups with research assets or connections, and leaders or decision-makers in the community, academia, industry, and government. A 2×2 Stakeholder Matrix can help with identifying who can and will do what. By plotting all persons and organizations in accordance with their influence and interest, the One Health project managers can better manage communications and expectations of stakeholders. It is important to recognize that as the project progresses, so do the stakeholders (Figure 14.2).

How Can We Work Together?

The mechanics of collaboration is a combination of information and resource sharing. Information sharing requires communication networks for discussing issues, ideas, options, planning, implementation, and progress updates. Resource sharing may involve funding, expertise, and facilities.

Information Sharing

Information comes in all shapes and sizes – for example, formal and informal, official and unofficial, cultural and traditional knowledge, and news and social media outlets. Managing and analyzing the multiple information streams on a project is challenging. In addition, it is highly likely that technical and non-technical information needs to be shared with collaborators that are not co-located in the same office, community, state, or even country. With today's collaborative software and management tools, collaborators can communicate, manage

Adobe LiveCycle	Group-Office	Simple Groupware
Basic Support for Cooperative Work	GroupWise	Skype for Business
Citadel	HCL Domino	Slack
Confluence	Horde	SOGo
Coogle	Kolab	Telligent Community
EGroupware	LogicalDOC	Tiki Wiki CMS Groupware
First Class	IceWarp	Tine 2.0
Flock	Microsoft Exchange Server	Trello
Google Docs	Microsoft Teams	WIMI
Google Sheets	Nuclino	Workplace
Google Slides	Oracle Beehive	Zimbra

FIGURE 14.3 Examples of online collaboration tools and platforms.

projects, update task boards, and share and store files while maintaining the flexibility of remote work environments and the advantage of connecting with faraway experts. Online collaboration tools range from online whiteboards and video conferencing to project management enterprise systems (Figure 14.3).

Resource Sharing

The benefits of sharing resources between collaborators are that the work gets done faster and smarter, which ultimately saves time and money. However, sharing resources is not necessarily easy, especially when resources are limited. Collaborators may be working for competing organizations. Even more challenging can be organizational guidelines or rules for resource sharing (e.g., confidential business information, expertise, facilities, assets, and funding). Generally, resource sharing must align with organizational mission and authorities and must be mutually beneficial to all parties. Table 14.1 provides examples of resources that could be shared across One Health sectors.

TABLE 14.1
Examples of Resources That Could Be Shared across One Health Sectors

Resources	Examples of Resources That Could Be Shared across One Health Sectors
Field Operations	Lodging/housing; vehicle/fuel; supply chains; PPE; data networks; field decontamination equipment
Field Sampling	Surface and fomite sampling kits and equipment; drinking water/wastewater sampling equipment; air sampling equipment
Diagnostics	Zoonotic pathogen test kits; soil diagnostic tests for pathogens and chemicals; water test kits
Lab Capacity	Lab Infrastructure (e.g., facilities; workforce, biosafety and biosecurity management); Lab equipment (freezers, PCR machines, gene sequencers, incubators, microscopes, pipettes, reagents, etc.); Bio-waste management

Are We Making a Difference?

The goal of One Health collaborations is to jointly make a positive difference in the health of people, animals, and their shared environment. Unlike business ventures that are focused on revenue, profits, and customers, One Health beneficiaries are communities and ecosystems. Given the varied endpoints in the health of humans, animals, plants, and the environment, there is no definitive set of indicators to measure progress on One Health projects. Examples of project performance measures that can be adapted for individual One Health projects to assess progress toward the overall objective are listed in Table 14.2.

CASE STUDY: LOCAL ENVIRONMENTAL OBSERVER

Arctic residents see first-hand the changes in the presence and health of wildlife and the environment that are the result of temperatures outside of normal and landscapes being altered. The Alaska Native Tribal Health Consortium (ANTHC) recognized the value of connecting with Arctic residents and tribal elders that held traditional ecological knowledge with topic experts around the world on environmental health issues impacting the health of animals and people in the circumpolar Arctic regions. In 2012, the ANTHC launched the Local Environmental Observer (LEO) Network (www.leonetwork.org) as a tool to help the tribal health system and local observers to share information about climate and other drivers of environmental change. The LEO strategy is to address the impacts of climate change, one event at a time. The LEO members submit observations of unusual animal, environment, and weather events to the network by submitting pictures and corresponding text to describe the event via a mobile application. All observations are reviewed by the LEO editorial team. Selected observations and data are published to the LEO database, allowing it to be viewed on the LEO map and searchable to the LEO community (Figure 14.4).

The LEO Network is a model participatory science program bridging traditional knowledge from local observers to the universe of topic experts in multiple disciplines. Given the remoteness and hard-to-reach areas, continuous long-term monitoring of ecosystems would not be possible without Arctic communities and Indigenous peoples collecting and reporting data that show trends in species, population numbers over time, unusual wildlife behaviors, and interactions between species. Beyond data reporting, the LEO Network strives to connect experts with community members to develop community-specific adaptation strategies.

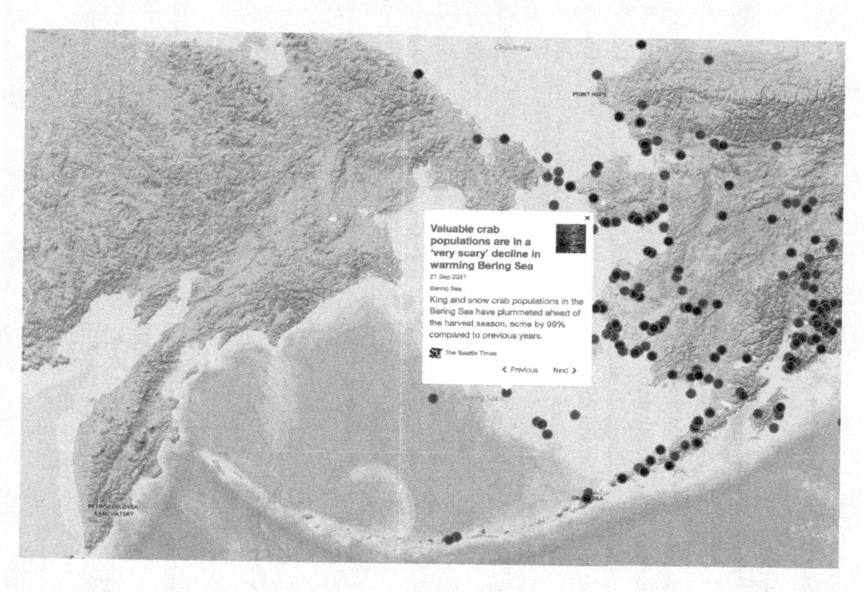

FIGURE 14.4 Local environmental observer network map example.

TABLE 14.2
Example Performance Measures for One Health Projects

	Collaboration – Systems and Stakeholders	Human Health	Animal Health	Plant Health	Environmental Health	Community Health
Output measures	• Stakeholder engagement sessions • Expert collaborators • Data networks shared • Facilities or assets shared	• Outbreak data • Demographic health data • Occupational health data • Biosurveillance data • Healthcare statistics • Self-help guidance for families, schools, businesses	• Local fauna, biodiversity, and habitat data • Outbreak data • Farmed animal health data • Companion animal health data • Wildlife animal health data • Captive animal health data • Invasive species data • Animal Permit data • Veterinary care statistics	• Local flora, biodiversity, and habitat data • Crop data • Farm to market data • Nursery data • Pest control data • Water usage data • Soil parameter data • Climate impact data	• Local flora, fauna, biodiversity, and habitat data • Water quality data • Air quality data • Soil quality data • Weather data • Built infrastructure data (water systems, transportation, etc.) • Climate impact data	• Cultural knowledge • Indigenous knowledge • Historical data • Land-use data • Economic data • Education data • Demographic data • Public health data • Community policies and laws
Efficiency measures	• Virtual vs in-person meetings • Shared networks	• Time to diagnosis • Diagnostic costs • Treatment costs • Intervention costs	• Time to diagnosis • Diagnostic costs • Treatment costs • Intervention costs	• Time to diagnosis • Diagnostic costs • Treatment costs • Intervention costs	• Time to sample environmental matrices • Time and cost to remediate • Intervention costs	• Time to communicate with community

(*Continued*)

TABLE 14.2 (*Continued*)
Example Performance Measures for One Health Projects

	Collaboration – Systems and Stakeholders	Human Health	Animal Health	Plant Health	Environmental Health	Community Health
Effectiveness measures	• Time to address issues • Collaborators contributing to online project management tools • Communication	• % reduction in illness and death • % reduction in diagnostic costs • % reduction in time and costs to treat • Communication with stakeholders	• % reduction in illness and death in target animals • % reduction in diagnostic costs • % reduction in time and costs to treat • % increase in intervention implementation • Communication with stakeholders	• % reduction in infection and crop loss • % reduction in diagnostic costs • % reduction in time and costs to treat • % increase in intervention implementation • Communication with stakeholders	• % reduction in environmental contamination • % reduction in time and costs to remediate • % reduction in secondary impacts • % increase in intervention implementation • Communication with stakeholders	• % economic recovery • % improvement in health of impacted populations • Culture-specific communication

(*Continued*)

TABLE 14.2 (*Continued*)
Example Performance Measures for One Health Projects

	Collaboration – Systems and Stakeholders	Human Health	Animal Health	Plant Health	Environmental Health	Community Health
Capability measures	• Expert networks established to address issues	• At-home kits for impacted community • Capability to diagnose and treat X number of patients in Y timeframe for Z cost	• Field test kits for animals (companion, production, and wildlife) • Capability to diagnose and treat X number of animals (species) in Y timeframe for Z cost • Capability to detect contamination in animal products before introduced into market or food supply	• Field test kits for farmers • Capability to diagnose and treat X number of plants (species) in Y timeframe for Z cost • Capability to detect contamination in animal products before introduced into market or food supply	• Field test kits for environmental hazard, contaminant, and pathogen • Capability to deploy sampling teams in timely manner • Capability to remediate in a timely manner	• Capability to distribute test kits • Capability to establish community-based testing facilities • Local school and business guidance • Critical infrastructure protected and available for use • Workforce protected

HARNESSING CONNECTEDNESS

One Health, by definition, recognizes the interconnection between people, animals, plants, and their shared environment through multisectoral and interdisciplinary collaboration. Connectedness (or lack of it) is often an invisible glue that separates success from failure with One Health projects and programs.

Connectedness is a concept that may provide insights into the advantages of social relationships, including the trust and strength that arises from relationships, yet definitions and measures lack consistency. Attributes of connectedness, as a concept, include intimacy, sense of belonging, caring, empathy, respect, reciprocity, and trust (Phillips - Salimi et al., 2012). One singular, authoritative definition of connectedness for the One Health framework does not yet exist. Conceptual ambiguity could limit the usefulness of this connectedness concept to some, yet general attributes exist.

Some find it helpful to think of connectedness as a concept, like leadership, which similarly lacks a singular definition. Both are nebulous yet important – arguably a lynchpin for success.

Like connectedness, leadership comes in many forms. Despite this variety, both connectedness and leadership have generalizable traits and attributes that may be taught, learned, and refined. Leadership styles vary, as Master of Business Administration (MBA) and education programs around the globe teach wide ranges of skills to millions of people – each with unique talents, behavior, and preferences. Somewhat similar to leadership, connectedness is essential for practitioners to refine and adapt to their own unique style.

With connectedness, like leadership, there is no right or wrong way. Every situation and interpersonal interaction offers new opportunity to connect; the "how" is not a one-size-fits-all formula or cookie-cutter approach. Connectedness will arise differently depending on the individuals – their unique context, orientations, preferences, intentions, and emergent properties from the individuals operating within the broader system. Discover what works best with your personal strengths and One Health community, given your unique context. Then, strengthen your individual skills, prioritize relationships, and remember the human dimensions.

Organizational Design to Foster Connectedness

Organizations and institutional networks have attributes, which foster or hinder connectedness. The best individual practitioners of One Health recognize – and operate – with real-time awareness of all the many nested scales that influence the community and decision-making within the systems where they operate.

One Health institutional approaches arise from Ostrom's contributions to managing the commons (Ostrom, 1990) and understanding institutional diversity (Ostrom, 2005) to maximize the chances of success through cooperative management, also known as co-management or co-created management. Ostrom's Institutional Analysis Design (IAD) framework identified eight IAD design principles that form the basis of every decentralized community that shares resources and underlies institutional actions – each arguably necessary for sustainable co-management, yet insufficient for connectedness.

Below are the eight IAD principles, adapted from Indiana University's Ostrom Workshop Toolkit (Indiana University, 2020).

1. **Defined Boundaries**: Clear boundaries can emerge from constitutive processes, competition among neighboring groups, and local resource knowledge.
2. **Effective Rules**: Rule-making will have wide participation if those suffering grievances have dispute resolutions processes available for redress.
3. **Rule-Making Process**: Long-term sustainability can't persist unless appropriation and maintenance rules become congruent with local conditions and values.
4. **Monitoring**: Unbiased monitoring done by monitors responsible to the core users will generate useful knowledge.
5. **Sanctions:** Sanctioning applied in a graduated fashion can reinforce shared community values.
6. **Conflict Resolution with Appellate Process**: Processes for the resolution of disputes that are widely available and operate at a reasonable cost in time and effort can also reinforce shared values.
7. **Self-Governance**: Constitutive processes that can be carried out relatively easily facilitate the establishment and operation of limited-task teams.
8. **Polycentricity**: Organizations established by legitimate constitutive processes at the local or regional and nested across scales, therefore interconnected (whether directly or indirectly) to systems at the state, national, and global scales, while retaining sufficient autonomy to make meaningful allocations of resources at the local level.

Elinor Ostrom
August 7, 1933–June 12, 2012
Human Environmental Sciences
Indiana University

As an anthropologist and political scientist, the late Ostrom is the only woman to win the Nobel Prize in Economics. In 2009, Ostrom jointly won the Nobel Prize in Economics for insights on economic governance and institutional incentives by demonstrating how ordinary people can create self-sustaining institutions and self-regulating rules, which allow for equitable management of sustainable resources when aligned to economic incentives (Lopez & Moran, 2016). Under certain conditions, natural resources may be both economically and ecologically sustainable.

In the 50 years since IAD publication, subsequent research highlights additional contributing factors for cooperative management and sustainability success. These additional attributes beget connectedness, as supplementary processes to strengthen community and improve outcomes.

- **Leadership**: Authentic leadership built on core values and shared goals.
- **Long-Term Commitment**: Commitment and shared concern for long-term outcomes.
- **Information Symmetry**: Shared access to timely information.
- **Trust and Reciprocity Norms**: Two-way interactions with mutual benefits over time.

Case studies to demonstrate the importance of connectedness span disciplines and sectors across the globe. Agriculture and sustainable forestry (Lopez & Moran, 2016; Romanelli & Boschi, 2019; Jenkins et al., 2020) exemplify the importance of connectedness and expanded IAD attributes to include the intangible, and often hard to quantify or measure human dimensions. Freshwater management examples also date back decades to Balinese water temple networks (Lansing, 2007; Helmreich, 1999) and groundwater management in California (Langridge & Ansell, 2018), where IAD principles endure through modern day. From the oceans and nearshore marine fisheries (Honey et al., 2010; Fujita et al., 2010), self-sustaining systems emerge with cooperative management in communities where transparent, committed, and long-term leadership combines with near-real-time information flows.

Across studies and diverse systems – terrestrial, freshwater, and marine – best practices for connectedness include:

- Effective leadership demonstrated in all settings;
- Long-term concerns incorporated in dispute resolution and other evaluative processes;
- Information available in a timely fashion for all monitoring and evaluative processes;
- Trust and reciprocity norms reinforced by participation in most or all of these processes.

One Health researchers, practitioners, and policy makers can incorporate cooperative design principles and connectedness best practices using IAD Toolkits (Indiana University, 2020). No matter which specific tools, always remember the multiple scales – ideally, nested scales – that interact across scales, as one dynamic system in time and space. One Health practitioners should expect non-linear dynamics, which will only be understood holistically by looking, planning, and evaluating across nested scales to drive outcomes and self-sustaining impact.

CASE STUDY: LYME INNOVATION

The Lyme Innovation movement exemplifies One Health impact originated by, for, and with a community of hyperlocal "bottom-up" action that used the U.S. government's open data (GSA, 2024a), open science (U.S. Department of Energy, 2024), and the international "Open Government" (GSA, 2024b; Open Government Partnership, 2024b) playbooks to transform national policy for Lyme and tick-borne diseases. The Lyme Innovation lynchpin provided connectedness.

After greater than 30 years of acrimonious "Lyme wars" (Stricker & Lautin, 2003; Tonks, 2007; Rupprecht et al., 2008) that stymied Lyme disease scientific progress, a handful of interdisciplinary scientists and doctors became Lyme patients themselves after tick bites that transmitted the bacteria causing Lyme disease. When their clinical and academic understanding from the peer-reviewed science journals failed to return full wellness, these physicians and scientists opened their minds to the collective wisdom of the Lyme disease advocacy community, effectively crowdsourced over millions of lived experiences over the decades. A handful of individuals used their first-hand experiences – radically different from the existing scientific paradigm – to augment understanding of Lyme disease. Both had merit. Both offered unique pieces to a complex puzzle.

What is Lyme disease? Lyme disease is a multistage and multisystemic disease transmitted by ticks that affect both physical and mental health, manifesting with diverse symptoms in humans (Rupprecht et al., 2008). According to the U.S. CDC, approximately 476,000 Americans are diagnosed and treated for Lyme disease annually (Centers for Disease Control and Prevention, 2023). This accounts for nearly 80% of all U.S. vector-borne disease cases with trends worsening (CDC, 2024; The U.S. Department of Health and Human Services, 2024). Lyme disease can debilitate when undiagnosed and not treated early, potentially leading to death due to acute carditis (Marx et al., 2020) or comorbidities. Late manifestations of Lyme disease are difficult and expensive to treat, so Lyme disease costs patients and the U.S. economy billions every year in direct medical expenses (Adrion et al., 2015) but the full economic costs-of-illness (including indirect costs) and societal impacts remain uncharacterized.

Why is Lyme so controversial? The bacteria (*Borrelia burgdorferi*, sensu lato) is an evolutionarily complex bacteria (e.g., neither gram-negative nor gram-positive) with multiple hosts, so the vector transmitting the bacteria involves multi-host, complex systems challenges. Complexity exists from the bacteria's biology, human (host)-pathogen interactions, and environmental factors including climate change, habitat loss, and land-use changes that affect arthropods (vectors) transmitting tick-borne diseases like Lyme disease. Many scientific unknowns and uncertainty exist, creating Lyme knowledge gaps in multiple disciplines.

Amidst these scientific unknowns and uncertainty, the Lyme disease problem compounded as a real-world threat to human health. The problem exploded on the ground, while Lyme science stagnated. Federally funded research and industry investments remained largely unchanged (until recently), despite the exponential growth of Lyme disease. Those experiencing Lyme disease first-hand saw a major problem, which remained largely obscured from scientists and policy makers due to the confluence of the disease's slow-moving biology, a lack of accurate diagnostic tests, and the time required to report cases and transform raw data into understanding. Without diagnostics to accurately test and detect all stages of Lyme disease, doctors and patients missed cases. This created a circular paradox because unreliable diagnostic tests resulted in the underreporting of Lyme disease cases that masked the growing problem, later estimated by the U.S. CDC to be two orders of magnitude larger than surveillance cases reported to government (Centers for Disease Control and Prevention, 2013). From these estimates, other sectors assumed that low numbers meant low disease incidence, hence small markets, so university and industry investments in Lyme science remained minimal for decades.

Bottom-up community catalyzes action and Lyme Innovation emerges. In 2013, the Lyme community petitioned the White House through new emerging technologies, including crowdsourcing campaigns such as We the People (The White House, President Barack Obama, 2016; Leland, 2013) and Change.gov (Childs, 2015) campaigns. Rather than waiting for government action, however, the Lyme community partnered with academic centers to find a diverse mix of problem solvers to propel the pace of research on Lyme disease.

The group called itself "Lyme Innovation" and the American Association for the Advancement of Science and Technology Policy Fellowship program empowered university affiliates, Lyme patients, and an eclectic group of volunteers to self-organize around shared goals. Lyme Innovation defined its core values, which rested on an "open" ethos for transparent information exchange (e.g., open data, open science, open-source code, and open innovation) as community standards. This empowered everyone, so that Lyme disease patients and those with lived experiences would work side-by-side, as co-equals, with physicians and scientists, and decision-makers in science and policy discussions.

Throughout 2015 and 2016, Lyme Innovation hosted co-creation events at universities and professional societies that created opportunities for diverse interdisciplinary teams to form and benefit from mentorship, share resources, and unlock information. They catalyzed a yearlong sprint of Lyme Innovation milestone events with everyone included and patients as equal partners, beyond conventional scientific workshops and conferences. Lyme Innovation partnered with an ever-growing number of academic and local organizations to invigorate the field of Lyme disease through user-centered exercises such as crowdsourcing, data "hackathons," and accelerators designed to bring new researchers into the field and strengthen existing teams. This bottom-up, community-driven, Lyme Innovation movement resulted in more than 25 new Lyme-related research projects and initiatives.

Top-down government meets bottom-up community. In 2018, the U.S. federal government formally joined the LymeX Innovation movement. The U.S. Department of Health and Human Services announced the HHS Lyme Innovation initiative, building upon the shared goals and core values identified by the community-driven Lyme Innovation movement (U.S. DHHS, 2024b). The community rested on shared values that mobilized action to address shared goals and solve real-world human needs through patient-centered, data-driven collaborations to accelerate science and improve government services. In five years, the public's demand for Lyme disease prioritization resulted in government action at the highest levels, including the White House under both Democratic (Office of the Press Secretary, The White House, 2016) and Republican (Open Government Partnership, 2019) administrations. The U.S. Congress also passed the Kay Hagan Tick Act in 2019, signaling new priority for Lyme and vector-borne disease.

In 2020, as part of HHS Lyme Innovation, a $25 million LymeX Innovation Accelerator (U.S. DHHS, 2024b) was launched – the world's largest public-private partnership for Lyme disease. The LymeX partnership sponsored almost 700 hours of human-centered design (HCD) research with Lyme patients, caregivers, and frontline practitioners, as synthesized in the 2021 Health+ Lyme Disease Human-Centered Design Report (Coforma, 2021), with a focus on how to help patients today amidst scientific unknowns. To address the diagnostic challenges important to the community, LymeX launched an innovative prize competition – the LymeX Diagnostics Prize – with up to $10,000,000 in cash prizes for the next generation of technologies (U.S. DHHS, 2024a). Consistent with its community-led Lyme Innovation roots, the LymeX Diagnostics Prize and partnership included patients in every step of identifying, developing, and implementing advancements for Lyme disease.

Transformative change endures by connecting the bottom-up and top-down. Impacted communities know their own problems and which needs to prioritize, as each individual or group is the expert in their own experience. As Lyme Innovation exemplified, the lived experiences of patients, caregivers, and frontline doctors offer a wealth of information – yet these individuals and bottom-up community groups often lack resources. People need resources to identify, implement, and scale solutions.

Government and large enterprises often have access to resources, while further removed from the frontline problems. A co-created understanding, which emerges by embedding those closest to the problem and those controlling decisions and resources (and vice versa), helps everyone to better know which questions to ask. What problems should government solve to address the community's most pressing needs? Given a well-defined problem or challenge statement, the full power of government and large enterprises can focus attention, convene groups, and unlock resources to expand the solution space.

This bottom-up and top-down combination, together, leads to transformative change that "sticks" for long-term impact. Lyme Innovation reveals how partnerships – transcending the Lyme wars through the connectedness of a few individuals – can unite a previously fractured group so that a new collective accelerates change faster than any one person or one sector could deliver.

Start small. Think big. Span silos. "Never underestimate the power of a small group of committed people to change the world. In fact, it is the only thing that ever has." ~ Margaret Mead

Co-creation Toolkit and Resources

One Health practitioners can look to co-creation standards and guidance with examples, best practices, templates, and information from the Open Government Partnership (OGP) National Handbook (Open Government Partnership, 2024a). The OGP co-creation toolkit aims to help reformers in government and civil society to improve government integrity, transparency, accountability, and responsiveness to citizens.

More generally, the design profession, including HCD, uses an iterative approach to problem-solving that includes service and system design, management, and engineering frameworks relevant to One Health. HCD professionals and designers co-create solutions by involving the human perspective and emotion at the core of the process of listening and learning with people at the center, using iterative development of concepts (e.g., Agile design).

HOW TO BUILD CONNECTEDNESS?

Often invisible or intangible, connectedness is an essential ingredient for One Health success. This requires long-term commitment, both in time and leadership, to build reciprocity and trust. Change happens at the speed of trust.

Trust

Trust can be defined as consistency over time. It is built fastest when consistent interactions, repeatedly over time, generate positive outcomes with each interaction. Trust takes active work and cannot be rushed, forced, or replaced (although it may be rebuilt, if lost).

At an individual level, people build trust by consistently showing up with authenticity and congruency through time where actions, verbal communication, and nonverbal communication match. When trust is built, it is the glue that binds everything together when unexpected challenges or complexities arise. Invest in trust as it is foundational for One Health research and solutions to "stick" – meaning that desired One Health solutions will be long-term and self-sustaining through voluntary engagement.

Given that trust can take years or decades to develop, One Health practitioners often partner with influencers already within a community – someone already known and trusted. Shared goals must align authentically with two-way information flow and reciprocity, not in any way forced or manipulated to achieve One Health goals before shared interests. When practitioners honor the unique orientation of individuals and the community within which they co-create, while harnessing their One Health expertise and resources to achieve partner goals, then knowledge flows with likely win-win outcomes.

Trust has nonlinear dynamics, which may be visualized as trust thermoclines. A thermocline is the transition layer, for example, when there is a sudden temperature change between warm surface water and cool deep water below. In a body of water, it is relatively easy to tell when you have reached the thermocline. The water thermocline is a noticeable hot/cold differential with abrupt temperature changes driven by physical science. The trust thermocline is an abrupt relationship change, often resulting from cumulative damage that erodes trust, driven by the human dimension. Once trust is lost and relationships breached, the partnership enters a different state. There is no going back. Project managers may call this the "Thermocline of Truth" that appears when Red-Amber-Green boards shift abruptly from amber to red.

One Health practitioners can never know where they exist on another's trust scale, yet you can control your own actions to be trustworthy. Embody unwavering integrity by being impeccable with your word and honoring commitments. Demonstrate with actions that match your words, repeatedly, for consistency over time. Show up, when promised. Deliver, as promised. Responsibly share information and volunteer insights or resources to help others, while continuously iterating and adapting as new information arises.

Transparency

One Health projects and practitioners work in complex environments, whether optimizing across multiple (often competing) objectives or balancing heterogeneous groups. Transparency can help.

Think of "open by default" as a toolkit or tactic for One Health practitioners to instantaneously communicate across multiple groups in near-real time. Resources for "open" programs and foundational principles include open data (e.g., Data.gov); open science (e.g., Science.gov); open innovation (e.g., Challenge.gov for crowdsourcing, citizen science, prizes, challenges); and collaboration in the public view. This is especially useful when groups are fractured or unable to travel or physically unite, given how open sharing will democratize information access and create a level playing field for everyone interested in participatory co-creation.

Successful open collaborations demystify processes, often revealing the people behind the scenes and how they operate. Humanize the process by pulling back the curtain on the organization, especially in large organizations with black box bureaucracies, so that the public may see how work is done and the individuals doing the work. Create safe and inclusive environments, where diverse groups may interact constructively (with enforced norms agreed by everyone) to foster relationships, collaborations, and connectedness through these human interactions in the open. Open events may be in-person or virtual. Working in full public view with transparency may take adjustment and "open by default" practice, yet this investment in consistent and transparent engagements (when done right with positive outcomes) will build trust with the public through time.

Thomas Kuhn
July 18, 1922–June 17, 1996
Scientific Historian and Philosopher
Massachusetts Institute of Technology

Trained as a physicist, Kuhn revolutionized science with the book, *The Structure of Scientific Revolutions* (1962) that brought the modern concept of scientific paradigms into the mainstream. Kuhn characterized the conditions in which science evolves over time and how new discoveries lead to paradigm shifts that have wide-reaching effects on the process of science and our understanding of the world.

Open Mindedness

Beyond the open collaboration strategies, an open and curious mind is essential for One Health practitioners. Think boldly and learn with open minds from people who think radically different from you. Solutions come from unexpected places, so lean into the discomfort, uncertainty, and unknowns. Embrace serendipity through constructive collisions by crossing cultural divides, disciplinary fields, party lines, and groups.

One Health practitioners will often operate at the transdisciplinary, spanning silos and connecting paradigms in novel ways. Transformative ideas often arise from other disciplines, so breakthrough science "discoveries" are often existing concepts applied to a new field in novel ways. This translation from one discipline, or sector, or party to another begets change and scientific revolutions – even though, nothing new was technically invented but something pre-existing was shared with a new audience and synthesized or applied in novel ways.

Interdisciplinary challenges are like a dichroic cube that looks like a cube of plain, clear glass – yet reflects different colors of light in different directions, so each side of the cube shows a different color. Like a disciplinary field of scientific research, each side of the cube has a special arrangement of filters and prisms for organizing inputs (e.g., light or information). This dichroic cube will show six different sides – each a different color – with its own way of refracting light. The vibrant rainbow colors from the cube, however, can only be seen and experienced through the holistic system view. All six sides and colors – each like its own disciplinary field – are essential and beautiful, yet the full rainbow of dynamic possibilities will only be seen through the amalgamated whole.

A One Health practitioner recognizes that no one discipline, or person, has all the answers. Therefore, embrace diversity in teams, diversity in disciplines, and diversity in context. Teams often have greater collective knowledge because they each shine a different color of an interdisciplinary dichroic cube.

Albert Einstein said it well, "We cannot solve our problems with the same thinking we used when we created them."

Connectedness – Cultivate the Karass

One Health practitioners know that vocabulary varies with keywords differing by disciplines and sectors, even when describing the same or similar concepts. One's intention behind a word (or action) arguably matters more than the word itself. When discussing connectedness or related concepts, some may use a different lexicon to describe the connectedness across our human dimensions. Potential substitutes may include: the matrix, network, influence, invisible hand, milieu, or even relatively new words in the English language like "karass."

Coined by author Kurt Vonnegut (1963), the karass (Vonnegut, 2006) is an invisible network between people that, unknown to them, are linked even in our unawareness, specifically to fulfill a higher or collective purpose. According to Vonnegut, and social scientists subsequently exploring this concept, this interconnected network of human experiences – the karass – is where life's beauty, wonder, and truth lie (Marvin, 2002). The karass exists to foster change through a shared matrix of collective connectedness, connecting individuals who are informally networked. This invisible influence rests on human relationships, where the whole is greater than the sum of its parts and emergent properties drive change.

Individually and collectively, One Health practitioners can cultivate the karass (Cultivate the Karass, 2024) for global sustainability and health. Everyone is connected through a shared matrix of collective connectedness.

CONCLUSION

Like collaborations and connectedness, data systems are linked networks with shared data vocabularies, rules, and norms that operate most effectively when intentionally linked through interoperable design. Networked linkages and influences occur through both formal and informal pathways. Whether explicitly acknowledged, or not, these nested scales influence outcomes, policy, research, and all of us every day.

No matter the application, One Health data and collaborations offer new pathways for co-creating solutions, especially when amplified through human connectedness. Nested scales of influence ensure dynamic change that One Health practitioners may co-create with and through. Whether a quantitative data network or a qualitative human connection, One Health involves nesting scales – individual, local, regional, state, national, and global – to combine information flows with diverse teams to transform data into understanding, decisions, and actions.

REFERENCES

Adrion, E. R., Aucott, J., Lemke, K. W., & Weiner, J. P. (2015). Health Care Costs, utilization and patterns of care following Lyme disease. *PLoS One*, *10*(2), e0116767. https://doi.org/10.1371/journal.pone.0116767

CDC. (2024, May 20). *Lyme disease surveillance and data*. Lyme Disease. https://www.cdc.gov/lyme/data-research/facts-stats/index.html

Centers for Disease Control and Prevention. (2010). *Understanding evidence: Evidence based decision-making summary*. CDC. https://vetoviolence.cdc.gov/apps/evidence/docs/EBDM_82412.pdf

Centers for Disease Control and Prevention. (2013, August 19). *CDC provides estimate of Americans diagnosed with Lyme disease each year*. CDC Online Newsroom - Press Release. https://archive.cdc.gov/#/details?url=https://www.cdc.gov/media/releases/2013/p0819-lyme-disease.html

Centers for Disease Control and Prevention. (2023, May 16). *Notifiable Infectious Disease Tables*. National Notifiable Diseases Surveillance System. https://www.cdc.gov/nndss/data-statistics/infectious-tables/index.html

Childs, A. (2015, March 29). *Lyme disease has reached epidemic proportions*. Change.Org. https://www.change.org/p/president-trump-and-congress-drain-the-cdc-swamp-and-legalize-lyme-disease

Coforma. (2021). *Health+ Lyme disease human-centered design report*. U. S. DHHS and Steven & Alexandra Cohen Foundation. https://www.hhs.gov/sites/default/files/healthplus-lyme-disease-hcd-report.pdf

Cultivate the Karass. (2024). *Home*. Cultivate the Karass. https://www.cultivatethekarass.org

Fujita, R. M., Honey, K. T., Morris, A., Wilson, J. R., & Russell, H. (2010). Cooperative strategies in fisheries management: Integrating across scales. *Bulletin of Marine Science*, *86*(2), 251–271.

GSA. (2024a). *Data.gov Home*. Data.Gov. https://data.gov/

GSA. (2024b). *U.S. open government initiatives | open.USA.gov*. https://open.usa.gov/

Helmreich, S. (1999). Digitizing 'development': Balinese water temples, complexity and the politics of simulation. *Critique of Anthropology*, *19*(3), 249–265. https://doi.org/10.1177/0308275X9901900303

Honey, K. T., Moxley, J. H., & Fujita, R. M. (2010). From rags to fishes: Data-poor methods for fishery managers. *Managing Data-Poor Fisheries: Cases Studies, Models & Solutions*, *1*, 159–184.

Indiana University. (2020). *IAD framework*. Ostrom Workshop. https://ostromworkshop.indiana.edu/courses-teaching/teaching-tools/iad-framework/index.html

Jenkins, M. E., Simmons, R., & Wardle, C. (2020). *The environmental optimism of Elinor Ostrom*. https://doi.org/10.22004/AG.ECON.307179

Knapp, R. (2022). *Wholesome design for wicked problems*. Public Sphere Project. https://www.publicsphereproject.org/content/wholesome-design-wicked-problems

Langridge, R., & Ansell, C. (2018). Comparative analysis of institutions to govern the groundwater commons in California. *Water Alternatives, 11*(3), 481–510.

Lansing, J. S. (2007). *Priests and programmers: Technologies of power in the engineered landscape of Bali*. Princeton University Press.

Leland, D. K. (2013, January 21). *Touched by Lyme: Squeaking under the wire for White House petition*. Lymedisease.Org. https://www.lymedisease.org/white-house-lyme-petition/

Lincoln, Y. S., & Guba, E. G. (1986). But is it rigorous? Trustworthiness and authenticity in naturalistic evaluation. *New Directions for Program Evaluation, 1986*(30), 73–84. https://doi.org/10.1002/ev.1427

Lopez, M. C., & Moran, E. F. (2016). The legacy of Elinor Ostrom and its relevance to issues of forest conservation. *Current Opinion in Environmental Sustainability, 19*, 47–56. https://doi.org/10.1016/j.cosust.2015.12.001

López-Pacheco, I. Y., Silva-Núñez, A., Salinas-Salazar, C., Arévalo-Gallegos, A., Lizarazo-Holguin, L. A., Barceló, D., Iqbal, H. M. N., & Parra-Saldívar, R. (2019). Anthropogenic contaminants of high concern: Existence in water resources and their adverse effects. *Science of The Total Environment, 690*, 1068–1088. https://doi.org/10.1016/j.scitotenv.2019.07.052

Martinez, S., Sullivan, A., Hagan, E., Goley, J., Epstein, J. H., Olival, K. J., Saylors, K., Euren, J., Bangura, J., Zikankuba, S., Mouiche, M. M. M., Camara, A. O., Desmond, J., Islam, A., Hughes, T., Wacharplusadee, S., Duong, V., Nga, N. T. T., Bird, B., ⋯ the PREDICT Consortium. (2022). Living safely with bats: Lessons in developing and sharing a global One Health educational resource. *Global Health: Science and Practice, 10*(6), e2200106. https://doi.org/10.9745/GHSP-D-22-00106

Marvin, T. F. (2002). *Kurt Vonnegut: A critical companion*. Greenwood Press.

Marx, G. E., Leikauskas, J., Lindstrom, K., Mann, E., Reagan-Steiner, S., Matkovic, E., Read, J. S., Kelso, P., Kwit, N. A., Hinckley, A. F., Levine, M. A., & Brown, C. (2020). Fatal Lyme carditis in New England: Two case reports. *Annals of Internal Medicine, 172*(3), 222. https://doi.org/10.7326/L19-0483

Office of the Press Secretary, The White House. (2016, September 28). *Fact sheet: Data by the people, for the people — Eight years of progress opening government data to spur innovation, opportunity, & economic growth*. Whitehouse.Gov. https://obamawhitehouse.archives.gov/the-press-office/2016/09/28/fact-sheet-data-people-people-eight-years-progress-opening-government

Open Government Partnership. (2019). *The open government partnership: Fourth open government national action plan for the United States of America*. https://open.usa.gov/assets/files/NAP4-fourth-open-government-national-action-plan.pdf

Open Government Partnership. (2024a). *OGP national handbook rules and guidance for participants* (Version 6). https://www.opengovpartnership.org/wp-content/uploads/2024/04/OGP-National-Handbook_2024.pdf

Open Government Partnership. (2024b, June 6). *Home*. Open Government Partnership. https://www.opengovpartnership.org/

Ostrom, E. (1990). *Governing the commons: The evolution of institutions for collective action*. Cambridge University Press.

Ostrom, E. (2005). *Understanding institutional diversity*. Princeton University Press.

Phillips - Salimi, C. R., Haase, J. E., & Kooken, W. C. (2012). Connectedness in the context of patient–provider relationships: A concept analysis. *Journal of Advanced Nursing, 68*(1), 230–245. https://doi.org/10.1111/j.1365-2648.2011.05763.x

Project Management Institute. (2017). *Agile practice Guide*. Agile Alliance.

Racaniello, V. (2013, September 6). *How many viruses on Earth? | Virology Blog*. https://virology.ws/2013/09/06/how-many-viruses-on-earth/

Risk Assessment Forum. (1998). *Guidelines for ecological risk assessment* (EPA/630/R-95/002F). U.S. Environmental Protection Agency. https://www.epa.gov/sites/default/files/2014-11/documents/eco_risk_assessment1998.pdf

Romanelli, J. P., & Boschi, R. S. (2019). The legacy of Elinor Ostrom on common forests research assessed through bibliometric analysis. *CERNE*, *25*(4), 332–346. https://doi.org/10.1590/01047760201925042658

Rupprecht, T. A., Koedel, U., Fingerle, V., & Pfister, H.-W. (2008). The pathogenesis of Lyme Neuroborreliosis: From infection to inflammation. *Molecular Medicine*, *14*(3–4), 205–212. https://doi.org/10.2119/2007-00091.Rupprecht

Sackett, D. L., Rosenberg, W. M. C., Gray, J. A. M., Haynes, R. B., & Richardson, W. S. (1996). Evidence based medicine: What it is and what it isn't. *BMJ*, *312*(7023), 71–72. https://doi.org/10.1136/bmj.312.7023.71

Stricker, R. B., & Lautin, A. (2003). The Lyme wars: Time to listen. *Expert Opinion on Investigational Drugs*, *12*(10), 1609–1614. https://doi.org/10.1517/13543784.12.10.1609

Sur, R., & Dahm, P. (2011). History of evidence-based medicine. *Indian Journal of Urology*, *27*(4), 487. https://doi.org/10.4103/0970-1591.91438

The U.S. Department of Health and Human Services. (2024). *The national Public Health strategy to prevent and control vector-borne diseases in people*. U.S. DHHS, CDC. https://www.cdc.gov/vector-borne-diseases/php/data-research/national-strategy/?CDC_AAref_Val=https://www.cdc.gov/ncezid/dvbd/framework.html

The White House, President Barack Obama. (2016). *Petitions under the Obama administration*. https://petitions.obamawhitehouse.archives.gov/responses

Tonks, A. (2007). Lyme wars. *BMJ*, *335*(7626), 910–912. https://doi.org/10.1136/bmj.39363.530961.AD

US Agency for International Development. (2023, July 11). *Emerging pandemic threats program | Global Health*. U.S. Agency for International Development. https://www.usaid.gov/emerging-pandemic-threats-program

U.S. Department of Energy. (2024). *Science.gov*. https://www.science.gov/

U.S. DHHS. (2024a). *Lyme X Diagnostics Prize*. https://www.lymexdiagnosticsprize.com/

U.S. DHHS. (2024b). *Lyme X Tick-borne disease innovation accelerator*. Lyme X. https://www.lyme-x.org/

US EPA, O. (2015, March 6). *EPA EcoBox (A toolbox for ecological risk assessors)* [Collections and Lists]. https://www.epa.gov/ecobox

Vonnegut, K. (2006). *Cat's cradle* (Dial Press trade paperback ed). Dial Press.

Woolhouse, M., Scott, F., Hudson, Z., Howey, R., & Chase-Topping, M. (2012). Human viruses: Discovery and emergence. *Philosophical Transactions of the Royal Society B: Biological Sciences*, *367*(1604), 2864–2871. https://doi.org/10.1098/rstb.2011.0354

Wu, K. J. (2020, April 15). *There are more viruses than stars in the universe. Why do only some infect us?* Science. https://www.nationalgeographic.com/science/article/factors-allow-viruses-infect-humans-coronavirus

15 Present and Future Climate Change Crisis – One Health Challenges and Solutions

Warren G. Lavey and Helena J. Chapman

INTRODUCTION

"Taking action on climate change" became part of the operational definition of One Health in a 2021 statement supported by four leading global organizations (One Health High-Level Expert Panel (OHHLEP) et al., 2022). This revision acknowledges that climate change affects the interrelated health of humans, animals, and ecosystems, and that One Health practitioners should increase the effectiveness of and support for actions to mitigate and adapt to climate change.

To implement this revised mission, climate change should be mainstreamed in One Health analyses and recommendations for actions. One Health's multidisciplinary, multisectoral approach should prioritize the pervasive, transformative impacts of climate change on the connected health systems. To promote climate change solutions, One Health practitioners should educate students and the public, and advocate with government and private sector decision-makers.

This chapter briefly describes (i) selected leading climate change impacts on human, animal, and ecosystem health; (ii) climate change impacts on One Health's focus areas of food safety, zoonoses, and antimicrobial resistance (AMR); (iii) utilizing innovative sources of data and expertise in the One Health approach to climate change; (iv) justice and ethics in One Health climate change actions; (v) One Health in recent climate change laws and policies; and (vi) action agenda for One Health on climate change.

SELECTED LEADING CLIMATE CHANGE IMPACTS ON HUMAN, ANIMAL, AND ECOSYSTEM HEALTH

The *2021–2022 Sixth Assessment Reports of the Intergovernmental Panel on Climate Change* (IPCC) found that climate change has strong adverse effects

DOI: 10.1201/9781003232223-15

on humans, animals, and ecosystems (Intergovernmental Panel on Climate Change (IPCC), 2023). The brief overview here is limited to a few highlights of the IPCC's extensive analyses and similar findings in other major reports (USGCRP, 2023).

On the physical science of climate change, the IPCC concluded: "It is unequivocal that human influence has warmed the atmosphere, ocean and land" (Intergovernmental Panel on Climate Change (IPCC), 2023). Reflecting on a thorough review of scientific studies, the scientific panel pointed to increasing atmospheric greenhouse gas (GHG) concentrations; carbon dioxide concentrations were higher in 2019 than at any time in at least two million years. One physical consequence of rapidly rising GHG was that each of the last four decades was warmer than any preceding decade since 1850. Hot extremes (including heat waves) have become more frequent and intense across most land regions since the 1950s. Other likely physical manifestations of elevated GHG include more frequent, intense, heavy precipitation events as well as droughts; increased major tropical cyclones; rising sea levels; and more acidic, warmer oceans.

Regarding human health impacts of climate change, adverse physical and mental outcomes include increased heat-related mortality and illnesses; food-, water-, and vector-borne diseases; malnutrition; trauma and related mental health challenges; and cardiovascular and respiratory distress. Additionally, climate change impairs food and water security for many people. Climate change impacts on human health are mediated through natural systems, such as increased exposure to wildfire smoke as well as emergence of zoonoses, aquatic pathogens, and toxic cyanobacteria in new areas. Climate-related health burdens fall disproportionately on economically and socially marginalized communities, especially women, children, ethnic minorities, poor communities, migrants or displaced persons, older populations, and those with underlying health conditions. As the World Health Organization (WHO) recognizes: "Climate change is the single biggest health threat facing humanity.... The climate crisis threatens to undo the last fifty years of progress in development, global health, and poverty reduction, and to further widen existing health inequalities between and within populations" (World Health Organization, 2023).

For land and marine animals and ecosystems, increased heat, wildfires, droughts, floods, and ocean acidification contribute to the rising number of animal and plant species that suffered extinction, local losses, increased diseases, mass mortality events, coral reef bleaching, ecosystem restructuring, and migration toward the global poles. Moreover, degradation by humans from deforestation, pollution, and overexploitation of land and other natural resources exacerbates ecosystem vulnerability to climate change. In contrast, people in some areas are applying nature-based strategies to mitigate and adapt to climate change. Mitigation strategies include storing atmospheric carbon dioxide through terrestrial and kelp forests, wetlands, and cover crops. Adaptation examples include expanding habitats like coastal mangroves and wetlands to protect against flooding, as well as reforesting to reduce desertification and recharge groundwater supplies. According to the UN Convention on Biological Diversity: "Biodiversity is declining at an unprecedented rate, and the pressures driving this decline are intensifying.... The time is right for change in our approach to the natural world ... and for the inextricable links between human well-being, climate change and biodiversity to be fully understood and acted upon" (Secretariat of the Convention on Biological Diversity, 2020).

CLIMATE CHANGE IMPACTS ON ONE HEALTH'S FOCUS AREAS OF FOOD SAFETY, ZOONOSES, AND AMR

A perspective on the important connections between climate change and One Health comes from recognizing that several key One Health focus areas are adversely affected by climate change. In 2017, the WHO observed that areas for which a One Health approach is particularly relevant include food safety, zoonoses, and AMR (World Health Organization, 2017). These three areas of work were again highlighted in the March 2022 Memorandum of Understanding on One Health cooperation among the UN Food and Agriculture Organization (FAO), World Organisation for Animal Health (WOAH), WHO, and UN Environment Programme (UNEP) (FAO et al., 2022).

Climate change worsens food safety, food security from both terrestrial and marine sources, and malnutrition (Mirzabaev et al., 2023). As for food safety, extreme weather events related to climate change increase the risk of disease-causing microorganisms in food, elevate various chemical hazards such as greater use of pesticides due to pest resistance, disrupt power for cooling and heating food, expand mold, and spur runoffs of human and animal wastes and other contaminants (Duchenne-Moutien & Neetoo, 2021). Regarding food production, rising temperatures and droughts decrease crop and livestock yields, in part by expanding the geographical distribution and incidence of pests and diseases (USGCRP, 2023). In the seas, climate change reduces marine fish capacity and capture through increases in water temperatures (associated with reduced oxygenation), acidification, and coral reef bleaching. Furthermore, elevated levels of carbon dioxide decrease by 10–15% the protein concentrations of wheat, barley, rice, and potato crops, contributing to malnutrition, particularly in children (Swinburn et al., 2019).

Next, climate change increases the transmission of several zoonotic diseases (Leal Filho et al., 2022). Rising temperatures and precipitation create favorable conditions for breeding mosquitoes, ticks, and other vectors; these conditions influence the incidence and geographical distribution of vector-borne diseases. In northern regions, early winters have increased wet conditions associated with climate change which amplify the intensity of pathogen transmission. Also, droughts and flooding in low- and middle-income countries can increase community risk of exposure to water contaminated with bacterial diseases like cholera or zoonotic water-borne diseases like schistosomiasis. Finally, opportunities for disease transmissions in animal-human contacts rise as migrations occur and habitats are disrupted due to extreme weather events, wildfires, sea level rise, ecosystem destruction, and other conditions related to climate change.

In a third priority area for One Health, climate change affects the proliferation and dissemination of AMR (Burnham, 2021). Higher temperatures and humidity are associated with increasing infection case numbers, colonization rates in animals, and AMR pathogens such as *Salmonella*. Flooding events worsen nitrogen fertilizer and sewage runoff, leading to eutrophication, polluted soils and waters with heavy metals, and spread of AMR. When extreme weather causes migration and crowding, conditions are conducive to acquiring AMR pathogens. Furthermore, the increased antibiotic prescribing practices by animal and human health practitioners may increase the risk of AMR.

This short overview of three priority areas for One Health demonstrates that climate change has become central to One Health's concerns and work. The Convention on Biological Diversity held in 2020 highlighted that both climate action and One Health are key transitions toward promoting the interrelated goals of healthy ecosystems, healthy people, and biodiversity, but that "overall considerably less attention has been paid through One Health approaches to broader aspects of human health beyond control of disease" (Secretariat of the Convention on Biological Diversity, 2020). One Health's analyses, programs, and recommendations should mainstream the pervasive, transformative impacts of climate change on the connected systems of human, animal, and ecosystem health.

UTILIZING INNOVATIVE SOURCES OF DATA AND EXPERTISE IN THE ONE HEALTH APPROACH TO CLIMATE CHANGE

Both the Joint Statement on One Health and the UN 2030 Sustainable Development Goals envision robust interdisciplinary, intersectoral collaborations, data sharing, and stakeholder engagements to promote well-being along social, economic, and environmental dimensions (United Nations Department of Economic and Social Affairs, 2015). By integrating various novel information sources, One Health practitioners can develop feasible, timely, and cost-effective interventions to address climate change as well as other human, animal, and environmental health risks. This section points to how innovative sources like Earth observations, citizen science applications, technological advancements and virtual platforms, local knowledge, and social sciences, can use the One Health approach to address climate change (Figure 15.1).

First, Earth observation data – collected from satellites, ground-based sensors, and aircraft – offer real-time information about the temporal and spatial patterns of many environmental parameters (Earth Science Data Systems, 2024). Each day, terabytes of Earth observation data relevant to health and climate change are collected and made available to the public, such as ambient GHG levels, wildfire emissions, extreme storms, land surface temperatures, sea levels, humidity, precipitation, and vegetation. Researchers can use these data to model climate change and other environmental hazards, monitor forest cover and crops, examine nutrition-rich and -poor areas, identify vector or disease hotspots, and other environmental applications. Potential climate-related actions to protect human and animal health through Earth observation data include early warning systems for heat waves, vector-borne diseases (e.g., malaria, West Nile virus, Rift Valley fever), harmful algal blooms, wildfires, and hurricanes. These data can also strengthen air quality monitoring and forecasts, such as for nitrogen oxides and fine particulate matter. Earth science experts can use these tools to support

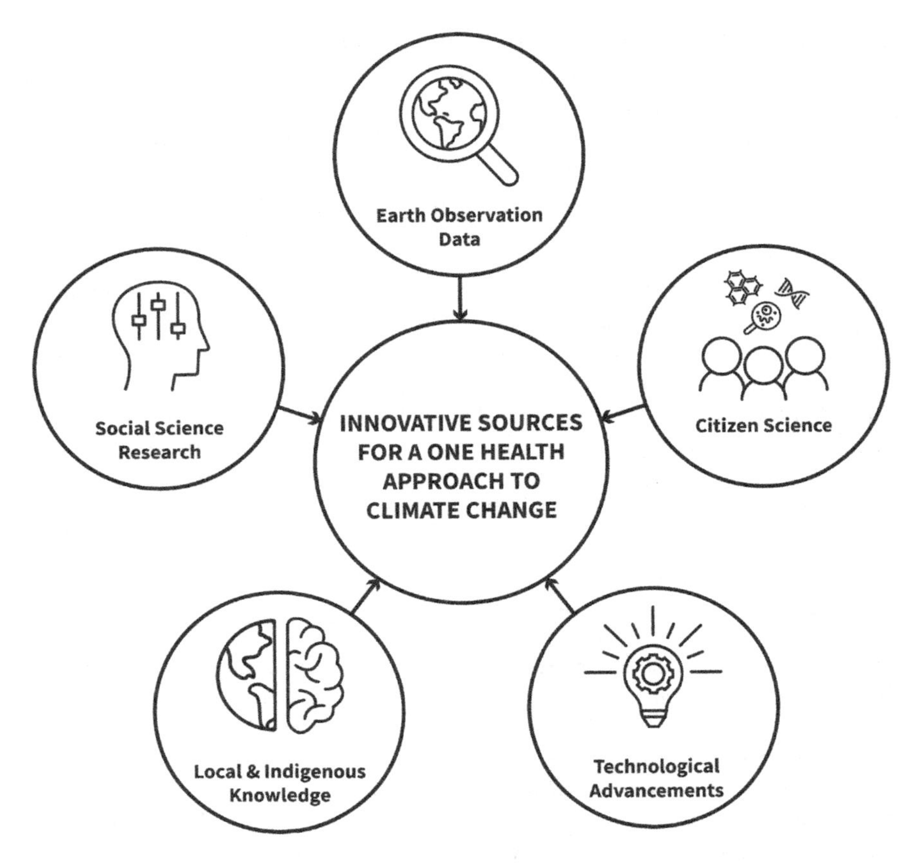

FIGURE 15.1 Innovative sources for promoting a One Health approach to address climate change

climate adaptation initiatives such as community heat management and mitigation plans and integrated vector control.

Next, 21st-century projects using citizen science applications increase learning and collaboration between scientific and non-scientific communities, resulting in a greater capacity to address climate change and health risks (Committee on Designing Citizen Science to Support Science Learning, 2018). Expertise of diverse disciplines, coupled with the use of novel information-gathering technologies such as smartphones, can enrich and expand scientific data collection. For example, the Global Learning and Observations to Benefit the Environment (GLOBE) Program – sponsored by the U.S. National Aeronautics and Space Administration (NASA), National Oceanic and Atmospheric Administration (NOAA), National Science Foundation (NSF), and Department of State – supports data collections by citizen scientists on cloud cover, dust, tree height, and land cover classification as well as mosquito habitats (GLOBE Program, NASA, 2024). Also, using the HABscope platform, citizen scientists collected short videos of water samples along Florida's Gulf of Mexico beaches, which helped expand sampling coverage and validate satellite data during *Karenia brevis* blooms, in order to produce more frequent and precise forecasts of respiratory hazards from aerosolized toxins (Hardison et al., 2019).

Third, the incorporation of technological advancements can propel cross-cutting approaches in One Health practice. Digital platforms like social media and smartphone apps are improving health efforts such as data collection, information dissemination, fitness programs, and medical consultation and follow-up (including medication adherence) (Animasahun et al., 2020; COVID-19 HPC Consortium, 2024; NASA, 2023; NASA Scientific Visualization, 2024). Health professionals can use podcasts, videos, data animations, blogs, and other media to educate the public on climate change and other environmental risks (NOAA & NIDIS, 2024; World Health Organization & Infodemic Management, 2021; World Health Organization & World Meteorological Organization, 2024). Global hackathon activities like NASA's International Space Apps Challenge can unite students and professionals to collaborate on real-world challenges and critically analyze novel approaches. Even high-performance computing resources, artificial intelligence, and machine learning techniques, led by data scientists, can add valuable insight for the analysis of big datasets such as for modeling climate change and environmental health. These approaches aid in the development of predictive models and risk maps for public health surveillance systems such as food insecurity (e.g., Famine Early Warning Systems Network) and disease outbreaks (e.g., cholera) (Usmani et al., 2023).

Another important source of information for One Health practitioners is Indigenous and other local knowledge (Dawson et al., 2021). Such expertise can shed light on significant temporal changes affecting ecosystems, animals, and humans, including marine and terrestrial food systems. They can also provide insights into human behaviors and community governance which shape the effectiveness of climate change interventions.

Finally, addressing the health risks of climate change requires utilizing the social sciences to better examine human behavior and socio-cultural factors (Degeling & Rock, 2020). Social science information can complement epidemiological data in assessing needs and influencing decision-making by the public, governments, and various practitioners. These research strategies, which often require an iterative process for data collection and analysis, can lead to robust research questions, findings, and policy recommendations.

In expanding the toolkit of information sources for One Health to tackle climate issues, workforce training, access to information technologies, community governance, and other capacity challenges must be considered (The White House, 2022). Collecting certain data and interdisciplinary collaborations with certain types of partners may be feasible in some communities but not others. Practitioners should consider various factors creating disconnects between generated knowledge and applications (e.g., "knowledge-action" gap) (Haines et al., 2004) and identifying best practices to enhance the application of such knowledge to practice.

JUSTICE AND ETHICS IN ONE HEALTH ACTIONS ON CLIMATE CHANGE

With the addition of climate change to the One Health mission, One Health should move toward strongly engaging with the critical challenges of justice and ethics. Climate change shows that these issues are not merely problems for low-income, minority human populations, but rather go to the heart of the linkages of human, animal, and ecosystem health which guide One Health (Ferdowsian, 2021; Tschakert et al., 2021).

The Joint Statement on One Health suggests, but does not clearly call out, that the mission of One Health encompasses pursuing justice for all people by remedying the disproportionate health burdens from environmental conditions borne by many disadvantaged communities and their exclusion from decision-making (United Nations Department of Economic and Social Affairs, 2015). In a related report, the One Health High-Level Expert Panel (OHHLEP) pointed to social justice and equal rights as a foundational principle in developing this new definition (One Health High-Level Expert Panel (OHHLEP) et al., 2022). Moreover, while the One Health definition does not refer to ethical actions toward animals and ecosystems, the OHHLEP stated another foundational principle of human stewardship, recognizing the importance of animal welfare and the integrity of the whole ecosystem.

Climate justice is a moral perspective with two parts (Canzi, 2015). First, disadvantaged human communities account for little of accumulated GHG and yet are more susceptible to the impacts of climate change on human health (Nicholas & Breakey, 2017). Climate justice analyses point to the greater burdens on low-income, marginalized people globally – in urban and rural areas as well as in high- and low-income nations – from heat stress, flooding, hurricanes, wildfires, and other effects of climate change. This thread of climate justice calls for reducing GHG emissions as rapidly as possible and meeting the needs of these communities in adaptation actions. Second, the transitions required to mitigate climate change should enable all people to realize the right to economic and social development. All people now and in the future deserve the right to sufficient health, energy, transportation, food, water, natural areas, and other resources for their development and well-being. Marginalized people should not be left behind so that others can thrive in the context of lower global GHG emissions.

The One Health approach is intrinsic to understanding and acting on several aspects of climate justice. Some of the impacts on disadvantaged people are tied to how climate change disrupts the health of animals and ecosystems. As illustrations, the warming and acidification of waters is bleaching coral reefs and forcing fish to migrate, thus harming people dependent on subsistence fishing; and the growing desertification of lands is killing the wildlife and livestock that feed Indigenous peoples, smallholder farmers (who are predominantly women), and nomadic herdsmen. From a different perspective, the pressures of climate change on food sources are forcing disadvantaged people around the world to engage in overexploitation and depletion of natural resources which destroys ecosystems and threatens species, including in fishing, hunting, and using water and forest resources. In the transition to practices aligned with greater climate stability, all people must have sufficient access to the ecosystem resources and services needed for human livelihood and development.

As for climate ethics, the UNESCO Declaration of Ethical Principles in relation to Climate Change was joined by 195 states in 2017 (UNESCO, 2017). These principles were intended to guide climate change policies and actions, complementing the Paris Agreement under the UN Framework Convention on Climate Change (United Nations, 2015) and the IPCC's scientific assessments. Most of these principles present a human-centered perspective on well-being, addressing issues like the welfare of present and future generations of people; solidarity and interdependence among peoples of different backgrounds; and meaningful involvement of all people in decision-making. Yet, importantly, several provisions reflect the One Health linkages of human health and well-being to other organisms as well as terrestrial and marine ecosystems (UNESCO, 2017). Furthermore, the One Health approach aligns with concerns about the ethical issues in how sea level rise and other effects of climate change are destroying species and ecosystems in locations which are spiritually and culturally significant for some human communities (Pantazatos, 2015). Also, advocates for the ethical rights of non-human species and natural systems are concerned about the impacts of climate change on biodiversity and ecosystems (Ikeke, 2021).

In summary, the One Health approach is closely linked to climate justice and climate ethics. Recognizing and acting on these linkages in mitigating and adapting to climate change can strengthen the elements in One Health associated with promoting social justice, partnering with groups of disadvantaged people, caring for animal welfare, and protecting ecosystem integrity.

ONE HEALTH IN RECENT CLIMATE CHANGE LAWS AND POLICIES

Many actions and plans on climate change focused on the important transitions away from fossil fuels in generating energy and transportation, primarily involving human-made structures, equipment, and technologies with little or no consideration for natural ecosystems and biodiversity (Guterres, 2022). The One Health approach to tackling climate change has penetrated some, but far from all, laws and policies. Four recent initiatives by the European Commission (EC), United Kingdom (UK), Commonwealth of Massachusetts, and State of Washington show a range of actions regarding One Health insights on climate change mitigation and adaptation.

a. On July 14, 2021, the EC adopted a series of legislative proposals designed to achieve climate neutrality in the European Union (EU) by 2050, including the intermediate target of reducing GHG emissions by at least 55% by 2030 (European Commission, 2021). Wisely, this framework prominently reflects One Health multisectoral climate change solutions.

The EC's agriculture strategy endorses the One Health perspective: "The link between healthy people, healthy societies and a healthy planet puts sustainable food systems at the heart of the European Green Deal, the EU's sustainable and inclusive growth strategy" (European Commission, 2024). Among the EU's goals are to "ensure food security in the face of climate change and biodiversity loss" and "reduce the environmental and climate footprint of the EU food system." The strategy embraces a Farm-to-Fork system for agriculture, fisheries, and aquaculture encompassing sustainable food production, sustainable food processing and distribution, sustainable food consumption, and food loss and waste prevention (European Commission, 2020).

Similarly, the EU Green Deal acts on biodiversity with a One Health perspective on challenges and solutions (Commission to the European Parliament et al., 2020):

> We humans are part of, and fully dependent on, this web of life.... The biodiversity crisis and the climate crisis are intrinsically linked.... Nature regulates the climate, and nature-based solutions, such as protecting and restoring wetlands, peatlands and coastal ecosystems, or sustainably managing marine areas, forests, grasslands and agricultural soils, will be essential for emission reduction and climate adaptation.

Specifically, the new EU Forest Strategy for 2030 calls for increasing forest coverage and improving forest resilience, observing that forests are essential for human health and well-being as well as the health of the planet, rich in biodiversity, and hugely important in fighting climate change (European Commission, Directorate General for Communication, 2021).

b. In its 2022 climate risk assessment, the UK Government explicitly committed to learning from the challenges in the climate-environment-health nexus and COVID-19 by taking a One Health approach for greater pandemic preparedness (HM Government, 2022). In one major sector, the UK's Agriculture Act of 2020 changed farm support payments by authorizing financial assistance to farmers for, among other goals, managing land or water in a way that protects or improves the environment, managing land, water, or livestock in a way that mitigates or adapts to climate change, protecting or improving the health or welfare of livestock, or protecting or improving the quality of soil (The National Archives, United Kingdom, 2020). The revised farming policy was centered around incentivizing environmentally sustainable farming practices, creating habitats for nature recovery, and supporting woodland and other ecosystem services to help tackle challenges like climate change (Coe & Finlay, 2020). Another manifestation of the One Health perspective is the UK's National Planning Policy Framework (Department for Levelling Up, Housing & Communities, 2023). Built around the objective of sustainable development encompassing social (including health), economic, and environmental dimensions, local governments' housing and other development plans should take proactive approaches to climate change mitigation and adaptation regarding communities and infrastructure as well as to biodiversity and landscapes.

c. Massachusetts in 2021 passed an Act Creating a Next-Generation Roadmap for Massachusetts Climate Policy which calls for comprehensive, clear, and specific plans to achieve net zero GHG emissions for the state by 2050 as well as interim targets (Commonwealth of Massachusetts, 2021). Three aspects of this law are especially noteworthy from the perspective of One Health policies. First, recognizing that inequitable environmental burdens on communities have related health consequences, the law defined environmental justice populations and changed how state agencies considered the needs of these people in determining the environmental impacts of construction projects. Next, the state must develop a plan for "natural and working lands" including setting goals for reducing GHG emissions and increasing carbon sequestration as well as outlining actions for land protection, management, and restoration (Lavey, 2021). However, the law does not address biodiversity, livestock practices, sustainable food consumption, or reducing food waste. Third, the law prioritizes emission reductions in six sectors with goals or sub-limits, but these do not cover natural and working lands despite the large GHG footprint of food systems (Kalibata, 2021).

d. Washington State's Climate Commitment Act of 2021 starts with One Health-aligned legislative findings that climate change is "an existential crisis with major negative impacts on environmental and human health" and requires "coordinated, comprehensive, and multisectoral implementation of policies, programs, and laws" (Washington State Department of Ecology, 2021). Unlike Massachusetts, the Washington program for GHG emission reduction metrics, goals, and compliance mechanisms does not cover the agricultural and natural land sectors. Nevertheless, some money from auctioning emissions allowances goes into a natural climate solutions account, leading to state investments intended to increase the resilience of waters, forests, and other vital ecosystems to climate change impacts, conserve working forests at risk of conversion, and increase carbon sequestration, storage, and overall system integrity. Another portion of the auction funds goes to address the disproportionate health effects from pollution on overburdened communities.

In summary, in various jurisdictions recent leading laws to tackle climate change challenges reflect several aspects of One Health's commitment to an integrated, multisectoral approach to human, animal, and ecosystem health. These climate change laws include actions on food security and consumption, zoonoses, livestock practices, soil health, environmental justice, natural lands, and other topics at the nexus of climate change, human health, animals, and ecosystems. On the other hand, many laws motivated by climate change ignore animals and ecosystems in aiming to reduce GHG emissions from electricity generators, transportation, buildings, manufacturing, equipment, and other human-made sources. While all these programs are important in climate change mitigation and adaptation, effective climate change solutions must also cover One Health systems and risks. Siloed approaches to climate change, which leave out animals and ecosystems, are handicapped in meeting goals for GHG emissions, atmospheric concentrations of GHG, climate justice and ethics, and sustainability.

ACTION AGENDA FOR ONE HEALTH ON CLIMATE CHANGE

Along with committing to take action on climate change, the Joint Statement anticipates developing a comprehensive Global Plan of Action for One Health which will "mainstream and operationalize One Health at global, regional, and national levels; support countries in establishing and achieving national targets and priorities for interventions; mobilize investment; promote a whole-of-society approach and enable collaboration, learning and exchange across regions, countries, and sectors" (FAO et al., 2022; One Health High-Level Expert Panel (OHHLEP) et al., 2022). A wide range of One Health initiatives should integrate actions on climate change, and actions on climate change should build on this framework for mainstreaming and operationalizing One Health. This section discusses some actions for promoting a One Health approach to examine the effects of climate change.

One Health's efforts to address challenges like food safety, zoonoses, and AMR should consider the pervasive, transformative impacts of climate change on the connected systems of humans, animals, and ecosystems. As UN Secretary General António Guterres bluntly said about the IPCC report released in February 2022, people and ecosystems "are getting clobbered by climate change" (Guterres, 2022). The direct and indirect effects of climate change as well as the influences of many actions aimed at addressing climate change should be major factors in One Health analyses and policy recommendations.

Climate change should be part of One Health actions targeting effective communications and learning throughout governments and societies. One Health practitioners should work globally with governments at all levels and private sector organizations to promote solutions to climate change, which integrate food systems, animals, ecosystems, land use, and other nature-based strategies. Moreover, understanding that human health is interrelated with animal and ecosystem health and that these systems are adversely affected by climate change should spur and guide important actions on GHG and resilience strategies. Laws and programs promoting clean energy and transportation are necessary climate actions but fall short of covering major sources of GHG emissions associated with food systems, biodiversity, and ecosystems. Many actions directed at the energy, transportation, building, and infrastructure sectors also do not encompass nature-based methods of GHG storage, ecosystem resilience needs and strategies, and climate justice/ethics issues. As described in the preceding section, some climate change laws and policies demonstrate an integration and leveraging of the One Health approach, but much more must be done to build these climate solutions through communications with policymakers, private organizations, and the public.

One Health efforts should target a wide range of students, professionals, and other workers for climate change education and advocacy. While the Joint Statement acknowledges that One Health requires a whole-of-society approach, this direction is especially applicable to actions on climate change. The existential threat of climate change demands, in the words of UN Secretary General Guterres, a "transformational" approach to all sectors of societies and economies: "If we don't urgently change our way of life, we jeopardize life itself" (Guterres, 2019). From its early mission to break through siloes separating human medicine, veterinary medicine, and public health, One Health should expand to become embedded in the training and work of many fields of engineering, agriculture, urban planning, biology, economics, business, political science, education, and more. A wide range of fields of expertise and work are affected by climate change and must be involved in One Health-informed climate solutions (Lavey, 2019).

Lastly, novel information, approaches, and applications will strengthen these efforts. The One Health interdisciplinary approach can foster the "4C's" (Communication, Coordination, Collaboration, and Capacity building) (World Health Organization, 2021) in climate change research and actions. The toolkit for applying One Health to climate change solutions can grow through university and continuing education courses, conferences and events, data sharing platforms, websites, reports, and publications as well as social media. Also, professional organizations along with communities of practice should provide global networks that share resources, communicate findings, and facilitate open discussions on an array of One Health approaches to climate change. Students interested in One Health and climate solutions can use their voices and learn through organizations like the Planetary Health Next Generation Network and Medical Students for a Sustainable Future (Medical Students for a Sustainable Future, 2024; Planetary Health Alliance, 2024).

As a guide for training students and other audiences to apply the One Health approach in understanding the impacts of climate change, Appendix 1 identifies and provides links to selected resources. Additionally, as an example of an active learning exercise to encourage thinking about the nexus between climate change and One Health, Appendix 2 outlines a case study on urban heat islands with discussion questions which include aspects of data collection and analysis, policy advocacy, climate justice, and natural solutions.

CONCLUSION

Climate change is worsening the health of humans, animals, and ecosystems. The One Health approach should be part of the path to more effective actions on mitigating and adapting to climate change. To pursue the recently revised mission of One Health, practitioners need new understandings and research directions, interdisciplinary collaborations, innovative information sources, and strong communications and advocacy actions. As Albert Einstein observed: "To raise new questions, new possibilities, to regard old problems from a new angle, requires creative imagination and marks real advance in science" (Einstein & Infeld, 1966).

Climate change is central to One Health priorities including food safety, zoonoses, and AMR. The One Health community should be a driving force for global, national, and local research, policies, and actions on climate change. Applying the One Health approach to tackling climate change will be crucial to advancing One Health goals and furthering the vision shared with the Sustainable Development Goals of multidisciplinary, multisectoral, and collaborative solutions. The perspective of linked human, animal, and ecosystem health can lead scientists and community practitioners to identify risk factors and develop strong approaches, interventions, and partnerships for societal benefits.

Innovative sources of data and expertise are advancing One Health efforts and should be applied in addressing climate change. Looking beyond scientific research, One Health practitioners need to consider the justice and ethical dimensions of climate change. While some recent legislation and government programs incorporate the One Health approach to climate change, much more needs to be done to educate and guide policymakers as well as the public.

For One Health practitioners, educating themselves on climate change and its impacts on health issues is just the start. Further actions are needed on communicating the One Health approach to climate change to students in all fields of study as well as throughout governments, private sector entities, diverse professionals, and the public. Collaborations across disciplines, sectors, and stakeholders can develop novel approaches to climate change and other determinants of health, and advance population health through prompt and effective interventions and policies.

APPENDIX 1: SELECTED RESOURCES FOR ONE HEALTH SUPPORTERS TO EXPAND THEIR KNOWLEDGE AND SKILLS TOOLKITS IN CLIMATE CHANGE

Category	Key Resources	Description
Organizations	One Health Initiative: https://onehealthinitiative.com/ One Health Commission: https://www.onehealthcommission.org/ Group on Earth Observations: https://earthobservations.org/	A list of key organizations that serve as knowledge hubs and professional networks
	Planetary Health Alliance's Planetary Health Next Generation Network: https://www.planetaryhealthalliance.org/next-gen Medical Students for a Sustainable Future: https://ms4sf.org/	A list of student-led organizations that offer scientific resources, workshops, mentorship, and networking opportunities
Definitions	One Health High-Level Expert Panel: https://www.who.int/news/item/01-12-2021-tripartite-and-unep-support-ohhlep-s-definition-of-one-health Centers for Disease Control and Prevention: https://www.cdc.gov/one-health/about/ World Health Organization: https://www.who.int/news-room/fact-sheets/detail/one-health (fact sheet) or https://www.who.int/health-topics/one-health	Definitions of One Health by selected institutions
Data Sources	NASA Earth Science Data: https://earthdata.nasa.gov/ Global Learning and Observations to Benefit the Environment (GLOBE) Observer: https://observer.globe.gov/ GLOBE Mission Mosquito: https://www.globe.gov/web/mission-mosquito GSA's Technology Transformation Services: https://data.gov/climate/ Global Climate Dashboard: https://www.climate.gov/climatedashboard European Union's Earth Observation Programme for Copernicus: https://www.copernicus.eu/en/access-data One Health Commission's One Health Tools and Toolkits: https://www.onehealthcommission.org/en/resources__services/one_health_tools__toolkits/	A list of relevant data sources and toolkits to examine the effects of climate change

(Continued)

Category	Key Resources	Description
Technological Advances	Comics, podcasts, videoblogs, YouTube videos, Instagram or TikTok videos Earth Information Center: https://earth.gov/ Data animations (e.g., NASA Scientific Visualization Studio: https://svs.gsfc.nasa.gov/) Global hackathon activities (e.g., NASA International Space Apps Challenge): https://www.spaceappschallenge.org/about/ High-performance computing resources (e.g., COVID-19 High-Performance Computing: https://covid19-hpc-consortium.org/)	Innovative technology that can help leverage expertise, share relevant information among networks, encourage collaborations on real-world case studies, and offer insight for big dataset analyses
Training Resources	Coursera: https://www.coursera.org/ NASA Applied Remote Sensing Training Program: https://appliedsciences.nasa.gov/what-we-do/capacity-building/arset National Institute of Environmental Health Sciences: Climate Change and Human Health Lesson Plans: https://www.niehs.nih.gov/health/scied/teachers/cchh Global Heat Health Information Network's Masterclasses: https://ghhin.org/masterclasses/ Global Consortium on Climate and Health Education (GCCHE) Courses: https://www.publichealth.columbia.edu/research/programs/global-consortium-climate-health-education/courses	Training opportunities to advance skill sets in One Health

APPENDIX 2: CASE STUDY ON URBAN HEAT ISLANDS

During heat waves, scientists have documented that some urban communities have experienced as much as a 5°F increase in air temperatures, when compared to surrounding rural areas. The density of built structures and paucity of green spaces contribute to this "urban heat island" effect. These conditions can endanger the health of humans, animals, and ecosystems in such urban areas, and are worsened by climate change (National Oceanic and Atmospheric Administration, 2023; Tong et al., 2021).

Reflections:

1. How do urban heat islands form, and what are the emerging risks from heat for the health of humans, animals, and ecosystems?
2. Where does the knowledge gap exist in research and practice activities?
3. Who are the relevant stakeholders, including vulnerable populations and policymakers?
4. What data sources can offer additional insights to the effects of climate change in urban areas?
5. How can we best disseminate our findings to the public as well as local decision-makers?
6. What is the local call to action to mapping and mitigating risk of urban heat islands to safeguard human, animal, and ecosystem health? What nature-based solutions should be considered?

REFERENCES

Animasahun, V. J., Chapman, H. J., & Oyewole, B. K. (2020). Social media to guide 'One Health' initiatives. *The Clinical Teacher*, *17*(2), 214–216. https://doi.org/10.1111/tct.13045

Burnham, J. P. (2021). Climate change and antibiotic resistance: A deadly combination. *Therapeutic Advances in Infectious Disease*, *8*, 204993612199137. https://doi.org/10.1177/2049936121991374

Canzi, G. (2015, August 4). *What is climate justice?* World Economic Forum. https://www.weforum.org/agenda/2015/08/what-is-climate-justice/

Coe, S., & Finlay, J. (2020). The Agriculture Act 2020. *House of Commons Library, Number CBP 8702.*

Commission to the European Parliament, The Council, The European Economic and Social Committee, & Committee of the Regions. (2020). *EU biodiversity strategy for 2030: Bringing nature back into our lives.* European Commission. https://eur-lex.europa.eu/resource.html?uri=cellar:a3c806a6-9ab3-11ea-9d2d-01aa75ed71a1.0001.02/DOC_1&format=PDF

Committee on Designing Citizen Science to Support Science Learning. (2018). *Learning through citizen science: Enhancing opportunities by design* (R. Pandya & K. A. Dibner, Eds.; p. 25183). National Academies Press. https://doi.org/10.17226/25183

Commonwealth of Massachusetts. (2021, March 26). *An act creating a next-generation Roadmap for Massachusetts climate policy*. Session Law - Acts of 2021 Chapter 8. https://malegislature.gov/Laws/SessionLaws/Acts/2021/Chapter8

COVID-19 HPC Consortium. (2024). *COVID-19 HPC consortium*. https://covid19-hpc-consortium.org/

Dawson, N. M., Coolsaet, B., Sterling, E. J., Loveridge, R., Gross-Camp, N. D., Wongbusarakum, S., Sangha, K. K., Scherl, L. M., Phan, H. P., Zafra-Calvo, N., Lavey, W. G., Byakagaba, P., Idrobo, C. J., Chenet, A., Bennett, N. J., Mansourian, S., & Rosado-May, F. J. (2021). The role of Indigenous peoples and local communities in effective and equitable conservation. *Ecology and Society*, *26*(3), art19. https://doi.org/10.5751/ES-12625-260319

Degeling, C., & Rock, M. (2020). Qualitative research for One Health: From methodological principles to impactful applications. *Frontiers in Veterinary Science*, *7*, 70. https://doi.org/10.3389/fvets.2020.00070

Department for Levelling Up, Housing & Communities. (2023). *National planning policy framework*. UK Ministry of Housing, Communities & Local Government. https://assets.publishing.service.gov.uk/government/uploads/system/uploads/attachment_data/file/1005759/NPPF_July_2021.pdf

Duchenne-Moutien, R. A., & Neetoo, H. (2021). Climate change and emerging food safety issues: A review. *Journal of Food Protection*, *84*(11), 1884–1897. https://doi.org/10.4315/JFP-21-141

Earth Science Data Systems, N. (2024, May 21). *Earthdata*. Earth Science Data Systems, NASA. https://www.earthdata.nasa.gov/

Einstein, A., & Infeld, L. (1966). *The evolution of physics: From early concepts to relativity and quanta*. Simon & Schuster.

European Commission. (2020). *Farm to fork strategy: For a fair, healthy and environmentally-friendly food systems*. https://ec.europa.eu/food/system/files/2020-05/f2f_action-plan_2020_strategy-info_en.pdf

European Commission. (2021, July 14). *The European green deal*. https://commission.europa.eu/strategy-and-policy/priorities-2019-2024/european-green-deal_en

European Commission. (2024). *Agriculture and the green deal*. https://commission.europa.eu/strategy-and-policy/priorities-2019-2024/european-green-deal/agriculture-and-green-deal_en

European Commission. Directorate General for Communication. (2021). *Making sustainable use of our natural resources*. Publications Office. https://data.europa.eu/doi/10.2775/706146

FAO, WOAH, WHO, & UNEP. (2022). *Memorandum of understanding regarding cooperation to combat health risks at the animal-human-ecosystems interface in the context of the "One Health" Approach and including antimicrobial resistance*. https://www.fao.org/3/cb9403en/cb9403en.pdf

Ferdowsian, H. R. (2021). Ecological justice and the right to health: An introduction. *Health and Human Rights*, *23*(2), 1–5.

GLOBE Program, NASA. (2024). *GLOBE Observer—GLOBE Observer—GLOBE.gov*. https://observer.globe.gov/

Guterres, A. (2019, December 2). *Remarks at opening ceremony of UN Climate Change Conference COP25*. United Nations Secretary General. https://www.un.org/sg/en/content/sg/speeches/2019-12-02/remarks-opening-ceremony-of-cop25

Guterres, A. (2022, February 28). *Message by António Guterres on the launch of IPCC climate report on impacts, adaptation and vulnerability | United Nations in Indonesia*. United Nations Indonesia. https://indonesia.un.org/en/173453-message-ant%C3%B3nio-guterres-launch-ipcc-climate-report-impacts-adaptation-and-vulnerability, https://indonesia.un.org/en/173453-message-ant%C3%B3nio-guterres-launch-ipcc-climate-report-impacts-adaptation-and-vulnerability

Haines, A., Kuruvilla, S., & Borchert, M. (2004). Bridging the implementation gap between knowledge and action for health. *Bulletin of the World Health Organization*, *82*(10), 724–732.

Hardison, D. R., Holland, W. C., Currier, R. D., Kirkpatrick, B., Stumpf, R., Fanara, T., Burris, D., Reich, A., Kirkpatrick, G. J., & Litaker, R. W. (2019). HABscope: A tool for use by citizen scientists to facilitate early warning of respiratory irritation caused by toxic blooms of Karenia brevis. *PLoS One*, *14*(6), e0218489. https://doi.org/10.1371/journal.pone.0218489

HM Government. (2022). *UK climate change risk assessment 2022—Presented to parliament pursuant to Section 56 of the Climate Change Act 2008*. https://assets.publishing.service.gov.uk/government/uploads/system/uploads/attachment_data/file/1047003/climate-change-risk-assessment-2022.pdf

Ikeke, M. O. (2021). The role of climate ethics in biodiversity conservation. *European Journal of Sustainable Development*, *10*(3), 205. https://doi.org/10.14207/ejsd.2021.v10n3p205

Intergovernmental Panel on Climate Change (IPCC). (2023). Summary for policymakers. In *Climate change 2022 – Impacts, adaptation and vulnerability: Working Group II contribution to the sixth assessment report of the intergovernmental panel on climate change* (1st ed.). Cambridge University Press. https://doi.org/10.1017/9781009325844

Kalibata, A. (2021, September 21). *The food systems summit- A new deal for people, planet and prosperity*. United Nations; United Nations. https://www.un.org/en/food-systems-summit/news/food-systems-summit-new-deal-people-planet-and-prosperity

Lavey, W. G. (2019). Teaching the health impacts of climate change in many American higher education programs. *International Journal of Sustainability in Higher Education*, *20*(1), 39–56. https://doi.org/10.1108/IJSHE-04-2018-0062

Lavey, W. G. (2021). Innovative U.S. state laws that help move land uses toward zero carbon emissions. *Carbon & Climate Law Review*, *15*(1), 92–110. https://doi.org/10.21552/cclr/2021/1/10

Leal Filho, W., Ternova, L., Parasnis, S. A., Kovaleva, M., & Nagy, G. J. (2022). Climate Change and zoonoses: A review of concepts, definitions, and bibliometrics. *International Journal of Environmental Research and Public Health*, *19*(2), 893. https://doi.org/10.3390/ijerph19020893

Medical Students for a Sustainable Future. (2024). *Medical students for a sustainable future*. Medical Students for a Sustainable Future. https://ms4sf.org/

Mirzabaev, A., Bezner Kerr, R., Hasegawa, T., Pradhan, P., Wreford, A., Cristina Tirado Von Der Pahlen, M., & Gurney-Smith, H. (2023). Severe climate change risks to food security and nutrition. *Climate Risk Management*, *39*, 100473. https://doi.org/10.1016/j.crm.2022.100473

NASA. (2023). *Space apps challenge*. https://www.spaceappschallenge.org/about/

NASA Scientific Visualization. (2024). *Scientific visualization studio*. NASA Scientific Visualization Studio. https://svs.gsfc.nasa.gov/

National Oceanic and Atmospheric Administration. (2023, April 4). *NOAA, communities to map heat inequities in 14 states, 1 international city*. https://www.noaa.gov/news-release/noaa-communities-to-map-heat-inequities-in-14-states-1-international-city

Nicholas, P. K., & Breakey, S. (2017). Climate Change, climate justice, and environmental health: Implications for the nursing profession. *Journal of Nursing Scholarship*, *49*(6), 606–616. https://doi.org/10.1111/jnu.12326

NOAA & NIDIS. (2024, June 6). *National drought status*. https://www.drought.gov/national

One Health High-Level Expert Panel (OHHLEP), Adisasmito, W. B., Almuhairi, S., Behravesh, C. B., Bilivogui, P., Bukachi, S. A., Casas, N., Cediel Becerra, N., Charron, D. F., Chaudhary, A., Ciacci Zanella, J. R., Cunningham, A. A., Dar, O., Debnath, N., Dungu, B., Farag, E., Gao, G. F., Hayman, D. T. S., Khaitsa, M., … Zhou, L. (2022). One Health: A new definition for a sustainable and healthy future. *PLOS Pathogens*, *18*(6), e1010537. https://doi.org/10.1371/journal.ppat.1010537

Pantazatos, A. (2015, November 26). *Cultural heritage rights and climate change*. Global Policy Journal. https://www.globalpolicyjournal.com/blog/26/11/2015/cultural-heritage-rights-and-climate-change

Planetary Health Alliance. (2024). *Next generation network*. Planetary Health Alliance. https://www.planetaryhealthalliance.org/next-gen

Secretariat of the Convention on Biological Diversity. (2020). *Global biodiversity Outlook 5*. https://www.cbd.int/gbo/gbo5/publication/gbo-5-en.pdf

Swinburn, B. A., Kraak, V. I., Allender, S., Atkins, V. J., Baker, P. I., Bogard, J. R., Brinsden, H., Calvillo, A., De Schutter, O., Devarajan, R., Ezzati, M., Friel, S., Goenka, S., Hammond, R. A., Hastings, G., Hawkes, C., Herrero, M., Hovmand, P. S., Howden, M., … Dietz, W. H. (2019). The global syndemic of obesity, undernutrition, and climate change: The Lancet commission report. *The Lancet*, *393*(10173), 791–846. https://doi.org/10.1016/S0140-6736(18)32822-8

The National Archives, United Kingdom. (2020). *Agriculture Act 2020* [Bill]. Statute Law Database. https://www.legislation.gov.uk/ukpga/2020/21/contents

The White House. (2022, May 11). *Fact Sheet: The Biden-Harris Administration Global Health Worker Initiative*. The White House. https://www.whitehouse.gov/briefing-room/statements-releases/2022/05/11/fact-sheet-the-biden-harris-administration-global-health-worker-initiative/

Tong, S., Prior, J., McGregor, G., Shi, X., & Kinney, P. (2021). Urban heat: An increasing threat to global health. *BMJ*, n2467. https://doi.org/10.1136/bmj.n2467

Tschakert, P., Schlosberg, D., Celermajer, D., Rickards, L., Winter, C., Thaler, M., Stewart-Harawira, M., & Verlie, B. (2021). Multispecies justice: Climate-just futures with, for and beyond humans. *WIREs Climate Change*, *12*(2), e699. https://doi.org/10.1002/wcc.699

UNESCO. (2017). *Declaration of ethical principles in relation to climate change*. https://www.unesco.org/en/ethics-science-technology/climate-change

United Nations. (2015). *Paris agreement*. https://unfccc.int/sites/default/files/english_paris_agreement.pdf

United Nations Department of Economic and Social Affairs. (2015). *The 17 goals: Sustainable development goals*. https://sdgs.un.org/goals

USGCRP. (2023). *Fifth national climate assessment*. U.S. Global Change Research Program. https://nca2023.globalchange.gov/

Usmani, M., Brumfield, K. D., Magers, B. M., Chaves-Gonzalez, J., Ticehurst, H., Barciela, R., McBean, F., Colwell, R. R., & Jutla, A. (2023). Combating cholera by building predictive capabilities for pathogenic Vibrio cholerae in Yemen. *Scientific Reports*, *13*(1), 2255. https://doi.org/10.1038/s41598-022-22946-y

Washington State Department of Ecology. (2021). *Climate Commitment Act 2021*. https://ecology.wa.gov/Air-Climate/Climate-Commitment-Act

World Health Organization. (2017, September 21). *One Health*. https://www.who.int/news-room/questions-and-answers/item/one-health

World Health Organization. (2021, December 1). *Tripartite and UNEP support OHHLEP's definition of "One Health."* https://www.who.int/news/item/01-12-2021-tripartite-and-unep-support-ohhlep-s-definition-of-one-health

World Health Organization. (2023, October 12). *Climate change*. https://www.who.int/news-room/fact-sheets/detail/climate-change-and-health

World Health Organization & Infodemic Management. (2021). *WHO competency framework: Building a response workforce to manage infodemics*. World Health Organization.

World Health Organization & World Meteorological Organization. (2024). *Heat health risks*. Global Heat Health Information Network. https://ghhin.org/

Acronyms

A4NH – CGIAR's Agriculture for Nutrition and Health
AAMC – Association of American Medical Colleges
AFROHUN – Africa One Health University Network
AGP – antimicrobial growth promoter
AHA – American Hospital Association
AM – antimicrobial
AMR – antimicrobial resistance
AMU – antimicrobial use
ANTHC – Alaska Native Tribal Health Consortium
APHA – American Public Health Association
APHIS – Animal and Plant Health Inspection Service
ARB – antimicrobial resistant bacteria
ARI – antimicrobial resistant infection
ARV – antiretroviral
ASF – Africa swine fever or animal-source foods
ASPIRE – African Science Partnership for Intervention Research Excellence
BRICS – Brazil, Russia, India, China, South Africa
CAESAR – Central Asian European Surveillance of Antimicrobial Resistance
CAPSCA – Collaborative Arrangement for the Prevention and Management of Public Health Events in Civil Aviation
CARB-X – Combating Antibiotic Resistant Bacteria Biopharmaceutical Accelerator
CBD – Convention on Biological Diversity
CBPP – contagious bovine pleuropneumonia
CDC – United States Centers for Disease Control and Prevention
CDSF – Capacity Development Support Facility
CE – continuing education
CHES – Certified Health Education Specialists
CIHR – Canadian Institutes of Health Research
CLASS – Calibrated, Lifelong, Adaptable, Scalable, and Sustainable
CLSI – Clinical and Laboratory Standards Institute
CME – Continuing Medical Education
COHEART – Kerala Veterinary and Animal Sciences University, Centre for One Health Education Advocacy Research & Training
COST – European Cooperation in Science and Technology
CPD – Continuing Professional Development
CRED – Centre for Research in the Epidemiology of Disasters
CSTE – Council of State and Territorial Epidemiologists
DALY – Disability-adjusted life-year

DANMAP – Danish Integrated Antimicrobial Resistance Monitoring and Research Program
DDD – defined daily dose
DNDi – Drugs for Neglected Diseases initiative
DRC – Democratic Republic of Congo
EARS -Net – European Antimicrobial Resistance Surveillance Network
EBM – Evidence-based medicine
EBS – Africa CDC's event-based surveillance
EC – European Commission
ECHO – European Commission Humanitarian Office
ECOWAS – Economic Community of West African States
EED – environmental enteropathy disorder
EID – emerging infectious disease
EPT – emerging pandemic threats
ESBL – extended-spectrum β-lactamase
ETC – education, training, and capacity building
EU – European Union
EUCAST – European Committee on Antimicrobial Susceptibility Testing
EVD – Ebola Virus Disease
EWEs – extreme weather events
FAO – Food and Agriculture Organization of the United Nations
FCDO – Foreign, Commonwealth & Development Office
FETP – Field Epidemiology Training Programs
GAP – global action plan
GARC – Global Alliance for Rabies Control
GARDP – Global Antibiotic Research and Development Partnership
GBF – Global Biodiversity Framework
GDP – gross domestic product
GHG – greenhouse gas
GHSA – Global Health Security Agenda
GLASS – Global Antimicrobial Resistance and Use Surveillance System
GLOBE – Global Learning and Observations to Benefit the Environment
GNI – Gross National Income
GOARN – Global Outbreak Alert and Response Network
HCD – human-centered design
HCWs – healthcare workers
HICs – high-income countries
HIPAA – Health Insurance Portability and Accountability Act
HLG – High-level group
IACG – United Nations Interagency Coordination Group on Antimicrobial Resistance
IAD – Institutional Analysis Design
IBCM – Integrated Bite Case Management
ICAO – International Civil Aviation Organization
ICARS – International Centre for Antimicrobial Resistance Solutions

ICC – intercultural competence
ICM – Saint Louis Zoo Institute for Conservation Medicine
IDRC – International Development Research Centre
IDSR – Integrated Disease Surveillance and Response
IFC – International Finance Corporation
IFIs – international financial institutions
IFRC – International Federation of Red Cross and Red Crescent Societies
IHR – International Health Regulations
IKS – Indigenous knowledge systems
INDDEX – International Dietary Data Expansion Project
InFARM – International FAO Antimicrobial Resistance Monitoring System
IPCC – Intergovernmental Panel on Climate Change
IUCN – International Union for the Conservation of Nature
IVSA – International Veterinary Student Association
IYC – infants and young children
JAMRAI – Joint Action on Antimicrobial Resistance and Healthcare-Associated Infections
JEE – Joint External Evaluation
JPIAMR – Joint Programming Initiative on Antimicrobial Resistance
KAP – knowledge, attitudes, and practices
LEO – local environmental observer
LMICs – low- and middle-income countries
M&E – monitoring and evaluation
MCHES – Master Certified Health Education Specialists
MDGs – Millennium Development Goals
MDR – multi-drug resistant
MERS – Middle East Respiratory Syndrome
MGHD – Masters in Global Health Delivery
MIA – medically important antimicrobial
MOOCs – massive open online courses
MOUs – memorandums of understanding
MPH – Master of Public Health
MRSA – methicillin-resistant *Staphylococcus aureus*
MS – Master of Science
NADPRP – National Animal Disease Preparedness and Response Program
NAP – national action plan
NARMS – National Antimicrobial Resistance Monitoring System
NASA – U.S. National Aeronautics and Space Administration
NbS – nature-based solutions
NCPD – Nursing Continued Professional Development
NEOH – Network for EcoHealth and One Health or Network for the Evaluation of One Health
NFELTP – Nigeria Field Epidemiology and Laboratory Training Program
NGO – non-governmental organization
NICE – National Institute for Clinical Excellence

NIH – U.S. National Institutes of Health
NOAA – U.S. National Oceanic and Atmospheric Administration
NSF – U.S. National Science Foundation
ODA – Official Development Assistance
OECD – Organisation for Economic Co-operation and Development
OFDA – Office of Foreign Disaster Assistance
OGP – Open Government Partnership
OH – One Health
OHCEA – One Health Central and Eastern Africa
OHEJP – One Health European Joint Programme
OHHLEP – One Health High-Level Expert Panel
OH JPA – One Health Joint Plan of Action
OHM – One Health Master's Degree
OHUNs – Country One Health University Networks
OHWA – One Health Workforce Academies
OHW-NG – One Health Workforce – Next-Generation project
OHZDP – One Health Zoonotic Disease Prioritization
OIHP – Office International d'Hygiène Publique
PEF – Pandemic Emergency Financing Facility
PEP – post-exposure prophylaxis
PEPFAR – President's Emergency Plan for AIDS Relief
PFGE – pulsed-field gel electrophoresis
PHEIC – Public Health Emergency of International Concern
PPPs – Public-Private Partnerships
PRISM – Performance of Routine Information System Management
ProMED – Program for Monitoring Emerging Diseases
PVS – Performance of Veterinary Services
QALY – Quality-adjusted life year
R&D – research and development
RA – regenerative agriculture
REDISSE – Regional Disease Surveillance Systems Enhancement
ReLAVRA – Red Latinoamericana de Vigilancia de la Resistencia a los Antimicrobianos
SARE – Stepwise Approach toward Rabies Elimination
SARS – Severe Acute Respiratory Syndrome
SCOH – Standing Committee on One Health
SDGs – Sustainable Development Goals
SDH – social determinants of health
SEAICRN – South East Asia Infectious Disease Clinical Research Network
SEAOHUN – Southeast Asia One Health University Network
SET – FAO Surveillance Evaluation Tool
SISOT – Tripartite Surveillance and Information Sharing Operational Tool
SNV – Sin Nombre virus
SPS – Sanitary and Phytosanitary Measures
STEM – science, technology, engineering, and mathematics

SUN – Scaling Up Nutrition
TATFAR – Transatlantic Taskforce on Antimicrobial Resistance
TB – tuberculosis
TD – transdisciplinary research
TOC – theory of change
ToT – Train-the-trainer
TrACSS – Tracking AMR Country Self-Assessment Survey
TRIPS – Trade-related aspects of intellectual property rights
UAF – University of Alaska at Fairbanks
UK – United Kingdom of Great Britain and Northern Ireland
UN – United Nations
UN-DHA – United Nations Department of Humanitarian Affairs
UNDP – United Nations Development Programme
UNDRR – UN Office for Disaster Risk Reduction
UNEP – United Nations Environment Programme
UNICEF – United Nations Children's Fund
U.S. – United States of America
USAID – United States Agency for International Development
USDA – United States Department of Agriculture
VOHUN – Vietnam One Health University Network
VRE – vancomycin-resistant enterococcus
VSL – Value of Statistical Life
WAHIS – World Animal Health Information System
WASH – water, sanitation, and hygiene
WGS – whole genome sequencing
WHO – World Health Organization
WOAH – World Organization for Animal Health
WPRACSS – Western Pacific Regional Antimicrobial Consumption Surveillance System
WRA – women of reproductive age
WTO – World Trade Organization
zTB – zoonotic tuberculosis

Index

Note: **Bold** page numbers refer to tables and *italic* page numbers refer to figures.